# TERZAGHI LECTURES 1974-1982

Geotechnical
Special Publication
No. 1

Published by the
American Society of Civil Engineers
345 East 47th Street
New York, New York 10017-2398

The material presented in this publication has been prepared in accordance with generally recognized engineering principles and practices, and is for general information only. This information should not be used without first securing competent advice with respect to its suitability for any general or specific application. The contents of this publication are not intended to be and should not be construed to be a standard of the American Society of Civil Engineers (ASCE) and are not intended for use as a reference in purchase specifications, contracts, regulations, statutes, or any other legal document.

No reference made in this publication to any specific method, product, process, or service constitutes or implies an endorsement, recommendation, or warranty thereof by ASCE.

ASCE makes no representation or warranty of any kind, whether express or implied, concerning the accuracy, completeness, suitability or utility of any information, apparatus, product, or process discussed in this publication, and assumes no liability therefor.

Anyone utilizing this information assumes all liability arising from such use, including but not limited to infringement of any patent or patents.

Copyright © 1986 by the American Society of Civil Engineers,
All Rights Reserved.
Library of Congress Catalog Card No.: 84-73061
ISBN 0-87262-532-X
Manufactured in the United States of America.

# PREFACE

This second volume of the collection of Terzaghi Lectures contains Terzaghi Lectures presented in 1974 through 1982, with the exception of the lecture presented in 1978 by the late Professor N. M. Newmark. As a consequence of Professor Newmark's untimely death on January 25, 1981, his Terzaghi Lecture has been irretrievably lost. The Geotechnical Engineering Community is fortunate indeed to have published 17 manuscripts of the first 18 Terzaghi Lectures. These manuscripts can be found in the first and second volumes of the Terzaghi Lecture collection. All of these have also appeared in the Journal of the Geotechnical Engineering Division or its predecessor, the Journal of the Soil Mechanics and Foundations Division.

The Executive Committee of the Geotechnical Engineering Division authorized the publication of the second volume which continues the effort toward easy accessibility of these lectures. It is well known that Dr. Karl Terzaghi had diverse interests in the fields of geotechnical engineering and engineering geology. His diversity of interests carried over in the lectures given in his honor under the Terzaghi Lectureship Program. These lectures presented in his honor and gathered in this volume attest to the fact that he exerted a broad and far-reaching influence on the fields of civil engineering and engineering geology. This influence still significantly affects the design and construction of all important foundation and earthwork projects.

W. F. MARCUSON III
Secretary, Executive Committee

JOHN T. CHRISTIAN
Chairman, Executive Committee

# CONTENTS

### LECTURE NO. 10

Some Effects of Dynamic Soil Properties on Soil-Structure
F. E. Richart, Jr. .................................................. 1
Introduction—Roy E. Olson ........................................... 2
Discussion—Michael J. Pender ........................................ 48
Closure ............................................................. 51

### LECTURE NO. 11

Bearing Capacity and Settlement of Pile Foundations
  George Geoffrey Meyerhof .......................................... 52
  Introduction—George F. Sowers ..................................... 53
  Discussion—Theodore K. Chaplin .................................... 86
            Jean Biarez and Pierre Foray ........................... 88
  Closure ........................................................... 91

### LECTURE NO. 12

Design and Construction of Drilled Shafts
  Lymon C. Reese ................................................... 94
  Introduction—Kenneth L. Lee ...................................... 95
  Discussion—Ibrahim H. Sulaiman .................................. 120
  Closure .......................................................... 122

### LECTURE NO. 13

Geology and Geotechnical Engineering
  Robert F. Legget ................................................ 124
  Introduction—Richard E. Gray .................................... 125
  Discussion—Richard J. Proctor ................................... 177
            Ronald J. Tannenbaum ................................. 179
  Closure .......................................................... 180

### LECTURE NO. 14

Observations on Stresses in Tunnel Linings
  Nathan M. Newmark ................................................. *

*Manuscript not available due to the death of Dr. Newmark

## LECTURE NO. 15
There Were Giants on the Earth in Those Days
George F. Sowers ........ 182
Introduction—William F. Swiger ........ 183
Discussion—N. J. Schmitter ........ 219

## LECTURE NO. 16
Investigation of Failures
Gerald A. Leonards ........ 222
Introduction—John A. Focht, Jr. ........ 223
Discussion—A. Isnard and T. J. Pilecki ........ 284
    G. E. Bratchell ........ 289
    H. Cambefort ........ 297
    Claudio A. Mascardi ........ 299
    Farrokh N. Screwvala ........ 300
Closure ........ 301

## LECTURE NO. 17
Evaluating Calculated Risk in Geotechnical Engineering
Robert V. Whitman ........ 312
Introduction—Robert L. Schuster ........ 313
Discussion—Herbert Klapperich, Ulrich Sturm and Stavros A. Savidis ........ 358
Closure ........ 359

## LECTURE NO. 18
Progress in Rockfill Dams
J. Barry Cooke ........ 361
Introduction—Ernest T. Selig ........ 362
Discussion—Ranji Casinader ........ 395
    W. L. Chadwick ........ 396
    Claude A. Fetzer ........ 397
    M. D. Fitzpatrick ........ 399
    E. M. Fucik, Jorge Hacelas and Carlos A. Ramirez ........ 401
    A. Clive Houlsby ........ 404
    A. Marulanda and C. S. Ospina ........ 405
    Bayardo Materon ........ 406
    A. H. Merritt ........ 407
    N. G. K. Murti ........ 409
    Ivor L. Pinkerton ........ 410
    Pietro De Porcellinis ........ 412
    C. F. Ripley ........ 414
    James L. Sherard ........ 418
    Arthur G. Strassburger ........ 420
    William F. Swiger ........ 423
    H. Taylor ........ 425
Closure ........ 427

**Subject Index** ........ 433
**Author Index** ........ 435

# JOURNAL OF THE GEOTECHNICAL ENGINEERING DIVISION

## THE TENTH TERZAGHI LECTURE

Presented at the American Society of Civil Engineers National Meeting on Water Resources Engineering, Los Angeles, California

January 24, 1974

F. E. RICHART, JR.

## INTRODUCTION OF TENTH TERZAGHI LECTURE

### By Roy E. Olson

President-elect Sangster, previous Terzaghi Lecturers, ladies and gentlemen, this afternoon we have the privilege of listening to the Tenth Terzaghi Lecture. This lecture series was established to honor, simultaneously, the father of our profession, and also its most gifted and reknowned practitioners. Terzaghi lecturers are selected not only for their expertise and their contributions to the profession over a period of years, but also for their continuing efforts and for their ability to communicate effectively. The high standards set for Terzaghi lecturers are well exemplified by a brief review of the names of previous Terzaghi lecturers. The first Terzaghi lecture was delivered by Ralph Peck in 1963 and successive lecturers have been:

| | |
|---|---|
| 1964, Arthur Casagrande | 1969, Stanley D. Wilson |
| 1966, Laurits Bjerrum | 1970, T. William Lambe |
| 1967, H. Bolton Seed | 1971, John Lowe III |
| 1968, Philip C. Rutledge | |

and the last lecture was delivered by Bramlette McClelland in 1973. Into this illustrious company we now welcome Professor F. E. "Bill" Richart.

Professor Richart has earned his way into such company with more than twenty years of dedicated service to his profession, his students, and to this society. Bill's entry into the Geotechnical Engineering field seems to have been partially an accident. His father was a distinguished Research Professor of Engineering Materials at the University of Illinois and Bill's undergraduate studies were in Mechanical Engineering and Aeronautical Engineering. He received his Ph.D. degree from the University of Illinois in Engineering Mechanics and Structural Engineering. In keeping with his outstanding academic performance, he was offered, and accepted, an appointment as an Assistant Professor at Harvard University where he worked with the renowned Professor Westergaard. When Professor Westergaard passed away shortly thereafter, Professor Richart was offered an opportunity to work in soil mechanics with Professor Terzaghi. Indeed, Professor Richart's first technical publication in Geotechnical Engineering was the paper "Stresses in Rock about Cavities," which was co-authored with Professor Terzaghi.

After spending four years at Harvard, Professor Richart accepted an Associate Professorship in Civil Engineering at the University of Florida, there to develop his own research program and, as rumor has it, to play golf twelve months out of the year. Bill lettered for three years in golf at the University of Illinois and remains an outstanding golfer as well as being a renowned engineer.

A continuation of his interests in Engineering Mechanics led to publication of a paper on the "Photoelastic Analogy for Non-Homogeneous Foundations," for which Bill was co-winner of his first Thomas A. Middlebrooks Award from the Soil Mechanics and Foundations Division of the American Society of Civil Engineers. Life in Florida inevitably led to an interest in coastal soils problems and three years after receiving his first Middlebrooks Award, Bill won another one, this time for a paper entitled "Analyses for Sheet-Pile Retaining Walls." During the summers of 1953, 1954, 1955, and 1957, Professor Richart developed his knowledge of practical aspects of soil mechanics and foundation engineering through employment with the firm of Moran, Proctor, Meuser, and Rutledge in New York City. A continuation of his Moran-Proctor work while back at the University of Florida led to several additional publications, one of which, entitled "Review of Theories for Sand Drains," led to a third Thomas A. Middlebrooks Award.

Professor Richart began his work in the field that has earned him worldwide acclaim, soil dynamics and foundation vibrations, in 1950, at the suggestion of Professor Terzaghi. His first of many papers on this topic, entitled "Foundation Vibrations," not only established his reputation in this emerging field but also earned him the Wellington Prize of the American Society of Civil Engineers in 1963. In the following years, he authored, or coauthored, numerous publications in soil dynamics and foundation vibrations, including a book that has become a standard throughout the Western World. One of these papers, entitled "Dynamic Response of Footings to Vertical Loading," made him a co-winner of a fourth Thomas A. Middlebrooks Award. His work also led to consulting assignments with the Waterways Experiment Station of the Corps of Engineers, the Air Force Special Weapons Laboratory, and NASA, and to numerous consulting assignments with industry.

During these years he continued his academic pursuits as well. In 1962 he returned to the University of Michigan as Professor and Chairman of the Department of Civil Engineering, a position he held for seven years. Bill's dedication to education and to his students is well known in our profession and is a model for the rest of us to emulate. His students have assumed prominent positions in the profession and have received numerous honors. Two have won the Norman Medal of the American Society of Civil Engineers and one is presently a member of the Executive Committee of the Geotechnical Engineering Division.

Professor Richart has been active in technical societies as well as in engineering and education. Besides serving on numerous committees in the Soil Mechanics and Foundations Division, he chaired the Division Committee on Session Programs, and was chairman of both the U.S. National Committee of the International Society of Soil Mechanics and Foundation Engineering, and the Executive Committee of our Division.

In addition to his honors from our Society, Professor Richart has received an honorary doctorate from the University of Florida and in 1969 he received the ultimate engineering honor of being elected to the National Academy of Engineering.

Like the man for whom this lecture was named, our lecturer today is truly a distinguished engineer, researcher, and educator. It is an honor and a privilege for me to introduce to you Professor F. E. "Bill" Richart, of the University of Michigan, who will deliver the Tenth Terzaghi Lecture on "Some Effects of Dynamic Soil Properties on Soil-Structure Interaction."

# SOME EFFECTS OF DYNAMIC SOIL PROPERTIES ON SOIL-STRUCTURE INTERACTION

### By F. E. Richart, Jr.,[1] F. ASCE

As the name implies, the Terzaghi Lectures were established to continue recognition of the great achievements in the field of soil mechanics by Dr. Karl Terzaghi. His influence extended into the studies of dynamic behavior of soils and foundations through his stimulating ideas and encouragement of younger colleagues. It was in the fall of 1951 that Professor Terzaghi said to me, "You should study vibrations of machine foundations—there is a need for knowledge in that field." Of course, since I was a junior assistant professor at Harvard then, I made a mental salute and began to study this problem—and have continued to this day. Fortunately, in the fall of 1951 a doctoral student, T. Y. Sung, was casting about for a dissertation topic, so Professor Terzaghi and I directed him toward analytical studies of foundations on the elastic half space. The study produced an excellent dissertation. During the intervening 23 years I have been most fortunate in finding a continuous string of brilliant and energetic students and colleagues who have contributed to studies of vibrations of foundations and special problems of soil dynamics. This lecture represents a progress report covering some studies we consider to be significant in the rapidly developing field of soil dynamics.

### INTRODUCTION

Dynamic loading may produce a wide range of deformations of soils. At the top end of the scale are nuclear explosions and blast effects on soils—topics that will not be treated in this lecture. In the intermediate range, soil deformations vary from small amplitude, nearly elastic, to plastic following earthquake, water wave, or severe machine-developed forces. Small amplitude deformations of soils are developed adjacent to foundations designed to sustain many stress repetitions without permanent settlements.

This lecture treats the influence of nonlinear soil behavior on the dynamic response of soil masses and the interaction between soils and structures. Linear elastic behavior is considered to be the reference conditions and variations introduced by nonlinear inelastic soil response are presented.

Seismic methods may be used to establish wave velocities and moduli corresponding to the "elastic" stress-strain conditions of soils in the field.

---

Note.—Discussion open until May 1, 1976. To extend the closing date one month, a written request must be filed with the Editor of Technical Publications, ASCE. This paper is part of the copyrighted Journal of the Geotechnical Engineering Division, Proceedings of the American Society of Civil Engineers, Vol. 101, No. GT12, December, 1975. Manuscript was submitted for review for possible publication on February 26, 1975.

[1] Prof. of Civ. Engrg., Univ. of Michigan, Ann Arbor, Mich.

Laboratory procedures for evaluating dynamic soil properties provide a means for checking the seismic field values at low amplitudes strains, and for extrapolating the low amplitude effects to probable field operating conditions. In this paper, special attention is directed to the effects of strain amplitude, numbers of repetitions at various strain levels, and duration of loading. These topics come under the general heading of *stress history effects*.

After determining the empirical stress-strain relations for a particular soil by laboratory and field tests, these relations can be approximated by analytical expressions. For the shearing stress-shearing strain curves described in this study, it was found that a hyperbolic or Ramberg-Osgood type curve could adequately represent the test data. Then the Ramberg-Osgood curves are introduced into the analytical method to simulate the influence of nonlinear soil behavior on the dynamic response of soil masses. Solutions by the method of characteristics include these nonlinear soil effects in studies of earthquake wave transmission through layered soils. Dynamic pore pressures have been evaluated also through application of the method of characteristics. Finally some examination of model tests on dynamically loaded foundations considers results of conventional tests and some preliminary results obtained by holographic interferometry.

## DYNAMIC STRESS-STRAIN RELATIONS FOR SOILS

**Constitutive Equations.**—Several stress-strain relationships may be developed for a given soil, depending upon the types of loading and restraining systems applied. It is sometimes convenient to express the general stress-strain relation in the form of a constitutive equation (Jackson, 1968, 1969)

$$\sigma_{ij} = Ke\delta_{ij} + 2G\left(\epsilon_{ij} - \frac{1}{3} e\delta_{ij}\right) \quad \ldots \ldots \ldots \ldots \ldots \ldots \ldots \ldots (1)$$

and a yield condition

$$\sqrt{J_2''} = f(\bar{\sigma}_o) \quad \ldots \ldots \ldots \ldots \ldots \ldots \ldots \ldots \ldots \ldots \ldots \ldots \ldots \ldots \ldots (2)$$

In Eq. 1, $\sigma_{ij}$ = total stress tensor; $\epsilon_{ij}$ = total strain tensor; $K$ = modulus of volume compressibility, or bulk modulus; $e = \epsilon_x + \epsilon_y + \epsilon_z$ = cubic dilatation, or volumetric strain; $G$ = shear modulus; and $\delta_{ij}$ = Kronecker delta function ($\delta_{ij} = 1$ when $i = j$, $\delta_{ij} = 0$ when $i \neq j$). In Eq. 2, $J_2''$ = second invariant of the stress deviation; and $\bar{\sigma}_o$ = average effective normal stress, or octahedral normal stress. Eq. 1 describes the stress that is developed by change in volume and change in shape, and Eq. 2 describes a yield condition (see Newmark, 1960 for an analysis of $J_2''$).

Constitutive equations similar to Eq. 1 are often used to describe soil properties for analyses of blast loadings on soils or soil-supported structures. For these conditions the soil is usually considered to deform in one-dimensional compression: to be compressed in the vertical direction only with no lateral deformation. Soils tested in one-dimensional compression (confined compression) exhibit a "strain hardening" behavior (Fig. 1, curve A) for stresses above about 200 psi (1,380 kN/m²). Below this stress level the stress-strain curve may exhibit strain hardening or "strain softening: (Fig. 1, curve C) depending on the type of soil, its initial relative density or degree of compaction, and the degree of saturation. In a summary of existing confined compression test data, Hadala (1973) has noted that nonlinearity of the strain-softening type is developed in many soils at low stress levels. For the tests reviewed by Hadala, a "low stress level" was about 50 psi (345 kN/m²) or less.

Because the primary purpose of this lecture is to examine the influence of the shearing stress-shearing strain relations (which are strain-softening) on the

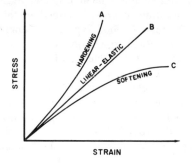

FIG. 1.—Types of Stress-Strain Curves

FIG. 2.—Basic Parameters for Hyperbolic Shearing Stress-Shearing Strain Curves

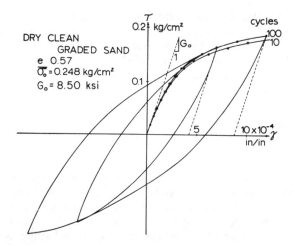

FIG. 3.—Stress-Strain Loop End Points at 10 and 100 Cycles of Loading (Hardin and Drnevich, 1972b)

dynamic behavior of soils and soil-supported structures, no further consideration will be given to the strain-hardening confined compression behavior. Where confined compression deformations are to be included in analytical studies, the stress-strain relations will be assumed as linearly elastic.

For shearing stress-strain conditions, Eq. 1 reduces to

$$\sigma_{ij} = 2G\epsilon_{ij} \quad \ldots \ldots \ldots \ldots \ldots \ldots \ldots \ldots \ldots \ldots \ldots \ldots \quad (3a)$$

or expressed for the $x$-$z$ plane in Cartesian coordinates:

$$\tau_{xz} = G\gamma_{xz} \quad \quad (3b)$$

Thus the relationship between shearing stress and shearing strain is governed by the shear modulus, $G$. Shearing stress-shearing strain relations are of the strain softening type.

**Laboratory Methods for Evaluating Shear Modulus and Damping.**—Laboratory devices may provide direct information about the shear modulus of soils by measuring the applied force and deformation developed in a sample. The shearing stress-strain curve thus developed establishes the initial, tangent, and secant values for the shear modulus. Generally, static tests provide this type of information. Dynamic tests provide an indirect evaluation of the shear modulus, $G$, through measurements of the shear wave velocity, $v_s$, developed in the sample and the equation

$$G = \rho v_s^2 \quad \quad (4)$$

in which $\rho$ = the mass density of the soil, defined as the unit weight of the soil divided by the acceleration of gravity. Thus, when $v_s$ is established by a dynamic test, the value of the corresponding shear modulus is easily determined.

*Resonant Column Tests.*—The resonant column method for determining wave propagation velocities in soil samples has been available for more than three decades, but major developments have occurred in the past fifteen years (see Iida, 1938, 1940, Shannon, Yamane, and Dietrich, 1959, Wilson and Dietrich, 1960, Hardin and Richart, 1963, Hall and Richart, 1963, Hardin and Black, 1966, 1968, Stevens, 1966 Drnevich, Hall and Richart, 1967, Humphries and Wahls, 1968, Afifi, 1970 and Anderson, 1974).

In the resonant column device, a cylindrical column of soil is contained within a rubber membrane, placed in a triaxial cell, and set into motion either in the longitudinal or torsional mode of vibration. The frequency of the input vibration is changed until the first mode resonant condition is determined. This resonant frequency, the geometry of the sample, and its conditions of end restraint provide the necessary information to calculate the velocity of wave propagation in the soil under the given testing conditions.

In the resonant column testing devices, an electromagnetic drive and pickup system provides excitation and response measurements for the soil column. The excitation and pickup mechanisms are arranged differently in the several types of equipment. The design by S. D. Wilson (see Wilson and Dietrich, 1960, Stevens, 1966) provides excitation in either torsional vibration or longitudinal vibration at the base of the sample and measurement of motion at the top of the sample (and at the base in the larger models). Thus the top of the sample is a free end and the base of the sample undergoes small motions, with the node point in the sample very close to the base. Hardin (1961) built a resonant column device in which the sample vibrated as a free-free rod. A separate driving device was needed for longitudinal and torsional motions. In Hall's (1962) device the base of the column was fixed and the excitation and measuring equipment was attached to the free upper end of the column. Hardin (see Hardin and Music, 1965) designed a special torsional vibration exciter which can be applied to standard-size triaxial specimens and which fits into standard triaxial equipment after slight modifications. This Hardin machine is now (1974) included as a part of the standard testing equipment at more than 22 different universities, consulting firms, and government laboratories in the United States, Canada, and Japan. The resonant column devices noted previously all provide for testing solid cylindrical samples of soil.

The strain developed in a solid cylinder under torsional stresses varies from zero at the center to a maximum at the periphery. However, in a hollow cylinder the shearing stresses on any transverse section do not differ significantly from an "average shearing stress." Drnevich (1967) developed a special torsional resonant column device which tested hollow cylindrical samples of soil 4 cm I.D., and either 1 cm or 0.5 cm wall thickness. The original device required specimens 30 cm long but recent modifications of this equipment accommodate shorter samples. Further details on resonant column equipment are given by Richart, Hall, and Woods (1970).

*Repeated Shear Tests.*—Hardin (1965) developed a large-amplitude low frequency ($f = 1/12$ Hz) torsional testing device which accommodated hollow samples of sand 9 in. (23 cm) long, 5 in. (13 cm) O.D., and 4 in. (10 cm) I.D. Hollow samples 2 in. (5.08 cm) long of undisturbed cohesive soils were prepared by trimming a 1.4-in. (3.56-cm) diam hole in the center of a 3-in. (7.62-cm) Shelby tube sample. The dead weight loading system applied a pure torque to the top end of the sample through a system of strings and levers. An improved model of this low-frequency high-amplitude device was developed by Hardin (1971) which incorporated electromagnetic loading. Approximately 60 kg-cm (589 N-cm) of torque could be applied by this system to samples of soils of dimensions noted previously. Both devices were connected to an $x$-$y$ plotter to record the shape of the shearing stress-shearing strain loop developed for reversal of each strain amplitude. With the improved model, shearing stress-strain loops were developed for shearing strain amplitudes as low as $10^{-5}$ and for strain amplitudes corresponding to 80% of the yield stress in the sample.

Yoshimi and Oh-Oka (1973) constructed a ring torsion device for testing samples 24 cm I.D. 2.4 cm wall thickness, and height varying from 2.0 cm. at the inside diameter to 2.4 cm at the outside diameter. The sample container consisted of stacks of aluminum or stainless steel disks with a dry lubricant on the surface of each disk. An electrodynamic vibration exciter provided either sinusoidal (0 Hz–80 Hz) or programmed input for the cyclic shear tests. Additional driving and restraining systems permitted either controlled shearing strain or controlled shearing stress tests. With this device the strength and deformation of soils in simple shear and plane strain could be evaluated for various loading conditions. Data were obtained for the shear modulus of a sand sample for shearing strains varying from about $3 \times 10^{-5}$ to $1 \times 10^{-2}$.

Simple shear testing devices were developed to evaluate shearing strength and volume changes during loading. The simple shear device designed and manufactured by the Norwegian Geotechnical Institute was modified for use in cyclic shear tests to give shearing stress-shearing strain data also. Descriptions of the equipment as well as significant test data were given by Silver and Seed (1971) and by Pyke (1973). In this device a cylindrical sample of soil 8 cm in diameter and 2 cm high is constrained within a wire-reinforced rubber membrane. The sample base is mounted on a sliding table and the cap is fixed to a cross head which is restrained from horizontal motion but is permitted to move vertically. Vertical loading of the cross head provides vertical stress on the sample. A lever and cam arrangement drives the sliding table horizontally to produce cyclic shearing strain in the sample ranging from 0.01%–0.5%.

**Effect of Test Variables on Shear Modulus and Damping.**—The shear modulus of soil is influenced by a number of quantities, which may be described as a functional relationship (Hardin and Black, 1968, Richart, Hall, and Woods, 1970).

$$G = f(\bar{\sigma}_o, e, H, S, \tau_o, C, A, f, t, \theta, T) \quad \quad \quad \quad \quad \quad \quad \quad \quad \quad \quad (5)$$

in which $\bar{\sigma}_o$ = average effective confining pressure; $e$ = void ratio; $H$ = ambient stress history and vibration history; $S$ = degree of saturation; $\tau_o$ = octahedral shearing stress; $C$ = grain characteristics; $A$ = amplitude of strain; $f$ = frequency of vibration; $t$ = secondary effects that are function of time and magnitude of stress increment; $\theta$ = soil structure; and $T$ = temperature, including effects of freezing. Of course, several of these quantities may be related (for example, $e$, $C$, and $\theta$).

Laboratory tests on clean sands at low shearing strain amplitudes (less than about $10^{-5}$) have demonstrated that the shear modulus is primarily a function of $e$ and $\bar{\sigma}_o$. Larger shearing strain amplitudes cause a reduction in the shear modulus. However, it has been found that grain size, grain shape, gradation, degree of saturation, and frequency of vibration introduced insignificant effects on values of the shear moduli of sands.

**Effects of Void Ratio and Confining Pressure.**—Resonant column tests on samples of clean cohesionless soils showed (Hardin, 1961, Hardin and Richart, 1963) that for shearing strain amplitudes less than about $10^{-5}$, the shear wave velocity depended essentially on the void ratio, $e$, and the average effective confining pressure, $\bar{\sigma}_o$. The test results were well represented by empirical equations

$$v_s = (170 - 78.2e) \, (\bar{\sigma}_o)^{0.25} \quad \quad \quad \quad \quad \quad \quad \quad \quad \quad (6a)$$

(fps) \quad \quad \quad \quad (lb/sq ft)

for *round-grained soils* ($e < 0.80$), and by

$$v_s = (159 - 53.5e) \, (\bar{\sigma}_o)^{0.25} \quad \quad \quad \quad \quad \quad \quad \quad \quad \quad (6b)$$

(fps) \quad \quad \quad \quad (lb/sq ft)

for *angular grained soils*. Values of shear modulus are then obtained by introducing $v_s$ and the mass density, $\rho$, into Eq. 4.

Hardin and Black (1968) found that Eq. 6b also gave reasonable results for *normally consolidated clays of low surface activity* at the end of *one-day* of pressure application. Thus Eq. 6b could be applied for a *first estimate* of the shear wave velocity for a cohesive soil, but such factors as duration of loading and stress history should be considered carefully.

**Effects of Shearing Strain Amplitude.**—Hardin and Drnevich (1972a) presented the results of an extensive series of reversed torsional tests on cohesionless and cohesive soils. With their test data and that information already available in the literature, they established that a modified hyperbolic curve satisfactorily represented the shearing stress-shearing strain relations throughout the range of strain amplitudes up to failure. Fig. 2 shows the factors that control the basic hyperbolic shearing stress-shearing strain curve. At zero shearing strain the tangent to the curve establishes the maximum value of shear modulus, $G_o$. The secant modulus corresponding to any point along the curve, for example point (A), is designated as $G$, and the limiting value of shearing stress as obtained in a simple shear test is designated as $\tau_{max}$. The horizontal line at the ordinate of $\tau_{max}$ is the second asymptote to the hyperbolic curve (see Kondner, 1963, for an analysis of hyperbolic stress-strain relations). Both $G_o$ and $\tau_{max}$ can be established by field or laboratory tests, or both.

Reversed torsional tests of soils produced hysteresis loops similar to those shown in Fig. 3. The black dots on the skeleton curves (0-10, 0-100) represent the extremities of a series of hysteresis loops. For each loop, the secant modulus passes through the dot and the origin. This figure illustrates the decrease in secant modulus with increasing shearing strain and the increase in width of the loop describes an increase in damping. Using test results of this type, Hardin and Drnevich (1972b) were able to develop basic relationships for the variation of $G/G_o$ and damping with shearing strain amplitude.

Instead of preparing diagrams of $G/G_o$ and damping as direct functions of the shearing strain, $\gamma$, Hardin and Drnevich (1972b) found it convenient to normalize the strains by use of the ratio $\gamma/\gamma_r$. The "reference strain," $\gamma_r$, represents the strain at the intersection of the initial tangent line, at the slope $G_o$, with the limiting shearing stress, $\tau_{max}$, as shown in Fig. 2. With this normalized representation of strain, the shearing stress-strain data collapsed into relatively narrow bands for cohesionless and cohesive soils. The test data approximated the hyperbolic relation, as indicated in Fig. 4, with the curve for clays falling below the basic hyperbola and that for sands falling above it. A further refinement for presentation of the test data resulted from a distortion of the strain scale to produce a "hyperbolic strain," defined by

$$\gamma_h = \frac{\gamma}{\gamma_r}\left[1 + a\exp\left(-b\frac{\gamma}{\gamma_r}\right)\right] \quad \dots \dots \dots \dots \dots \dots \dots \dots \dots \dots (7)$$

in which $a$ and $b$ are empirical quantities. For example, $a = 0.5$ and $b = 0.16$ when $\gamma_h$ is needed to evaluate the shear modulus versus shearing strain relations for clean dry sands. The $a$ and $b$ terms include consideration of the number of cycles of stress repetitions for damping of dry and saturated sands and for the modulus of saturated sands. As an example, Fig. 3 shows a slight shift of the skeleton curve when the number of stress cycles, $N$, increases from 10 to 100.

After establishing the hyperbolic strain, a single curve was developed which illustrated the relation of the shear modulus ratio, $G/G_o$, and damping ratio $D/D_{max}$ to $\gamma_h$ as shown in Fig. 5. Additional curves were presented by Hardin and Drnevich (1972b) to show the influence of the various test parameters on the relationship between $\gamma_h$ and $\gamma/\gamma_r$.

*Effects of Time of Loading.*—In tests of clean sands Hardin (1961) found that the shear wave velocity, measured in the resonant column device, reached a stable value at 15 min–30 min after pressure increments were applied. However, tests on dry and saturated samples of fine-grained crushed quartz (100% passing No. 200 and 80% passing No. 400 sieve) and dry samples of Novaculite No. 1250 (100% passing No. 400 sieve) showed that the shear wave velocity of these materials continued to change with time after application of a pressure increment. For example, the shear wave velocity increased from 726 fps (221.3 m/s) at 0.1 hr to 792 fps (241.4 m/s) at 63.3 hr. Hall (see Richart, 1961)

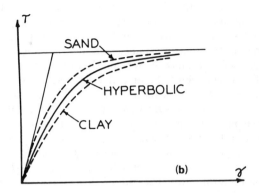

FIG. 4.—Hyperbolic Shearing Stress-Strain Curves for Sand and Clay (Hardin and Drnevich, 1972b)

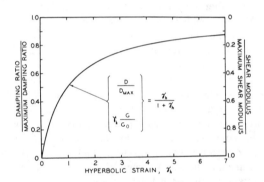

FIG. 5.—Normalized Shear Modulus and Normalized Damping Ratio for All Soils Versus Hyperbolic Strain (Hardin and Drnevich, 1972b)

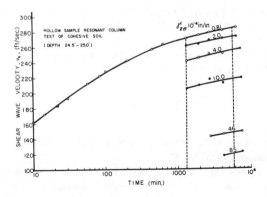

FIG. 6.—Effect of Time and Amplitude on Shear Wave Velocity (Anderson, 1974)

found from resonant column tests of undisturbed samples of Vicksburg loess that $v_s$ increased from 469 fps (143 m/s) after 20 min to 517 fps (157.6 m/s) after 20 hr, following a change of confining pressure from 360 psf (17.2 kN/m²) to 1,320 psf (63.2 kN/m²). Thus, it has been recognized for some time that resonant column test results of fine-grained soils involved *time effects*, but similar tests of coarse-grained soils showed no significant time effects.

To establish the magnitude of the time effect for samples of dry soils, Afifi (1970) ran resonant column tests on a variety of soils in the Hall device. In each test the confining air pressure was applied to the sample and the shear wave velocity was measured periodically by short intervals of vibration of the sample. A total of 3,039 days of testing time was accumulated while studying samples of seven types of air-dry soils ranging from Ottawa (30-50) sand to air-dry kaolinite. From these tests (see Afifi and Woods, 1971, and Afifi and Richart, 1973) he found that the time-rate of shear wave velocity increase was relatively unimportant for dry soils having $D_{50}$ larger than 0.04 mm, but for finer-grained dry soils this increase was significant.

Following the studies on air-dry soils, evaluations of time effects on saturated "undisturbed" and laboratory remolded samples of cohesive soils have been conducted (see Afifi and Richart, 1973, Stokoe and Richart, 1973, and Anderson, 1974). Fig. 6 shows the shear wave velocity-time relationship often observed while the first increment of pressure was applied to a sample of cohesive soil. Initially the sample undergoes primary consolidation which is indicated in the resonant column test by a relatively rapid increase in shear wave velocity with time. After 100 min–1,000 min of continuous pressure application, the increase of $v_s$ settles down to a rate represented by a straight line on the semi-log plot of shear wave velocity versus log time. This linear increase of $v_s$ with log time has been considered to represent a "secondary" time effect. In Fig. 6 this straight line increase continued to 10,000 min, at which time the test was terminated. Earlier tests (see Afifi and Woods, 1971) have shown this straight line relationship to continue for at least $10^6$ min.

The straight line increase of shear wave velocity with logarithm of time in this secondary zone can be designated as "$\Delta v_s$ per log time cycle," or simply as $\Delta v_s$. Because $\Delta v_s$ is dependent on the value of $v_s$, this rate of increase has been normalized by expressing it as a percentage of the shear wave velocity at 1,000 min duration of testing ($v_{s_{1,000}}$). Empirical expressions were developed from tests on nine cohesive soils (see Anderson, 1974, or Anderson and Woods, 1975) to relate the ratio $\Delta v_s / v_{s_{1,000}}$ to various physical parameters of the soils. The best fit of data was obtained by

$$\left(\frac{\Delta v_s}{v_{s_{1,000}}}\right) \text{(percentage)} = \exp(2.0 - 0.46\, S_u + 0.25\, e_o) \quad \ldots \ldots \ldots \ldots (8)$$

in which $S_u$, in kilograms per square centimeter, is the undrained shearing strength; and $e_o$ is the original void ratio.

Fig. 6 also shows the time-rate of secondary shear wave velocity increase corresponding to increased shearing strain amplitudes. The shearing strain amplitude of $0.8 \times 10^{-4}$ was used as the reference or control amplitude, because at that level insignificant strain history effects were introduced. After the low amplitude wave velocity was definitely into the secondary time range, the strain

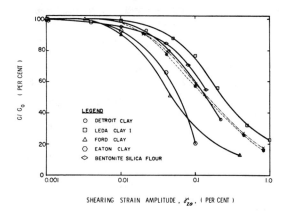

FIG. 7.—Effect of Shearing Strain Amplitude on Shear Modulus (Anderson, 1974)

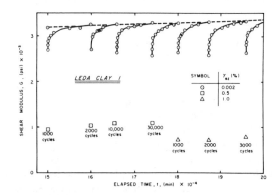

FIG. 8.—Time-Dependent Regain in Low Amplitude Shear Modulus after High Amplitude Cycling of Leda Clay I (Anderson, 1974)

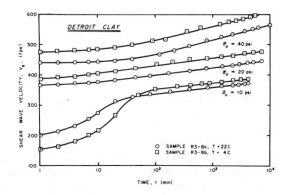

FIG. 9.—Effects of Temperature and Time on Shear Wave Velocity (Anderson and Richart, 1974)

amplitude was increased to $2.0 \times 10^{-4}$ and held at that level only long enough to take a reading, then again reduced to $0.8 \times 10^{-4}$. This process was repeated by stepping up to the next higher strain amplitude and back to the low amplitude until all the chosen amplitudes had been applied. Amplitudes of $2.0 \times 10^{-4}$, $4.0 \times 10^{-4}$, and $10.0 \times 10^{-4}$ were applied at two time intervals, then higher amplitudes of $46 \times 10^{-4}$ and $85 \times 10^{-4}$ were added to the pattern. Fig. 6 shows that the secondary rate of shear wave velocity increase with log time was essentially constant for all strain levels.

To determine the effect of shearing strain amplitude on $v_s$, it was necessary to compare values at a common time within the secondary range. Values of $v_s$ were taken from Fig. 6 at times of 1 day (1,440 min) and at 4 days (5,760 min) for each value of shearing strain. Then the shear modulus corresponding to each strain level was calculated and expressed in terms of the low amplitude shear modulus, $G_o$. Fig. 7 shows the two curves for the $G/G_o$ versus $\gamma_{z\theta}$ relationship for the 1-day and 4-day test data as dashed lines. Note that these curves are almost identical. Fig. 7 also includes test curves for samples of five other cohesive soils for which the data were taken at the end of a 1-day test.

It is important to note from the curves in Fig. 7 how rapidly the secant shear modulus, $G$, decreases as the shearing strain amplitude increases. All curves would indicate that $G$ is 20% or less of the small amplitude value, $G_o$, at a shearing strain of 1%, and even at a shearing strain of 0.1% $G$ may be reduced to 20%–40% of $G_o$. Therefore, the shearing strain amplitude developed in the soil during dynamic soil-structure interaction has a *very important* influence on the dynamic response of the system.

*Effects of Stress History.*—The term stress-history relates to the time-dependent pattern of static and dynamic stresses which may be applied to soils. In resonant column tests it is possible to change the confining pressure, $\bar{\sigma}_o$, on the sample and to measure the changes in low amplitude wave velocities. Generally, little effect of previous stress history has been observed in tests on sands or cohesionless soils having mean grain diameter greater than about 0.04 mm. Finer grained soils, even those tested in the dry condition (Hardin and Richart, 1963, Afifi and Richart, 1973), have shown an influence of previous higher confining pressures. However, in each case, the increase in wave velocity was associated with the reduction in void ratio caused by the high pressure. Thus stress-history effects caused by variations in confining pressure are related principally to void ratio changes, and in the case of saturated cohesive soils the time required for consolidation becomes an important factor.

Stress-history effects caused by changes in the level of repeated dynamic stresses are of a different nature. These depend upon the type of soil, the static and dynamic stress levels, the number of cycles of repeated dynamic stresses, and length and frequency of the no-load or "rest" periods between loadings. Torsional resonant column tests of dry sands have demonstrated a lower limit of shearing strain amplitude effect at about $10^{-4}$. Repeated vibrations at this strain amplitude or lower introduce no strain history effects. Repeated straining at a given higher amplitude causes the dynamic modulus and damping to increase (see Fig. 3), and may also increase the wave velocity in subsequent low amplitude vibrations (see Drnevich, 1967, and Drnevich and Richart, 1970). These beneficial effects of prestraining may be caused by a "wearing in" of particle contacts during repeated motions, because some of these effects may be destroyed by a few cycles of shearing strain large enough to cause particle rearrangement.

Saturated cohesive soils show different patterns of dynamic response after prestraining than do sands. There is again a "threshold" shearing strain of about $10^{-5}$ for stress-history effects. However, repeated cycling at a given shearing strain amplitude above this level will cause a slight, but continued, decrease in shear wave velocity. When the strain amplitude is returned to the low value after prestraining at a higher amplitude, the low amplitude shear wave velocity is lower than it was originally, depending upon the magnitude and number of repetitions of the high amplitude straining. However, if the sample is allowed to "rest" under the static confining pressure, the low amplitude shear wave velocity will gradually increase with time to its original value. This thixotropic regain of low amplitude wave velocity (in terms of shear modulus) is shown in Fig. 8 for Leda clay I (Anderson, 1974). In this test the low amplitude shearing strain was 0.002%. After the test had proceeded for $15 \times 10^4$ min, 1,000 repetitions of a shearing strain at a high amplitude of 0.5% were applied to the sample, and then the regain in the low amplitude shear modulus was evaluated as a function of time. The low amplitude modulus returned to the original value in about $10^4$ min (about 1 week). Subsequent overstrainings produced similar effects, as shown in Fig. 8. The "rest" time required to regain the original low amplitude modulus depends on the type of cohesive soil and the level and number of repetitions of overstrain. In the tests reported by Anderson (1974) this recovery time varied from a few hours to about 1 week. However, it is very important to note that this time-dependent recovery of the dynamic shear modulus *does occur*. It could be anticipated that during an earthquake, for example, that soil-structure interactions could cause large shearing strains in the soil, with a temporary reduction of the shear modulus (and ability to transmit large stresses). However after the earthquake motion terminated, it could be anticipated that the original dynamic response of the system would be restored.

*Effect of Temperature.*—Because the time effects noted in resonant column tests of cohesive soils exhibited what might be considered a secondary time effect and perhaps some evidence of secondary compression, the possibility of temperature effects was investigated. It was thought that the dynamic behavior might be appreciably different for cohesive soils in situ at perhaps 4° C and in the laboratory at about 22° C. Consequently a series of tests was run with a temperature-controlled chamber surrounding the resonant column sample (see Anderson and Richart, 1974).

Fig. 9 shows the shear wave velocity versus time data for tests of Detroit clay at 4° C and 22° C, and at three levels of confining pressure. During the first pressure increment (10 psi) the typical primary and secondary increases developed, with the primary portion being completed in a shorter time at 22° C, as might be anticipated. Note that in the secondary zone the numerical value of shear wave velocity was larger for the test at 4° C, at each time, than for the test at 22° C. Again, this might be anticipated. However, the curves in Fig. 9 emphasize the importance of evaluating test data *after* the *primary time effect is completed* and the tests are well into the secondary zone. If the tests had been continued for only 20 min and wave velocities determined then, the conclusion would be that the shear wave velocities are higher in Detroit clay at 22° C than they are at 4° C. Tests of seven cohesive soils (total testing time of about 1 yr) showed that the shear wave velocity at 4° C was from 0%–12.5% higher than when the samples were tested at 22° C (see Fig. 10). Consequently, the effect of testing temperature is relatively unimportant, but the wave velocity will be slightly higher at lower temperatures.

In the constant temperature tests, the temperature did not vary more than a maximum of 2° C over the time of the test (about 2 weeks for each test shown in Fig. 9). However, temperature fluctuations caused immediate variations

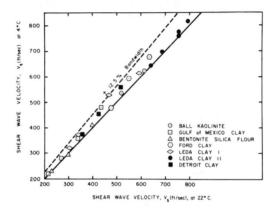

FIG. 10.—Relationship Between $v_{s1,000}$ at 4° C and $v_{s1,000}$ at 22° C (Anderson and Richart, 1974)

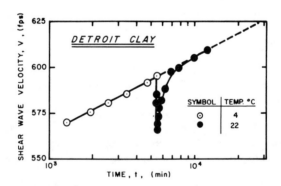

FIG. 11.—Effect of Rapid Temperature Change on $v_s$ (Anderson and Richart, 1974)

in the shear wave velocity, evidently because of pore pressure changes (and change in $\bar{\sigma}_o$) within the sample. This effect was discovered by accident following failure of the air-conditioning system one summer day. Subsequently a series of controlled temperature-change tests were run to evaluate this effect. Fig. 11 shows the influence of temperature change for a sample of Detroit clay. At the end of 4 days' testing time at 4° C, the temperature in the water bath was raised to 22° C in about 10 min. As the sample warmed, a rapid decrease in $v_s$ occurred, as noted in Fig. 11. Simultaneously, the axial length of the sample increased, and water began to flow out of the drainage line. As drainage occurred and the sample warmed to 22° C, the value of $v_s$ returned to its original projected value, as shown by the black circles in Fig. 11.

These studies of temperature effects emphasize the importance of maintaining a *constant* temperature condition throughout resonant column tests, but demonstrate that the level of this chosen temperature is of less significance.

**Field Methods for Evaluating Shear Wave Velocity.**—Seismic methods have been developed for measuring the shear wave velocity of soils in-situ. These methods develop small shearing strains in the soil and the values of $v_s$ measured permit calculation of the low-amplitude shear modulus, $G_o$, through application of Eq. 4.

Preliminary surveys covering large areas can be carried out by refraction surveys (Schwarz and Musser, 1972) or steady-state vibrations (Fry, 1963, Ballard,

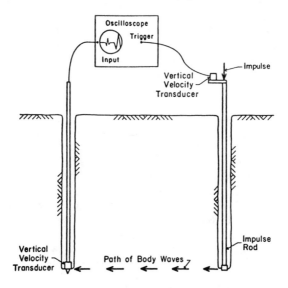

FIG. 12.—Schematic of Cross-Hole Seismic Survey Technique

1964). Both procedures involve excitation and measurements on the ground surface.

The down-hole method involves an energy source at the surface with sensors located at various depths down a borehole. This method averages the effects of layering and may be less effective if it is necessary to use a cased hole. Detailed information can be obtained by transmitting shear waves horizontally between two vertical bore holes spaced at 5 ft–25 ft (1.53 m–7.63 m) (Stokoe and Woods, 1972) to establish the shear wave velocity. This method, called the "cross-hole method," is effective and relatively inexpensive. The refraction, steady-state vibration, and down-hole methods are standard seismic procedures, but the recent developments and applications of the cross-hole method deserve a few comments.

The general arrangement of equipment for the cross-hole test is shown in Fig. 12. Two bore holes are required, one for the impulse and one for the sensor. As shown in Fig. 12, the impulse rod is struck at the top end and the impulse travels down the rod, then is transmitted to the soil at the bottom. It has been found that an open-ended tube or standard sampling spoon transmits shear energy to the soil more efficiently than does a flat-ended rod. The shear waves are transmitted horizontally to the vertical velocity transducer located in the second bore hole. The time needed for the wave to traverse the known distance between the bore holes is determined from the trace on the oscilloscope. As shown in Fig. 12, the transducer at the top of the impulse rod acts as a trigger to start the oscilloscope trace. Thus the time from initial triggering to arrival of the shear wave at the pick-up includes the time required for the wave energy to travel the length of the impulse rod. This time is calculated for each installation and a correction is then made to the recorded data. If the impulse rod is long [i.e., 200 ft (61 m)] then it has been found expedient to locate the impulse transducer at the bottom of the impulse rod (Hall, 1973). However, for the usual shallow surveys the correction procedure is quite adequate. One improvement to the triggering procedure was included in 1973 (see Stokoe and Woods, 1973) by replacing the impulse transducer by a capacitive electrical circuit which is activated by the contact of the hammer on the impulse rod. This system develops the full impulse signal within a few microseconds.

Fig. 13($a$) shows a typical travel time record for a cross-hole test. Traces from repeated impacts are recorded on a storage oscilloscope, then photographed with a Polaroid camera. The first arrival noted in Fig. 13($a$) is the compression wave, with $v_p$ = 800 fps (244 m/s). The second arrival is the shear wave, which for this test was found to have a velocity of $v_s$ = 416 fps (127 m/s). The shear wave arrival can be identified more precisely if the direction of impact is reversed. One trace is recorded for a downward impact, then a second trace is produced by an upward impact on the impulse rod. Ideally, the two traces should diverge at the point of arrival of the shear wave, as shown in Fig. 13($b$) (Schwarz and Musser, 1971). Test results from cross-hole tests with a 10-ft (3.05-m) spacing in a Detroit clay are shown in Fig. 13($c$) (Anderson and Woods, 1975). This procedure is recommended for enhancing the information from cross-hole tests.

For shallow surveys by the cross-hole method the distance between the bottom of the impulse and pick-up rods can be determined by projecting the surface measurements of distance between the rods and slope. For depths greater than perhaps 50 ft, some positive record obtained by a Slope Indicator or similar device is required to record the wanderings of the bore holes. For deep evaluations by the cross-hole method, special impact and recording devices (Lundgren, 1973) have been developed which may be lowered in cased boreholes, placed in contact with the hole periphery by expanding jacks, and readings taken throughout the lengths of a pair of bore holes.

**Evaluation of Resonant Column and Cross-Hole Methods for Determining Shear Wave Velocity in Soils.**—Either or both methods may be employed to determine the shear wave velocity of soils. It is useful to review briefly the advantages and disadvantages of each method.

*Resonant Column Test.*—The principle advantage of the resonant column test is its flexibility. The testing environment may be changed to evaluate the effects

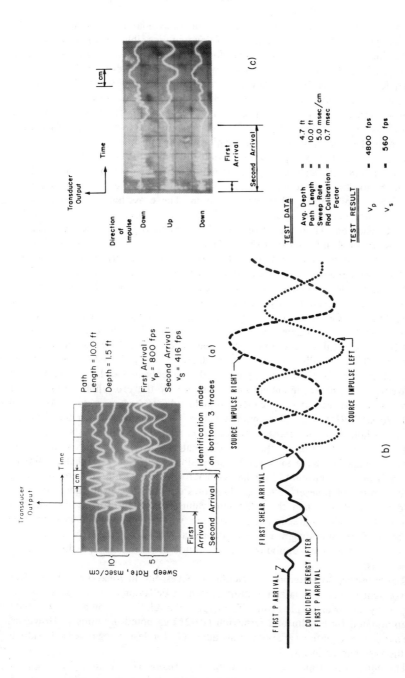

FIG. 13.—Time-Amplitude Traces from Cross-Hole Tests: (a) Under Laboratory Conditions; (b) Effect of Reversing Impact Direction; (c) Typical Set of Traces from Field Test

of changes in confining pressures, large shearing strain amplitudes, stress-histories, and duration of loading.

Disadvantages of the method include the unavoidable disturbances of the soil sample caused by unloading and reloading, possible disturbances caused by handling, and the problem of reproducing the in-situ confining pressures.

*Cross-Hole Method.*—The principle advantage of the cross-hole method is that it determines the shear wave velocity in undisturbed soil. It permits evaluation of $v_s$ in layered soils, and is relatively inexpensive and easy to use.

Disadvantages of the cross-hole method are primarily that the shear wave velocity is determined only for small shearing strains and for the stress conditions existing at the time of the test. Modifications of the confining pressure conditions by construction, for example, would modify the shear wave velocity.

*Comparison.*—From this brief examination it is evident that both methods should be used in connection with any significant effort to evaluate the dynamic response of soils or a soil-structure system. The cross-hole method provides the reference value for $v_s$ at low amplitude strains throughout the soil profile being considered. Then the resonant column method permits extrapolation of this low amplitude information to higher strain amplitude conditions, or to different stress or stress-history conditions. The assumption that the resonant column method provides a satisfactory extrapolation technique implies that both methods give comparable results at low amplitude shearing strains. For cohesionless soils comparable results have usually been obtained by both methods when correct values of $\bar{\sigma}_o$ were applied in the laboratory. On the other hand, significant differences have sometimes been found between $v_s$ determined by laboratory and field tests (Cunny and Fry, 1973, Seed and Idriss, 1970, Durgunoglu and Tezcan, 1974). Usually the value of $v_s$ determined by the laboratory method was lower than the field value, particularly for cohesive soils. The difference between $v_s$ by the resonant column tests and by the cross-hole tests can be minimized if the secondary time effect is included (see Anderson, 1974, Anderson and Woods, 1974). Fig. 14 compares values of $v_s$ determined in the field by the cross-hole test with those obtained from resonant column tests from eight different sites. At four of the sites the material was clay, and at the other sites the materials were silty sand, sandy silt, clayey silt, and shale. Samples for each site were brought to the laboratory for resonant column tests, including an evaluation of time effects. Fig. 14 shows the comparison of laboratory and field values of $v_s$ when the 1,000-min value from the resonant column test was used, and when the secondary time effect was included with an extrapolation to 20 yr. From this diagram it is evident that the secondary time effect must be incorporated when applying low amplitude resonant column data to field conditions.

**Ramberg-Osgood Equation for Nonlinear Stress-Strain Relations.**—It is often convenient to approximate the strain-softening behavior of soils (as shown in Fig. 7) by analytical expressions. The hyperbolic relations shown in Fig. 2 have been adopted by Hardin and Drnevich (1972b) as noted previously. However, the hyperbolic equations are not readily adaptable for describing reversed loadings of the type shown in Fig. 3.

The equations proposed by Ramberg and Osgood (1943), have been applied to structural problems by Jennings (1964) and Berg (1965), and to soils by Streeter, Wylie, and Richart (1974). In terms of the shearing stress-shearing strain relations, these equations are

$$\frac{\gamma}{\gamma_y} = \frac{\tau}{\tau_y}\left[1 + \alpha \left|\frac{\tau}{\tau_y}\right|^{R-1}\right] \quad \ldots \ldots \ldots \ldots \ldots \ldots \ldots \ldots \ldots \ldots \ldots (9)$$

for the skeletal curve which develops for initial loading, and by

$$\frac{\gamma - \gamma_1}{\gamma_y} = \frac{\tau - \tau_1}{\tau_y}\left[1 + \alpha \left|\frac{\tau - \tau_1}{2\tau_y}\right|^{R-1}\right] \quad \ldots \ldots \ldots \ldots \ldots \ldots \ldots (10)$$

for the unloading or reloading curves. The term $\tau_y$ represents the "yield" shearing stress, and $\tau_1$, $\gamma_1$ represent the last point of reversal of stress. The factor

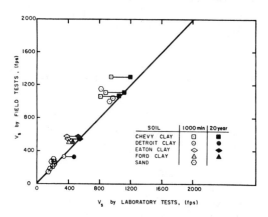

**FIG. 14.—Comparison of Lab and Field Shear Wave Velocities (Anderson and Woods, 1974)**

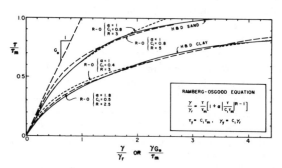

**FIG. 15.—Fit of Ramberg-Osgood Curves to Soil Data**

$\alpha$ can be varied to adjust the position of the curve along the strain axis, and the exponent $R$ controls the curvature. Note that for $R = 1$, Eq. 9 describes a linear relationship between shearing stress and shearing strain. For each chosen value of $\alpha$, the family of loading curves (Eq. 9) obtained by introducing different values of $R$, pass through the points $\tau = 0$, $\gamma = 0$, and $\tau = \tau_y$, $\gamma = (\tau_y/G_o)$ $(1 + \alpha)$. Upon unloading from any point $\tau_1$, $\gamma_1$, the initial slope of the unloading curve is always equal to the low amplitude shear modulus $G_o$.

Fig. 15 shows how the Ramberg-Osgood (R-O) curves can be adjusted to fit soil data. The Hardin-Drnevich curves for sand and for saturated clays at $N = 1,000$ cycles are shown as solid lines, and two R-O curves are shown which approximate each curve. For the Hardin-Drnevich curve representing sand, the factor $C_1$ was chosen as 0.8 (i.e. $\tau_y = 0.8\ \tau_m$). Therefore the two R-O curves for $R = 3$ and $R = 5$, with $\alpha = 1.0$ and $C_1 = 0.8$, intersect at the point $\tau/\tau_m = 0.8$, $\gamma G_o/\tau_m = 0.8(1 + \alpha) = 1.6$. The R-O curve for $R = 3$ fits the H-D (sand) curve quite well up to $\gamma G_o/\tau_m \approx 2.0$, then it diverges. Above values of $\gamma G_o/\tau_m \approx 1.6$ the R-O curve for $R = 5$ fits the H-D (sand) curve well, but the fit is fairly poor for lower values. For the H-D (clay) curve, a reasonable fit was found using $\alpha = 1.8$, $C_1 = 0.5$, and $R = 2.5$. Anderson (1974) found that a R-O curve based on $\alpha = 1.0$, $C_1 = 0.4$, and $R = 3$ fit his empirical data for six clay samples better than did the H-D (clay) curve. This curve is also shown in Fig. 15.

## ANALYSIS OF WAVE PROPAGATION IN SOILS BY METHOD OF CHARACTERISTICS

The preceding portion of this lecture has been devoted to a description of the dynamic shearing stress-shearing strain properties of soils and particularly to the factors related to inelastic behavior. Soil properties are then introduced into some theoretical procedure to evaluate the dynamic behavior of a mass or to determine soil-structure interaction. When soils are treated as elastic materials it may be possible to find an exact solution by the theory of elasticity or to develop an approximate solution by a numerical method such as the finite element method. Viscoelastic behavior may also be treated by exact theories or by the finite element method. However, when problems include permanent plastic deformations of soils, the number of available methods for analysis is limited. This section considers the application of the method of characteristics to a few simple problems in soil dynamics.

The method of characteristics is certainly not a new method for obtaining solutions of dynamic motions of materials. It was proposed by Westergaard (1933) for finding horizontal shears in buildings idealized as shear beams with distinct segments along the height and was found useful if the input ground motion was relatively simple. A description of the method of characteristics applied to one-dimensional wave transmission in soils and in reservoirs was given in the excellent book by Newmark and Rosenblueth (1971), and the method has long been applied to problems of hydrodynamics (Streeter and Wylie, 1967). The application of the method of characteristics to one and two-dimensional problems of wave propagations in soils, including nonlinear soil response, was given by Streeter, Wylie and Richart (1974).

Consider first the one-dimensional propagation of shear waves through an elastic layer of soil. Excitation is provided at the base and the top surface of the layer is unloaded. Fig. 16(a) shows the simple shear deformation of a soil element, and Fig. 16(b) shows the layer and section designations. The element deforms by displacement $u$ in the horizontal $x$ direction, with displacements $w$ in the vertical $z$ direction equal to zero. For this condition the shearing strain, $\gamma_{zx}$ is described by $\partial u/\partial z$, and the particle velocity, $V$, is denoted by $\partial u/\partial t$. For dynamic motions, the equation of equilibrium of the element is

$$\frac{\partial \tau}{\partial z} - \rho \frac{\partial^2 u}{\partial t^2} = \frac{\partial \tau}{\partial z} - \rho \frac{\partial V}{\partial t} = 0 \quad \ldots \ldots \ldots \ldots \ldots \ldots \quad (11)$$

in which $\rho$ = the mass density of the soil.

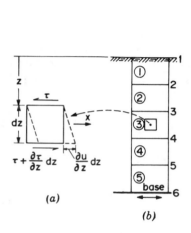

FIG. 16.—One-Dimensional Shear Wave Transmission Model: (a) One-Dimensional Element; (b) Layer and Section Designation

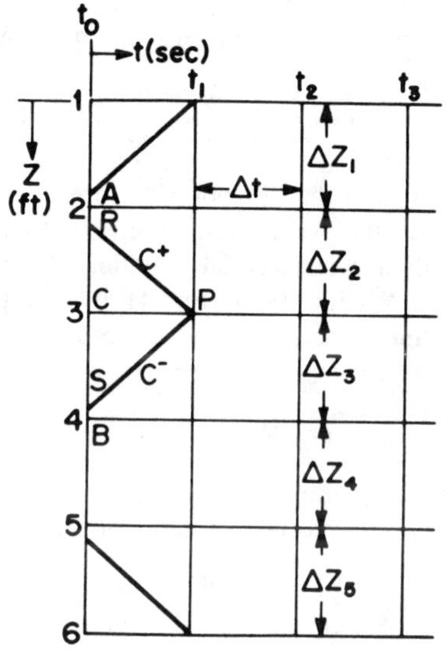

FIG. 17.—$z$-$t$ Diagram for Five Soil Layers

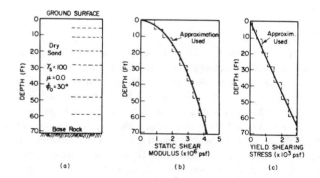

FIG. 18.—Layer of Dry Sand: (a) Eight Subdivisions of Thickness; (b) Shear Modulus Proportional to Square Root of Depth; (c) Yield Shearing Strength Proportional to Depth

The dynamic shearing stress-shearing strain relations for a viscoelastic material can be expressed by (see Kolsky, 1963)

$$\tau = G\gamma + \mu\frac{\partial\gamma}{\partial t} = G\frac{\partial u}{\partial z} + \mu\frac{\partial^2 u}{\partial z \partial t} \quad \ldots \ldots \ldots \ldots \ldots \ldots \ldots \ldots \quad (12)$$

in which $\mu$ = the coefficient of viscosity. If Eq. 12 is differentiated with respect to $t$ it can be represented in terms of particle velocities, $V$. Furthermore, the term involving viscous stresses may be represented by a finite difference approximation. The equation developed from these operations is

$$\frac{\partial\tau}{\partial t} - \left(G + \frac{\mu}{\Delta t}\right)\frac{\partial V}{\partial z} + \frac{\mu}{\Delta t}V^* = 0 \quad \ldots \ldots \ldots \ldots \ldots \ldots \ldots \quad (13)$$

in which the symbol $V^*$ represents the value of $(\partial V/\partial z)$ determined at point C on the $z$-$t$ diagram (Fig. 17). Eqs. 11 and 13 are now transformed into four ordinary differential equations by the method of characteristics (see Streeter and Wylie, 1967). Eq. 11 is multiplied by a factor $\theta$ and added to Eq. 13; terms are grouped to form ordinary derivatives if

$$\theta = \frac{dz}{dt} = \pm\sqrt{\frac{G}{\rho} + \frac{\mu}{\rho\Delta t}} = \pm v_s \quad \ldots \ldots \ldots \ldots \ldots \ldots \ldots \quad (14)$$

The equations are designated $C^+$ when the plus sign is used for $\theta$, and $C^-$ when the negative sign is introduced, to give

$$C^+ \begin{cases} \dfrac{d\tau}{dt} - \rho v_s \dfrac{dV}{dt} + \dfrac{\mu}{\Delta t} V^* = 0 & \ldots \ldots \ldots \ldots \ldots \quad (15) \\[1em] \dfrac{dz}{dt} = v_s & \ldots \ldots \ldots \ldots \ldots \quad (16) \end{cases}$$

$$C^- \begin{cases} \dfrac{d\tau}{dt} + \rho v_s \dfrac{dV}{dt} + \dfrac{\mu}{\Delta t} V^* = 0 & \ldots \ldots \ldots \ldots \ldots \quad (17) \\[1em] \dfrac{dz}{dt} = -v_s & \ldots \ldots \ldots \ldots \ldots \quad (18) \end{cases}$$

Eq. 15 is valid only when Eq. 16 is satisfied, and Eq. 17 is valid only when Eq. 18 is satisfied. The fact that this transformation of Eqs. 11 and 13 has been possible is a result of the hyperbolic character of the equations.

Solutions to the $C^+$ and $C^-$ equations can be readily obtained after they have been expressed in terms of finite differences. A convenient time interval, $\Delta t$, is chosen, and Eq. 14 then establishes the depth interval, $\Delta z$, associated with each small amplitude shear modulus, $G_o$ (i.e., substitute $\Delta t$ for $dt$, $\Delta z$ for $dz$, and $G_o$ for $G$ in Eq. 14. Although $G_o$ usually changes with depth below the surface, it is considered constant within each layer and is calculated at mid-thickness of each $\Delta z$ interval. The shear modulus, $G_o$, represents the "elastic" condition and is evaluated from low amplitude laboratory or field tests. Because the shearing stress-strain relations are strain softening, this determines that $G \leq G_o$ in each layer and that $v_s \Delta t \leq \Delta z$ at all times, a condition required to satisfy stability criteria of the method of characteristics. This insures that the characteristic lines through P intersect the previous time lines at points R and S which lie within the space interval A to B, as shown in Fig. 17.

A solution may be followed from the $z$-$t$ diagram shown in Fig. 17. It is assumed that the shearing stresses, $\tau$, and particle velocities, $V$, are known at time $t_o$, and these values permit calculation of $\tau$ and $V$ at point P through application of the difference equation equivalents of Eqs. 15 and 17. Within the distance interval AC ($\Delta z_2$) the average value of shearing stress, 0.5 ($\tau_A$ + $\tau_C$), is evaluated then the *tangent modulus* ($G = d\tau/d\gamma$) is determined at that stress level by differentiating Eq. 9 (or Eq. 10 if unloading occurs). Then $(v_s)_2$ is found by introducing this value of $G$ into Eq. 14. In a similar manner, $(v_s)_3$ is found for the interval $\Delta z_3$. When $G < G_o$ interpolation formulas are needed to evaluate $V_R$, $V_S$, $\tau_R$, and $\tau_S$ from the values of $V_A$, $V_B$, $V_C$, $\tau_A$, $\tau_B$, and $\tau_C$. Boundary values of $\tau$ and $V$ at the upper surface, and base of the soil layer permit evaluations at intermediate space points for each time interval.

Basically the one-dimensional solution by the method of characteristics involves an elastic wave propagation over each time interval. However, the elastic shear wave velocity, $v_s$, in this time interval is determined from the tangent modulus of a nonlinear shearing stress-strain curve, calculated at the average initial level of the shearing stress. Thus, nonlinear soil behavior is introduced by a sequence of elastic behaviors in each time step, during both loading and unloading of the soil. Additional details on application of this method of characteristics are given by Streeter, Wylie, and Richart (1974), Streeter and Richart, (1974), Papadakis (1973), and a computer program for calculating shearing stresses, displacements, velocities, and accelerations throughout a layer of nonlinear soil material is available as CHARSOIL (1974).

**Wave Propagation in Layer of Cohesionless Soil.**—To illustrate the use of the method of characteristics to study wave propagation in cohesionless soil layers, consider a layer of dry sand 70.4 ft (21.5 m) thick which rests on rock. The sand has a unit weight of $\gamma_s$ = 100 pcf (15.72 kN/m³) throughout and the static shear modulus, $G_o$, was determined at the midpoint of each layer at depth $z_i$ from

$$G_o = 50,000\, (\gamma_s z_i)^{0.5} \quad\quad\quad\quad\quad\quad\quad\quad\quad\quad\quad\quad\quad\quad\quad\quad (19)$$

Note that Eq. 19 was developed from Eqs. 4 and 6a. The soil viscosity, $\mu$, was assumed to be zero. Then the 70.4-ft (21.5-m) thick soil layer was subdivided into eight layers having thicknesses of 5.0 ft, 6.8 ft, 7.9 ft, 8.8 ft, 9.6 ft, 10.2 ft, 10.8 ft, and 11.3 ft (1.52 m, 2.07 m, 2.40 m, 2.68 m, 2.93 m, 3.11 m, 3.29 m, and 3.45 m) in accordance with Eq. 14 and the chosen value of $\Delta t$ = 0.01 sec. The nonlinear shearing stress-strain behavior was described by the R-O equations, Eqs. 9 and 10, using parameters of $\alpha$ = 1.0, $C_1$ = 0.8, and $R$ = 5, and the shearing strength at any depth was taken as $\tau_m = \gamma_s z_i \tan 30°$. Layer thicknesses, and distributions of $G_o$ and $\tau_m$ with depth are shown in Fig. 18.

This layer was excited by horizontal motions at the rock base which developed the velocity-time pattern shown in Fig. 19($a$). Also shown in Fig. 19($a$) are the velocity-time values computed for the surface of the layer. In Fig 19($b$) the shearing stress-strain relations developed at a depth of 8.4 ft (2.56 m) are given for the time interval of about 1.8 sec-2.8 sec. This diagram illustrates the nonlinear behavior of the soil when the maximum shearing stress was only about 0.9 $\tau_y$ (or 0.9 × 0.8 $\tau_m$ = 0.72 $\tau_m$). This nonlinear soil behavior absorbed energy and reduced the motions developed at the surface to values lower than expected for an elastic material.

**Base Motion Synthesis Method.**—In the preceding example, the excitation was provided at the base of the soil layer and motions were calculated throughout the soil mass and at the surface. However, in many instances, earthquake records have been obtained at the surface and it was desirable to determine what base motions caused these surface motions. A procedure using the method of characteristics for such computations was reported by Papadakis, Streeter, and Wylie (1974).

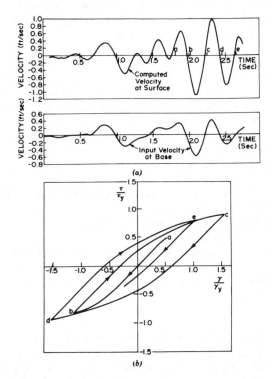

FIG. 19.—Seismic Shear Wave Transmission in Dry Sand Layer: (a) Applied Seismic Velocity at Base and Computed Surface Velocity; (b) Stress-Strain Diagram Computed at 8.4-ft Depth

This base motion synthesis method requires that soil properties be known throughout the depth of the layer, and a $z$-$t$ diagram is constructed. For a chosen value of $\Delta t$ the intervals $\Delta z$ must conform to average values of $v_s$ within that interval, according to Eq. 14. Then for this procedure, the surface velocities and shearing stresses ($\tau = 0$ for a free surface) are known throughout the entire time interval to be studied. The stresses and velocities are then calculated at the first depth interval, $\Delta z_1$ (Fig. 20) for each time interval. For example, values at point P in Fig. 20 are calculated by proceeding from point A along the $C^+$ characteristic and from point B along the $C^-$ characteristic. These two equations include two unknown quantities, $\tau_P$ and $V_P$, and are solved directly.

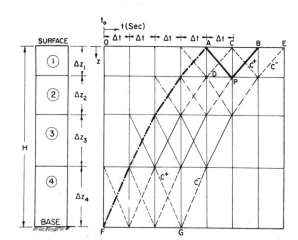

**FIG. 20.**—Layer Designation and $z\text{-}t$ Diagram for Base Motion Synthesis Method (Papadakis, Streeter, and Wylie, 1974)

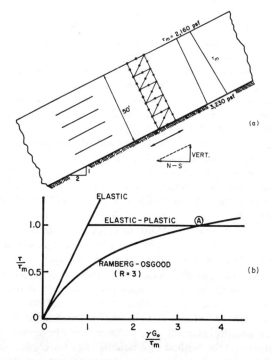

**FIG. 21.**—Cohesive Soil Layer on 2:1 Slope: (*a*) Layer and Shearing Strength Distribution; (*b*) Shearing Stress-Strain Relations Assumed for Soil

When stress levels were high enough that nonlinear stress-strain relations developed, an interpolation procedure was introduced. A study of the example shown in Fig. 18 was included by Papadakis, et al. (1974) wherein the surface velocity-time pattern of Fig. 19($a$) was taken as the starting point and the base rock velocity-time pattern was computed. The result was almost identical with the base rock velocity-time pattern, shown as the lower curve in Fig. 19($a$). The computed shearing stress-strain curves were nearly identical with those shown in Fig. 19($b$), with the outside loop displaced perhaps up to 5% on the strain scale. It was concluded that this method of synthesizing base motions starting from surface motions is an accurate and effective method, even when nonlinear soil properties are introduced.

**Downhill Movement of Cohesive Soil Layer on Slope.**—Fig. 21($a$) shows a 50-ft (15.2-m) thick layer of cohesive soil resting on an inclined rock base with a 2:1 slope. The shearing strength of the soil varied linearly from $\tau_m = 2,160$ psf (103.5 kN/m$^2$) at the upper surface to $\tau_m = 3,230$ psf (154.7 kN/m$^2$) at the soil-rock interface. This layer was excited by rock motions acting parallel to the soil-rock interface which were developed by superposing components of the N-S and vertical motions from the 1940 El Centro earthquake. The combined motions perpendicular to the rock surface was ignored. Thus one-dimensional shear waves were propagated in the soil layer along lines perpendicular to the rock surface. The dynamic shearing stresses developed by the soil motions were added to the shearing stresses required to hold the layer in static equilibrium.

This problem was solved by Finn and Miller (1973) who applied the finite element method. The element heights were 10 ft (3.05 m) and were arranged as shown in Fig. 21($a$). They assumed shearing stress-strain relations for the soil which were elastic, viscoelastic, and elastic-plastic with the elastic modulus varying throughout the layer according to the expression $G_o = 200 \, \tau_m$. The unit weight was $\gamma_s = 120$ pcf (18.9 kN/m$^2$). The relative displacement-time curve determined by Finn and Miller for the elastic-plastic case is shown in Fig. 22($a$) by the dash-dot line.

This problem was also treated by the method of characteristics (Streeter and Richart, 1974) using the same basic soil properties, but considering the shearing stress-strain relations to be elastic, elastic-plastic, and Ramberg-Osgood with slip [Fig. 21($b$)]. For the characteristics solutions represented in Fig. 22($a$) slip was allowed only at the soil-rock interface. For example, the Ramberg-Osgood curve shown in Fig. 21($b$) ($\alpha = 1$, $C_1 = 0.8$, $R = 3$) reached the maximum shearing stress, $\tau_m$, at point A. Thus, in calculating shearing stresses at the soil-rock interface, no stress larger than $\tau_m$ was permitted, but stresses greater than $\tau_m$ could be developed at other points in the layer. (Note: CHARSOIL now includes a limit of $\tau_m$ at any point in the layer, if this condition is desired).

In Fig. 22($a$) the response of an elastic layer consists of oscillations about the zero displacement line at the natural frequency of the layer. Note that in Fig. 22($a$) the *relative displacements* are shown. These represent the difference between the surface displacement and the base rock displacement at any given time. The solutions by the method of characteristics for elastic-plastic soil with slip, and Ramberg-Osgood soil with slip show a permanent downhill motion. As seen from Fig. 22($a$), these relative displacement-time patterns are quite similar to the one determined from the finite element method by Finn and Miller (1973).

Fig. 22(b) shows the motion of the rock base as developed by the 1940 El Centro tangential components of motions. It was considered of interest to evaluate the influence of the direction of the initial major impulse from this excitation. Consequently, the direction of the input was reversed to develop an "uphill" earthquake. Fig. 23 shows the difference in surface displacement developed for an "uphill" and a "downhill" earthquake. Consequently, it should be of some interest to evaluate the influence of direction of earthquake motions when studying the dynamic response of soils on slopes.

Fig. 23 also includes a curve that shows the increased surface deformation developed when a rigid slab weighing 400 psf (19.2 kN/m$^2$) was attached to the upper surface of the soil layer. The slab introduced a dynamic shearing

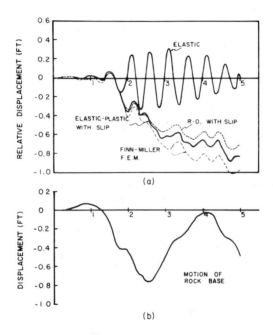

**FIG. 22.**—Displacement-Time Behavior of 50-ft Thick Cohesive Soil Layer on 2:1 Slope Excited by Tangential Components of 1940 El Centro Earthquake Acting at Rock Base: (a) Motion at Surface; (b) Motion at Rock Base

stress at the soil surface of $(W/g)\ddot{u}$ and a static shearing stress proportional to its weight, $W$. This slab represents a one-dimensional approximation to a building on the slope, and shows how the downhill motion of the soil mass is increased by the added surface weight following this type of earthquake excitation.

An advantage of the method of characteristics for this type of study is its low cost. The computations required for this example, including the Ramberg-Osgood nonlinear behavior and slip, required 6.5 sec of running time on the IBM 360/67 for 5 sec of earthquake excitation of the five-segment layer. For 12 sec of excitation the computer running time increased to 12 sec.

**Evaluation of Dynamic Pore Pressures.**—Two examples of dynamic pore

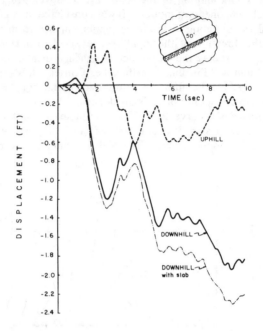

**FIG. 23.—Total Displacement at Surface of Soil Layer on 2:1 Slope Caused by Tangential Components of El Centro Earthquake**

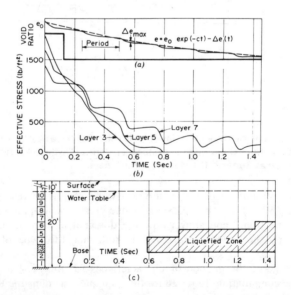

**FIG. 24.—Potential Liquefaction in Cohesionless Soil: (a) Forced Void Ratio Variation; (b) Computed Effective Stresses in Selected Layers; (c) Time-Dependent Development of Potential Liquefaction (Streeter, et al., 1974)**

pressure changes in cohesionless soils following external excitation were treated by Streeter, Wylie, and Richart (1974), using the method of characteristics. In the first study, a horizontal layer of saturated sand consisted of 10 layers, nine of these having an initial void ratio, $e_o = 0.66$, and coefficient of permeability, $k = 0.02$ ft/min. The tenth layer represented loose sand with $e_o = 1.0$ and $k = 0.03$ ft/min. Periodic decreases of void ratio were applied to both materials according to the pattern shown in Fig. 24(a). For the nine layers, $c = 0.0001$ and $\Delta e_{max} = 10^{-5}$ and for the loose layer, $c = 0.001$ and $\Delta e_{max} = 2 \times 10^{-5}$. Both pulses acted at a period of 0.25 sec. Fig. 24(b) indicates how the effective stresses in layers 3, 5, and 7 reduced with time as the pore pressures built up. Layer 3 was the loose layer which compacted at the higher rate, and reached the liquefied condition [Fig. 24(c)] first. This type of problem illustrates the importance of the time rate of void ratio change, $\Delta e(t)$, in saturated soils as compared to the permeability. If the "pump," $\Delta e(t)$, creates pressures at a greater rate than these can be dissipated by water flow, eventually the pore pressure will build up to a value equal to the overburden stress and the condition of "liquefaction" occurs.

The second example included a simplified treatment of the dynamic pore pressures developed in the upstream pervious section of an earthfill dam. The dam was subjected to a vertical sinusoidal motion with a single amplitude displacement of 1/8 in. (3.17 mm) at a frequency of 8 cps. Pore pressure transients were developed by vertical displacements of the dam into the reservoir, and these pressures at the dam-reservoir interface were dissipated by flow through a two-dimensional network of flow elements (representing the pervious dam material) as shown in Fig. 25(a). As a result of the pore pressures developed by this excitation, temporary conditions of liquefaction were noted during each cycle. These zones wherein the transient pore pressure reached values equal to the effective overburden pressure, therefore potential liquefaction, moved down the face of the dam as shown in Fig. 25(b).

Further studies of earthquake induced wave propagation in a water-saturated layer of sand (Papadakis and Richart, 1974) have shown the possibility of developing high pore pressures, by compression wave propagation in the water, at the same time high values of dynamic shearing stresses are developed in the soil structure. Analytical studies are continuing to evaluate one and two-dimensional wave propagations in water-saturated soils, including the nonlinear shearing stress-strain relations for the soil skeleton.

**Two-Dimensional Representation of Shear and Pressure Wave Transmission by Latticework Method.**—For calculation of two and three-dimensional transients, Streeter and Wylie (1968) have shown that the continuum may be replaced by a latticework of one-dimensional elements. The one-dimensional element of the latticework transmits shear and pressure waves. At interior nodes, as shown in Fig. 26, an imaginary transfer element receives and transmits all shear and pressure waves. The node element is considered to be rigid, weightless, and free to move in the $x$, $y$, and $z$ directions, but not to rotate. For the two-dimensional systems the method of characteristics involves five linear equations in five unknowns for each coordinate direction at each node point.

Boundary conditions at the edge of the latticework are handled as in the one-dimensional flow case, because the boundaries are considered to be made up of horizontal and vertical steps.

An improvement on the two-dimensional representation of the continuum by the latticework system was obtained by introducing a deformable transfer element (Papadakis, 1973). Use of this element required solution of eight equations for each of the two coordinate directions. Thus 16 explicit difference equations were developed for each node through the method of characteristics, and their solution was carried out readily by the computer.

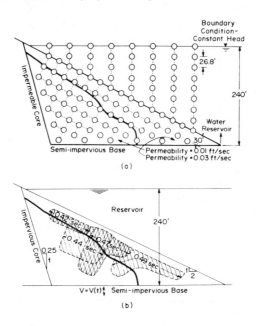

**FIG. 25.—Transient Liquefaction of Upstream Pervious Section of Earth Dam: (a) Elements Used to Model Pore Pressure Transmission; (b) Zones of Potential Liquefaction (Streeter, et al., 1974)**

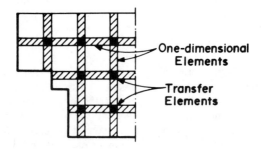

**FIG. 26.—Two-Dimension Latticework with Nodal Elements**

Papadakis (1973) treated several two-dimensional examples by this "16-Latticework" method, including the simplified representation of the Palos Grandes, Caracas, area [Fig. 27(c)]. He used constant values of shear modulus, Poisson's ratio, viscosity, unit weight of the soil, and the soil was treated as an elastic material. Two latticeworks were used to represent the geometry of the valley section. Latticework I had 330 linear elements of length 150 ft (45.75 m) and 133 transfer elements. The corresponding time increment was $\Delta t = 0.030$ sec. Latticework II had 500 linear elements of length 126 ft (38.4 m), 210 transfer

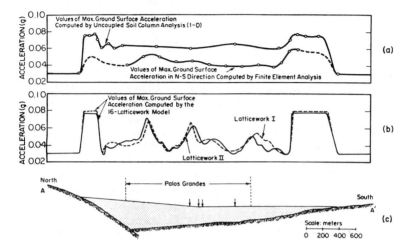

FIG. 27.—Comparison of Computed Values of Maximum Ground Surface Acceleration for Section A-A' through Palos Grandes, Caracas, Venezuela (Papadakis, 1973)

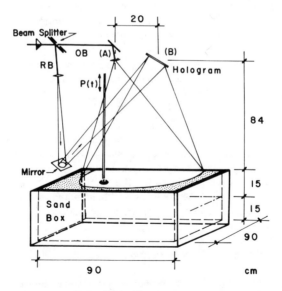

FIG. 28.—Optical Components of Interferometer and Half-Space Model (Woods and Saegesser, 1973)

elements, and a corresponding time increment of $\Delta t = 0.025$ sec. Values of the maximum ground surface accelerations computed by the Latticework Method are shown in Fig. 27(b) for excitation provided by the S21°W component of the Taft earthquake accelerogram, scaled by multiplying the amplitude by 0.166 and time by 0.9 (see Seed, et al., 1970). Quite similar results were obtained from the two latticeworks. It is of interest to note that the computer time for only executing the program was approximately 80 sec for Latticework I and was approximately 200 sec for Latticework II.

Fig. 27($a$) shows the ground surface acceleration obtained by Seed, et al. (1970) using the finite element method and more realistic soil properties which included variation of the shear modulus with overburden and strain-dependent values of both shear modulus and damping. Thus it might be expected that the peaks of ground surface accelerations obtained by the finite element method would be lower and less pronounced than those obtained by considering the soil to be an elastic material. Values obtained by the one-dimensional soil column analysis using the method of characteristics are also shown in Fig. 27($a$) for comparison.

**Summary of Method of Characteristics Solutions for Soil Dynamics.**—Solutions are now available for one-dimensional shear wave propagation through soils having elastic, viscoelastic, or nonlinear (strain softening) stress-strain properties. The total thickness may be composed of layers of any given thicknesses and soil properties. For excitation at the base of the soil mass, the computer program CHARSOIL may be used directly, whereas for calculations based on surface records the program for the Base Motion Synthesis Method was given by Papadakis (1973).

Wave propagation in the fluids contained in saturated soils and simultaneous propagation of shear waves in the soil structure have been developed, which represent a simplified approach to the problem. It is anticipated that solutions for two-dimensional wave propagation in water saturated soils (the Biot problem) will be obtained using the method of characteristics. The development of time-dependent changes in pore pressures caused by stress-dependent void ratio changes will contribute to understanding of the liquefaction process. Such studies are also underway.

The two-dimensional solutions by the latticework method are now (January, 1974) being developed and the examples presented here represent simplified systems. It is of particular importance to include the nonlinear (Ramberg-Osgood) shearing stress-strain conditions, including slip, into the latticework.

Finally, the primary reason for considering these recently developed methods for solving soil dynamics problems by the method of characteristics is that they are inexpensive. Even if the nonlinear soil behavior is included, the computer solution for a one-dimensional system including multiple layers seldom costs more than a few dollars.

### Selected Studies of Soil-Structure Interaction

**Model Tests.**—Model tests are often performed to check on theoretical studies, or to provide a relatively inexpensive evaluation of changes in design parameters before the prototype design is completed. This section describes two types of laboratory tests using models of dynamically loaded soil-structure systems.

*Holographic Interferometry.*—A new tool for laboratory studies in soil dynamics has recently been developed [see Woods and Saegesser, (1973), Woods, (1973), and Woods, Barnett, and Saegesser, (1974)] which permits evaluations of footing

FIG. 29.—Interferogram from Static Loading of Model Footing (Woods, Barnett, and Saegesser, 1974)

FIG. 30.—Model Half-Space and Dynamic Holography Apparatus (Woods, Barnett, and Saegesser, 1974)

motions and wave patterns on the surface of soils. Stroboscopic double exposure holographic interferometry can be employed to get "stopped motion" records of traveling waves.

To make a hologram, a coherent light beam produced by a laser is divided into an object beam (OB) and a reference beam (RB) by a "beam splitter" (see Fig. 28). The object beam is reflected from the object soil surface onto a photographic plate where it intersects the reference beam and a standing wave interference pattern is recorded. The developed photographic plate is the hologram. When a hologram is subsequently illuminated only by the reference beam, the hologram will reconstruct a virtual image of the object, even if the object is gone. However, if the object is left in place but deformed slightly, the deformed object will not coincide exactly with the image on the hologram and interference occurs. Fig. 29 shows the interference pattern developed by superposing the image of an unloaded footing on a statically loaded footing-soil system. Each black fringe in Fig. 29 represents a contour line of equal elevation. Between each fringe the elevation changes by vertical steps of 0.00013 mm (i.e., one-half wave length of the laser light). It is evident from the parallel fringes across the footing that the loading was not exactly centered and that the footing tipped slightly as it displaced vertically.

Dynamic problems were studied for steady-state vertical excitation of a circular footing resting on sand. The test setup (see Fig. 30) included a sand box 100 cm × 100 cm × 30 cm, an electromagnetic oscillator to provide vertical motions to the footing, and a laser system for recording surface motions. By pulsing the laser beam at the same frequency as the footing was vibrated vertically, the stroboscopic effect produced a "stopped motion" which could be recorded in a hologram. Using a double-exposure technique, the static hologram was superposed on the dynamic "stopped motion" hologram and the interference pattern was obtained on one film. Woods, et al. (1974) first applied this holographic interferometry technique to model studies of interception of surface waves by trenches, open cylindrical holes, model piles, and wall-type barriers. Fig. 31 shows the surface wave patterns developed in the vicinity of a vibrating footing when a portion of these propagating waves were intercepted by two rows of open holes. It is obvious from this picture that the pattern of the holes *was* effective in shielding the zone beyond them from the incoming waves. Scaling to prototype situations was based on the length of the Rayleigh waves, $\lambda_R$, i.e., all geometrical test parameters were determined as a function of $\lambda_R$.

By introducing an optical wedge into the object beam, Woods (1973) found that the static interferogram consisted of a pattern of parallel lines, but that the dynamic interferogram produced a three-dimensional illusion as shown in Fig. 32. This type of interferogram is useful for visualizing the shape and distributions of waves propagating from a vibrating footing, and the modifications of these waves by adjacent footings or other wave barriers.

The technique of holographic interferometry has potential value in evaluating the response of odd-shaped, or multiple foundations when excited by incoming wave energy. A particular advantage of the holographic technique is that the object to be studied may have appreciable initial contours (as, for example, an automobile tire) and need not be a continuum. Thus holograms may be obtained for sand surfaces that are not originally flat, and as shown in Fig. 29 useful information was obtained from the sand surface and the upper surface of the footing which projected approximately 3/8 in. (0.95 cm) above the sand surface. Techniques have also been devised to determine the nearly horizontal component of vibratory motion by a reorientation of the optical system (Woods, 1973).

*Model Tests of Embedded Footings.*—Scale model tests represent a standard procedure for evaluating the dynamic response of foundation-soil systems. The soil is seldom modeled, but may be prepared in large containers or bins for

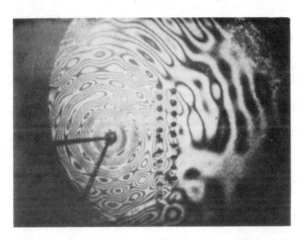

FIG. 31.—Interference Patterns for Interception of Traveling Wave by Two Rows of Cylindrical Holes (Woods, et al., 1974)

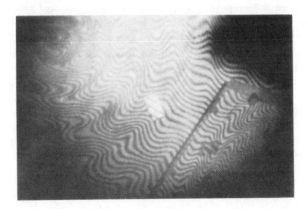

FIG. 32.—Dynamic Interferogram with Three-Dimensional Illusion (Woods, 1973)

laboratory tests, or natural soils may be used directly in outdoor tests. It is necessary to evaluate the shear wave velocity, $v_s$, the mass density, $\rho$ (from in-situ unit weight), and an effective Poisson's ratio for the soil. Then with the scaled dimensions of the foundation, the experimental results may be compared with theoretical predictions.

Careful tests of model footings resting on the surface of soils were reported by Fry (1963), and an evaluation of the results by Richart and Whitman (1967) indicated good agreement with theories treating footings resting on an elastic half space. However, it was noted that the dynamic response of these model footings demonstrated a nonlinear behavior (strain softening) of the supporting soil.

Studies of the dynamic behavior of embedded footings introduce additional variables into the problem. In theoretical treatments of rocking motions, of embedded footings (see Beredugo and Novak, 1972, Krizek, Gupta, and Parmelee, 1972, Urlich and Kuhlemeyer, 1973, Novak, 1973, for example) it has been assumed that full contact has been maintained by the soil against the vertical surfaces of the embedded portion of the foundation, and that the material was elastic. The solutions by Beredugo and Novak, (1972) and Novak (1973) permitted different elastic properties for the soil against the vertical face of the foundation and for the soil beneath the base.

Although there have been many model tests of dynamically loaded footings resting on or embedded in soils, the series of tests by Stokoe (see Stokoe, 1972, and Stokoe and Richart, 1974) will be described briefly because of the importance of a few test details. Cylindrical footings 8 in. (3.15 cm) to 12 in. (4.72 cm) in diameter were embedded in sand to depths ranging from 0 to 1.6 times the footing radius, and were given a transient excitation. A static force was applied to the upper surface of the footing by a nylon cord and transient motions of the footing developed after the cord was cut. The dynamic response depended upon the type and direction of the static forces applied, the geometrical properties of the footing, the dynamic soil properties, and the contact between the footing surface and the soil.

Footings resting on the surface were formed by pouring concrete into a steel cylinder which penetrated about 1/16 in. (1.6 mm) into the sand. Pouring was accomplished in several lifts, but all pouring was completed within 1/2 hr. Therefore, the static pressure distribution at the contact surface was between a uniformly distributed pressure and that developed by a rigid base on an elastic medium. Because no footing was subjected to dynamic tests within 3 weeks after pouring, it was anticipated that the dynamic response would correspond to that of a rigid footing. The embedded footings were constructed by first digging a hole the diameter of the steel shell (which was used to form the portion of the footing above the sand surface) to the required depth. Then the steel shell was embedded about 1/16 in. (1.6 mm) into the sand surface and the footing was poured, in lifts, within a period of about 1 hr. By this procedure, direct contact was developed between the vertical surface of the concrete and the soil.

Fig. 33 shows the influence of embedment on the damping ratio, $D$, and the damped natural frequency, $f_d$, for one model footing when tested in the coupled rocking and sliding mode of vibration. The black symbols denote full soil contact against the vertical surface of the footing, whereas the open symbols designate results obtained after sand was excavated around the periphery of the footing. This diagram shows that good soil-footing contact will increase the damped natural frequency and damping ratio for this system as the depth of embedment increases. However, if soil contact was removed from the vertical surface of the footing, the embedment depth exerted no influence on these dynamic properties of the system.

From these model tests, and from tests on larger models by Berdugo and Novak (1972) it is evident that construction procedures and stress-history can have a significant influence on the dynamic response of embedded footings. Cast-in-place concrete footings subsequently loaded only by low-amplitude dynamic loads might be expected to benefit as much from embedment as indicated by the black symbols in Fig. 33. Finally, if the vibration stress history was such that the soil along the vertical faces of the footing was compacted to the point where an open void developed, then the condition of no lateral support would exist.

**Tests of Embedded Machine Foundations.**—Descriptions of corrective measures to improve the dynamic behavior of existing foundations have been reported

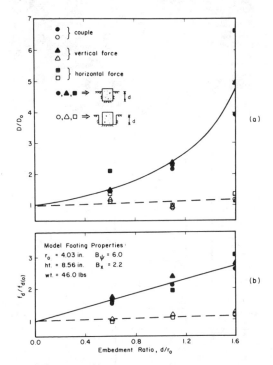

**FIG. 33.—Variation of Damping Ratio and Natural Frequency of Model Footings with Depth of Embedment (Stokoe and Richart, 1974)**

by Gnaedinger (1961), Tschebotarioff (1964), and McNeill (1969). Margason, McNeill, and Babcock (1968) presented five case studies involving predicted behaviors of dynamically loaded foundations. Other studies are available in the literature, but most of these give an incomplete description of the dynamic soil properties, evaluated at strain levels anticipated in the prototype situation, and give little or no evaluation of the effectiveness of soil-structure interaction.

Careful studies of two types of machine foundations were described by Stokoe and Richart (1974), which included evaluations of the shear wave velocities in situ for each installation. One foundation block, 87 in. (221 cm) × 55 in. (140 cm) × 48 in. (122 cm) deep was embedded into a clayey silt soil (CL-ML) and supported a two-cylinder air compressor. The air compressor, shown in Fig. 34(a), produced forces that rocked the foundation block, shown in Fig. 34(b), in a N-S direction. An analysis, based on the results of field evaluations of the soil, geometries of the machine and foundation, model tests of equivalent embedment ratios, and theory, was compared with measured dynamic motions

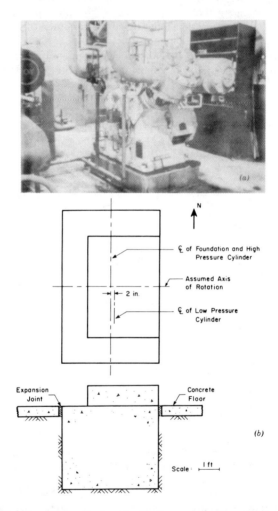

**FIG. 34.—Machine-Foundation System in Air Compressor Case Study: (a) Air Compressor; (b) Foundation**

of the system. The dynamic motions were developed by steady-state operation of the machine, and by a special "plucking" test developed by a quick release of a horizontal static force applied to the foundation. From this study it was found that the machine-foundation systems exhibited a dynamic response comparable to that for a system resting on the *surface* of the soil. No significant effect of embedment was developed in this case. It was concluded that this lack of horizontal restraint could have been caused by: (1) Inadequate compaction of the soil during backfilling; (2) drying of the soil adjacent to the foundation block because of the elevated temperature of the boiler room above; or (3) permanent deformation of the backfill because of previous dynamic motions of the foundation. The result of any or all of these actions was to cause a separation of the soil and foundation face over an appreciable portion of the vertical surface.

The second installation consisted of a pair of horizontal single-cylinder vacuum pumps, each mounted on foundation blocks 189 in. (480 cm) × 60 in. (152 cm) maximum width, to 38 in. (97 cm) minimum width × 90 in. (229 cm) embedment depth. The foundation blocks were parallel with the long axes oriented in the N-S direction. They were supported by and backfilled with a brown silty find sand. Again the dynamic motions from steady-state operation and

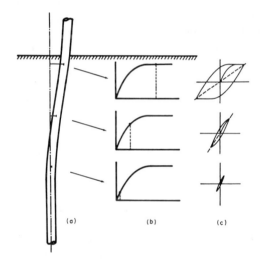

FIG. 35.—Soil-Structure Interaction for Laterally Loaded Pile: (a) Pile Displacement; (b) Lateral Pressure Versus Deflection for First Loading; (c) Lateral Pressure Versus Deflection for Reversed Loading

from "plucking" tests were measured and evaluated. For these installations, the dynamic response was clearly influenced by the depth of embedment, which showed that the soil was in contact with the vertical faces of the foundation block and provided dynamic restraint.

These two case studies dealt with low-amplitude dynamic motions of embedded foundation blocks. Therefore, no strain amplitude effects were noticed. However the significant conclusion was that for certain types of soils and operating conditions, soil-foundation contact may be reduced or eliminated completely from the vertical faces of embedded foundations.

**Dynamic Lateral Loading of Piles.**—Piles required to resist dynamic loads range from those used to stiffen small machine foundations to deep penetration piles for offshore structures. No attempt will be made here to summarize the literature on the topic of laterally loaded piles but state-of-the-art summaries have been presented recently by Broms (1972) and McClelland (1974), for example.

Many static tests and a few dynamic tests have demonstrated that the pressure versus displacement relations developed as a pile face is forced laterally against the soil is nonlinear. This $p$-$y$ relation is strain-softening (Fig. 1, curve C) and has been approximated by hyperbolic curves similar to that shown in Fig. 2 (Kondner and Cunningham, 1963, Parker and Reese, 1971) or by a power function which relates the pressure developed by $y^n$, with $n < 1.0$. Yegian and Wright (1973) incorporated hyperbolic stress-strain soil behavior into a finite element study of a pile section being forced into soil. The load-displacement curves thus developed were, of course, also nonlinear. It would be expected that these various nonlinear load-displacement curves could also be approximated by a Ramberg-Osgood type curve.

The determination of a suitable pressure versus displacement curve for a particular pile-soil system depends upon the dimensions of the pile, the properties of the soil, and the operating environment. Particular attention should be directed to the displacement-history pattern to evaluate special conditions of separation of pile from the soil at the rear face of the pile, or of the pile being pushed through yielding soil. If neither extreme condition exists then a nonlinear pressure-displacement relation can be developed at each depth along the pile. Fig. 35($b$) shows the strain energy developed at three elevations along a flexible pile [Fig. 35($a$)] being forced horizontally into a soil mass. Fig. 35($c$) indicates the hysteresis loops formed if the deformation is repeatedly reversed. From this diagram it is evident that under vibratory loads the hysteretic damping varies along the pile. In addition, the strain-dependent soil reaction delivers a stiffness that varies along the pile.

The dynamic response of cylinders or piles have been studied by Tajimi (1969) and Novak (1974), for example, by considering the soil as an elastic medium. Penzien (1970) used a lumped mass model which made it possible to incorporate nonlinear behavior of soils into the study of dynamic response of piles. It is anticipated that future analytical studies will include dynamic nonlinear soil-pile interactions, and will lead to improved design procedures.

### SUMMARY AND CONCLUSIONS

The central theme of this lecture has been an examination of the nonlinear shearing stress-shearing strain behavior of soils, as this behavior changed with strain from nearly elastic to nearly plastic conditions. Considerations were given to laboratory and field methods for evaluating the "elastic" shear modulus, $G_o$, and to procedures for estimating the secant modulus at higher strains. Experimental shearing stress-strain data were approximated by a Ramberg-Osgood mathematical curve, which may be incorporated into analytical procedures.

The method of characteristics was presented as one analytical method which permitted studies of dynamic behavior of soil masses. It was found possible to treat nonlinear behavior of layered soils, and to include inelastic slip, when the layer was excited by arbitrary dynamic loadings acting either at the base or at the surface. The method of characteristics can be adopted for the study of transient pore pressures in water-saturated soil masses, and these studies form the basis for investigations of transient liquefaction which may develop during dynamic loading. Finally, the method of characteristics can be extended to treat two and three-dimensional problems by the latticework method. It is anticipated that improvements of the two-dimensional latticework will soon (after January, 1974) be developed.

Several types of model studies involving soil-structure interaction were presented to illustrate the new method of holographic interferometry for study of soil dynamics problems, and to present some comments on the importance of test and construction details on the effectiveness of soil structure interaction. The importance of developing and maintaining positive contact between the foundation block and soil along the vertical faces of embedded footings was noted.

In conclusion, the process of improving design procedures for interaction of soils and structures must be based on: (1) A thorough evaluation of the dynamic soil properties at the field location; (2) an adequate theory which includes the dynamic loading, soil properties, and soil-foundation geometrical parameters; and (3) field measurements from the completed prototype to check on the reliability of 1 and 2. Therefore, I conclude by continuing to recommend field measurements on the prototype, whenever possible, to provide a basis for improvement of our design procedures.

### Acknowledgments

I wish to acknowledge with thanks the continuous support provided by the Department of Civil Engineering of the University of Michigan and by the National Science Foundation for much of the work described in this lecture. Special thanks go to the group of young colleagues, noted in the outer two columns of the list of names herein for their many contributions to our group effort in soil dynamics, and to my education during the past 15 years. Dr. J. R. Hall, Jr. developed our Soil Dynamics Laboratory at Michigan, and Professor R. D. Woods has continued to improve it as a research facility. The central column of names below lists my present colleagues at Michigan who are participating actively in our Soil Dynamics studies. I recognize them with my continuing thanks:

| S. S. Afifi | D. G. Anderson | B. O. Hardin |
| Y. S. Chae | C. N. Papadakis | J. Lysmer |
| V. P. Drnevich | V. L. Streeter | J. G. Jackson, Jr. |
| J. R. Hall, Jr. | R. D. Woods | K. H. Stokoe, II |
| | E. B. Wylie | |

And finally, my special thanks to Betty, who makes everything worthwhile!

### Appendix.—References

1. Afifi, S. S., "Effects of Stress History on the Shear Modulus of Soils," thesis presented to The University of Michigan, at Ann Arbor, Mich., in 1970, in partial fulfillment of the requirements for the degree of Doctor of Philosophy.
2. Afifi, S. S., and Richart, F. E., Jr., "Stress-History Effects on Shear Modulus of Soils," *Soils and Foundations* (J. JSSMFE), Vol. 13, No. 1, Mar., 1973, pp. 77-95.
3. Afifi, S. S., and Woods, R. D., "Long-Term Pressure Effects on Shear Modulus of Soils," *Journal of the Soil Mechanics and Foundations Division*, ASCE, Vol.

97, No. SM10, Proc. Paper 8475, Oct., 1971, pp. 1445-1460.
4. Anderson, D. G., "Dynamic Modulus of Cohesive Soils," thesis presented to The University of Michigan, at Ann Arbor, Mich., in 1974, in partial fulfillment of the requirements for the degree of Doctor of Philosophy.
5. Anderson, D. G., and Richart, F. E., Jr., "Temperature Effect on Shear Wave Velocity in Clays," *Journal of the Geotechnical Engineering Division*, ASCE, Vol. 100, No. GT12, Proc. Paper 10981, Dec., 1974, pp. 1316-1320.
6. Anderson, D. G., and Woods, R. D., "Comparison of Field and Laboratory Shear Modulus," *Proceedings of the ASCE Conference on In-Situ Measurement of Soil Properties*, Raleigh, N.C., Vol. I, June, 1975, pp. 69-92.
7. Ballard, R. F., Jr., "Determination of Soil Shear Moduli at Depth by In-Situ Vibratory Techniques," U.S. Army Engineer Waterways Experiment Station, Vicksburg, Miss., *Misc. Paper No. 4-691*, Dec., 1964.
8. Beredugo, Y. O., and Novak, M., "Coupled Horizontal and Rocking Vibration of Embedded Footings," *Canadian Geotechnical Journal*, Vol. 9, No. 4, Nov., 1972, pp. 477-497.
9. Berg, G. V., "A Study of the Earthquake Response of Inelastic Systems," *Proceedings of the Structural Engineering Association of California*, Oct., 1965.
10. Broms, B. B., "Stability of Flexible Structures (Piles and Pile Groups)," General Report, Fifth European Conference on Soil Mechanics and Foundation Engineering, Vol. 2, 1972, pp. 239-269.
11. CHARSOIL, "Characteristics Method Applied to Soils" programmed by V. L. Streeter, E. B. Wylie, and F. E. Richart, Jr., University of Michigan, Ann Arbor, Mich., can be ordered from NISEE, 729 Davis Hall, University of California, Berkeley, Calif., 94720.
12. Cunny, R. W., and Fry, Z. B., "Vibratory In-Situ and Laboratory Soil Moduli Compared" *Journal Soil Mechanics and Foundations Division*, ASCE, Vol. 99, No. SM12, Proc. Paper 10219, Dec., 1973, pp. 1055-1076.
13. Drnevich, V. P., "Effect of Strain History on the Dynamic Properties of Sand," thesis presented to The University of Michigan, at Ann Arbor, Mich., in 1967, in partial fulfillment of the requirements for the degree of Doctor of Philosophy.
14. Drnevich, V. P., Hall, J. R., Jr., and Richart, F. E., Jr., "Effects of Amplitude of Vibration on the Shear Modulus of Sand," *Proceedings*, International Symposium on Wave Propagation and Dynamic Properties of Earth Materials, Albuquerque, N.M., Aug., 1967.
15. Drnevich, V. P., and Richart, F. E., Jr., "Dynamic Prestraining of Dry Sand," *Journal of the Soil Mechanics and Foundations Division*, ASCE, Vol. 96, No. SM2, Proc. Paper 7160, Mar., 1970, pp. 453-469.
16. Durgunoglu, H. T., and Tezcan, S. S., discussion of "Vibratory In-Situ and Laboratory Soil Moduli Compared," by R. W. Cunny and Z. B. Fry, *Journal of the Geotechnical Engineering Division*, ASCE, Vol. 100, No. GT12, Proc. Paper 10973, Dec., 1974, pp. 1303-1304.
17. Finn, W. D. L., and Miller, R. I. S., "Dynamic Analysis of Plane Non-Linear Earth Structures," *Proceedings of the Fifth World Conference on Environmental Engineering*, Paper No. 42, Session 1D, 1973.
18. Dry, Z. B., "Development and Evaluation of Soil Bearing Capacity. Foundations of Structures. Field Vibratory Tests Data," U.S. Army Engineer Waterways Experiment Station, Corps of Engineers, Vicksburg, Miss., *Technical Report No. 3-632, Report No. 1*, July, 1963.
19. Gnaedinger, J. P., "Symposium on Grouting: Grouting to Prevent Vibration of Machinery Foundations," *Journal of the Soil Mechanics and Foundations Division*, ASCE, Vol. 87, No. SM2, Proc. Paper 2793, Apr., 1961, pp. 43-54.
20. Hadala, P. F., "Effect of Constitutive Properties of Earth Media on Outrunning Ground Shock from Large Explosions," U.S. Army Engineer Waterways Experiment Station, Corps of Engineers, Vicksburg, Miss., *Technical Report S-73-6*, 1973.
21. Hall, J. R., Jr., "Effect of Amplitude on Damping and Wave Propagation in Granular Materials," thesis presented to the University of Florida, at Gainesville, Fla., in 1962, in partial fulfillment of the requirements for the degree of Doctor of Philosophy.
22. Hall, J. R., Jr., and Richart, F. E., Jr., "Dissipation of Elastic Wave Energy in Granular Soils," *Journal of the Soil Mechanics and Foundations Division*, ASCE,

Vol. 89, No. SM6, Proc. Paper 3698, Nov., 1963, pp. 27-56.
23. Hardin, B. O., "Study of Elastic Wave Propagations and Damping in Saturated Granular Materials," thesis presented to the University of Florida, at Gainesville, Fla., in 1961, in partial fulfillment of the requirements for the degree of Doctor of Philosophy.
24. Hardin, B. O., "Dynamic Versus Static Shear Modulus for Dry Sands," *Materials Research and Standards*, American Society for Testing and Materials, Vol. 5, No. 5, May, 1965, pp. 232-235.
25. Hardin, B. O., "Constitutive Relations for Airfield Subgrade and Base Course Materials. 1. Characterization and Use of Shear Stress-Strain Relation," University of Kentucky, Lexington, Ky., *Technical Report UKY 32-71-CE5, Soil Mechanics Series No. 4*, Contract No. F29601-70-C-0040, with U.S. Air Force Weapons Lab., Albuquerque, N.M., 1971.
26. Hardin, B. O., and Black, W. L., "Sand Stiffness Under Various Triaxial Stresses," *Journal of the Soil Mechanics and Foundations Division*, ASCE, Vol. 92, No. SM2, Proc. Paper 4712, Mar., 1966, pp. 27-42.
27. Hardin, B. O., and Black, W. L., "Vibration Modulus of Normally Consolidated Clay," *Journal of the Soil Mechanics and Foundations Division*, ASCE, Vol. 94, No. SM2, Proc. Paper 5833, Mar., 1968, pp. 353-369.
28. Hardin, B. O., and Drnevich, V. P., "Shear Modulus and Damping in Soils: Measurement and Parameter Effects," *Journal of the Soil Mechanics and Foundations Division*, ASCE, Vol. 98, No. SM6, Proc. Paper 8977, June, 1972a, pp. 603-624.
29. Hardin, B. O., and Drnevich, V. P., "Shear Modulus and Damping in Soils: Design Equations and Curves," *Journal of the Soil Mechanics and Foundations Division*, ASCE, Vo. 98, No. SM7, Proc. Paper 9006, July, 1972b, pp. 667-692.
30. Hardin, B. O., and Music, J., "Apparatus for Vibration During the Triaxial Test," Symposium on Instrumentation and Apparatus for Soils and Rocks, American Society for Testing and Materials, *Special Technical Publication No. 392*, 1965.
31. Hardin, K. G., and Richart, F. E. Jr., "Elastic Wave Velocities in Granular Soils," *Journal of the Soil Mechanics and Foundations Division*, ASCE, Vol. 89, No. SM1, Proc. Paper 3407, Feb., 1963, pp. 33-65.
32. Humphries, W. K., and Wahls, H. E., "Stress History Effects on Dynamic Modulus of Clay," *Journal of the Soil Mechanics and Foundations Division*, ASCE, Vol. 94, No. SM2, Proc. Paper 5834, Mar., 1968, pp. 371-389.
33. Iida, K., "The Velocity of Elastic Waves in Sand," Tokyo Imperial University, *Bulletin of the Earthquake Research Institute*, Vol. 16, 1938, pp. 131-144.
34. Iida, K., "On the Elastic Properties of Soil Particularly in Relation to its Water Content," *Bulletin of the Earthquake Research Institute*, Vol. 18, 1940, pp. 675-690.
35. Jackson, J. G., Jr., "Factors that Influence the Development of Soil Constitutive Relations," U.S. Army Engineer Waterways Experiment Station, Corps of Engineers, Vicksburg, Miss., *Misc. Paper No. 4-980*, 1968.
36. Jackson, J. G., Jr., "Analysis of Laboratory Test Data to Derive Soil Constitutive Properties," U.S. Army Engineer Waterways Experiment Station, Corps of Engineers, Vicksburg, Miss., *Misc. Paper S-69-16*, 1969.
37. Jennings, P. C., "Periodic Response of a General Yielding Structure," *Journal of the Engineering Mechanics Division*, ASCE, Vol. 90, No. EM2, Proc. 3871, Apr., 1964, pp. 131-166.
38. Kolsky, H., *Stress Waves in Solids*, Dover Publications, Inc., New York, N.Y., 1963.
39. Kondner, R. L., "Hyperbolic Stress-Strain Response: Cohesive Soils," *Journal of the Soil Mechanics and Foundations Division*, ASCE, Vol. 89, No. SM1, Proc. Paper 3429, Feb., 1963, pp. 115-143.
40. Kondner, R. L., and Cunningham, J. A., "Lateral Stability of Rigid Poles Partially Embedded in Sand," *Highway Research Record No. 39*, Washington, D.C., 1963, pp. 49-67.
41. Krizek, R. J., Gupta, D. C., and Parmelee, R. A., "Sliding and Rocking of Embedded Foundations," *Journal of the Soil Mechanics and Foundations Division*, ASCE, Vol. 98, No. SM12, Proc. Paper 9421, Dec., 1972, pp. 1347-1358.
42. Margason, B. E., NcNeill, R. L., and Babcock, F. M., "Case Histories in Foundation Vibrations," *Special Technical Publication 450*, American Society for Testing and Materials, June, 1968, pp. 167-196.

43. McClelland, B., "Design of Deep Penetration Piles For Ocean Structures," The Ninth Terzaghi Lecture, *Journal of the Geotechnical Engineering Division*, ASCE, Vol. 100, No. GT7, Proc. Paper 10665, July, 1974, pp. 709-747.
44. McNeill, R. L., "Machine Foundations," *Soil Dynamics Specialty Session, Proceedings*, Seventh International Conference on Soil Mechanics and Foundation Engineering, Mexico City, Mexico, August.
45. Newmark, N. M., "Failure Hypotheses for Soils," *Proceedings*, ASCE Research Conference on Shear Strength of Cohesive Soils, Boulder, Colo., 1960, pp. 17-32.
46. Newmark, N. M., and Rosenblueth, E., *Fundamentals of Earthquake Engineering*, Prentice-Hall, Inc., Englewood Cliffs, N.J., 1971.
47. Novak, M., "Vibrations of Embedded Footings and Structures," presented at the April 9-13, 1973, ASCE National Structural Engineering Meeting, held at San Francisco, Calif. (Preprint No. 2029).
48. Novak, M., "Dynamic Stiffness and Damping of Piles," *Soil Mechanics Research Report SM-1-74*, Faculty of Engineering Science, The University of Western Ontario, London, Ontario, Canada, Jan., 1974.
49. Papadakis, C. N., "Soil Transients by Characteristics Method," thesis presented to The University of Michigan, at Ann Arbor, Mich., in 1973, in partial fulfillment of the requirements for the degree of Doctor of Philosophy.
50. Papadakis, C. N., and Richart, F. E., Jr., "Earthquake Wave Transmission Through Saturated Soil," *Proceedings of the Conference on Analysis and Design in Geotechnical Engineering*, ASCE, Austin, Tex., June 9-12, 1974, Vol. 1, pp. 1-32.
51. Papadakis, C. N., Streeter, V. L., and Wylie, E. B., "Bedrock Motions Computed from Surface Seismograms," *Journal of the Geotechnical Engineering Division*, ASCE, Vol. 100, No. GT10, Proc. Paper 10853, Oct., 1974, pp. 1091-1106.
52. Parker, F., Jr., and Reese, L. C., "Lateral Pile-Soil Interaction Curves for Sand," *Proceedings of the International Symposium on Engineering Properties of Sea-Floor Soils and Their Geophysical Identification*, Seattle, Wash., July 25, 1971, pp. 212-223.
53. Penzien, J., "Soil-Pile Foundation Interaction," *Earthquake Engineering*, R. L. Wiegel, coord. ed., Prentice-Hall, Inc., Englewood Cliffs, N.J., 1970.
54. Pyke, R. M., "Settlement and Liquefaction of Sands under Multi-Directional Loading," thesis presented to the University of California, at Berkeley, Calif., in 1973 in partial fulfillment of the requirements for the degree of Doctor of Philosophy.
55. Ramberg, W., and Osgood, W. T., "Description of Stress-Strain Curves by Three Parameters," *Tech. Note 902*, National Advisory Committee for Aeronautics, 1943.
56. Richart, F. E., Jr., closure to "Foundation Vibrations," *Journal of the Soil Mechanics and Foundations Division*, ASCE, Vol. 87, No. SM4, Proc. Paper 2917, Aug., 1961.
57. Richart, F. E., Jr., Hall, J. R., Jr., and Woods, R. D., *Vibrations of Soils and Foundations*, Prentice-Hall, Inc., Englewood, Cliffs, N.J., 1970.
58. Richart, F. E., Jr., and Whitman, R. V., "Comparison of Footing Vibration Tests with Theory," *Journal of the Soil Mechanics and Foundations Division*, ASCE, Vol. 93, No. SM6, Proc. Paper 5568, Nov., 1967, pp. 143-168.
59. Schwarz, S. D., and Musser, J. M., Jr., "Various Techniques for Making In-Situ Shear Wave Velocity Measurements—A Description and Evaluation," *Proceedings of the International Conference on Microzonation for Safer Construction, Research and Application*, Seattle, Wash., Vol. II, 1972, pp. 593-608.
60. Seed, H. B., and Idriss, I. M., "Soil Moduli and Damping Factors for Dynamic Response Analysis," Earthquake Engineering Research Center, College of Engineering, University of California, Berkeley, Calif., *Report No. EERC 70-10*, Dec., 1970.
61. Seed, H. B., Idriss, I. M., and Dezfulian, H., "Relationships Between Soil Conditions and Building Damage in the Caracas Earthquake of July 29, 1967," Earthquake Engineering Research Center, University of California, Berkeley, Calif., *Report No. EERC 70-2*, Feb., 1970.
62. Shannon, W. L., Yamane, G., and Dietrich, R. J., "Dynamic Triaxial Tests on Sand," *Proceedings of the First Panamerican Conference on Soil Mechanics and Foundation Engineering*, Mexico City, Mexico, Vol. 1, 1959, pp. 473-486.
63. Silver, M. L., and Seed, H. B., "Deformation Characteristics of Sands under Cyclic Loading," *Journal of the Soil Mechanics and Foundations Division*, ASCE, Vol. 97, No. SM8, Proc. Paper 8334, Aug., 1971, pp. 1081-1098.
64. Stevens, H. W., "Measurement of the Complex Moduli and Damping of Soils under

Dynamic Loads—Laboratory Test Apparatus-Procedure and Analysis," U.S. Army Material Command, Cold Regions Research and Engineering Laboratory, Hanover, N.H., *Technical Report 173*, Apr., 1966.
65. Stokoe, K. H., II, "Dynamic Response of Embedded Foundations," thesis presented to The University of Michigan, at Ann Arbor, Mich., in 1972, in partial fulfillment of the requirements for the degree of Doctor of Philosophy.
66. Stokoe, K. H., II, and Richart, F. E., Jr., "In-Situ and Laboratory Shear Wave Velocities," *Proceedings of the VIII International Conference on Soil Mechanics and Foundation Engineering*, Moscow, U.S.S.R., Vol. 1.2, 1973, pp. 403-409.
67. Stokoe, K. H., II, and Richart, F. E., Jr., "Dynamic Response of Embedded Machine Foundations," *Journal of the Geotechnical Engineering Division*, ASCE, Vol. 100, No. GT4, Proc. Paper 10499, Apr., 1974, pp. 427-447.
68. Stokoe, K. H., II, and Woods, R. D., "In-Situ Shear Wave Velocity by Cross-Hole Method," *Journal of the Soil Mechanics and Foundations Division*, ASCE, Vol. 98, No. SM5, Proc. Paper 8904, May, 1972, pp. 443-460.
69. Stokoe, K. H., II, and Woods, R. D., closure to "In-Situ Shear Wave Velocity by Cross-Hole Method," *Journal of the Soil Mechanics and Foundations Division*, ASCE, Vol. 99, No. SM11, Proc. Paper 10112, Nov., 1973, pp. 1014-1016.
70. Streeter, V. L., and Richart, F. E., Jr., "One-Dimensional Soil Transients" (submitted for publication).
71. Streeter, V. L., and Wylie, E. B., *Hydraulic Transients*, McGraw-Hill Book Co., Inc., New York, N.Y., 1967.
72. Streeter, V. L., and Wylie, E. B., "Two- and Three-Dimensional Transients," *Transactions*, American Society of Mechanical Engineers, *Journal of Basic Engineering*, Vol. 90, Series D, No. 4, Dec., 1968, pp. 501-510.
73. Streeter, V. L., Wylie, E. B., and Richart, F. E., Jr., "Soil Motion Computations by Characteristic Method," *Journal of the Geotechnical Engineering Division*, ASCE, Vol. 100, No. GT3, Proc. Paper 10410, Mar., 1974, pp. 247-263.
74. Tajimi, H., "Dynamic Analysis of a Structure Embedded in an Elastic Stratum," *Proceedings of the 4 World Conference on Earthquake Engineering*, Santiago, Chile, Vol. III, 1969, pp. A-6, 53-69.
75. Tschebotarioff, G. P., "Vibration Controlled by Chemical Grouting," *Civil Engineering*, ASCE, Vol. 34, No. 5, May, 1964, pp. 34-35.
76. Urlich, C. M., and Kuhlemeyer, R. L., "Coupled Rocking and Lateral Vibrations of Embedded Footings," *Canadian Geotechnical Journal*, Vol. 10, No. 2, May, 1973, pp. 145-160.
77. Yoshimi, Y., and Oh-Oka, H., "A Ring Torsion Apparatus for Simple Shear Tests," *Proceedings of the VIII International Conference on Soil Mechanics and Foundation Engineering*, Moscow, U.S.S.R., Vol. 1.2, 1973, pp. 501-506.
78. Westergaard, H. M., "Earthquake-Shock Transmission in Tall Buildings," *Engineering News-Record*, Vol. 111, 1933, pp. 654-656.
79. Wilson, S. D., and Dietrich, R. J., "Effect of Consolidation Pressure on Elastic-and-Strength Properties of Clay," *Proceedings ASCE Research Conference on Shear Strength of Cohesive Soils*, Boulder, Colo., June 1960.
80. Woods, R. D., "Holographic Interferometry in Soil Dynamics," Communication at Specialty Session No. 8, *VIII International Conference on Soil Mechanics and Foundation Engineering*, Moscow, U.S.S.R., 1973.
81. Woods, R. D., Barnett, N. E., and Saegesser, R., "Holography—A New Tool for Soil dynamics," *Journal of the Geotechnical Engineering Division*, ASCE, Vol. 100, No. GT11, Proc. Paper 10949, Nov., 1974, pp. 1231-1247.
82. Woods, R. D., and Saegesser, R., "Holographic Interferometry in Soil Dynamics," *Proceedings of the VIII International Conference on Soil Mechanics and Foundation Engineering*, Moscow, U.S.S.R., Part 1.2, 1973, pp. 481-486.
83. Yegian, M., and Wright, S. G., "Lateral Soil Resistance-Displacement Relationships for Pile Foundations in Soft Clay," 1973 Offshore Technology Conference, Vol. II, 1973, pp. 663-676.
84. Yoshimi, Y., and Oh-Oka, H., "A Ring Torsion Appartus for Simple Shear Tests," *Proceedings of the VIII International conference on Soil Mechanics and Foundation Engineering*, Moscow, U.S.S.R., Vol. 1.2, 1973, pp. 501-506.

## SOME EFFECTS OF DYNAMIC SOIL PROPERTIES ON SOIL-STRUCTURE INTERACTION[a]

**Discussion by Michael J. Pender**[2]

One of the topics Richart has discussed in his wide-ranging and most interesting Terzaghi Lecture is the relationship between strain and shear modulus for soil. For the monotonic loading case the widely used hyperbolic representation is considered. For the reversed loading case, which is more important in earthquake engineering applications, the Ramberg-Osgood representation of the nonlinear stress-strain characteristics of the soil is used to considerable advantage. The purpose of this discussion is to set out briefly an alternative approach to modeling

---

[a] December, 1975, by F. E. Richart, Jr. (Proc. Paper 11764).
[2] Geomechanics Engr., Ministry of Works and Development, Central Labs., Lower Hutt, New Zealand.

the effect of shear strain amplitude on apparent shear modulus.

The writer has developed a stress-strain model for the behavior of overconsolidated soils (85) which is capable of giving a very good qualitative picture of the whole range of overconsolidated soil behavior. The model is based on a constitutive relationship for a work-hardening plastic material. It is based on the group of concepts known as critical state soil mechanics (86) and has the great advantage that only four material properties are needed to characterize a particular soil.

The general expressions for the strain increments (both volumetric and shear strains) consequent on a given stress increment are such that the stress-strain behavior can generally be evaluated only by numerical integration. However,

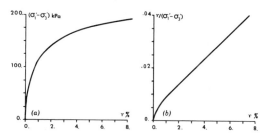

FIG. 36.—(a) Stress-Strain Response for Monotonic Loading; (b) Hyperbolic Representation of Fig. 36(a)

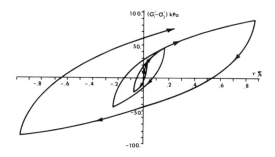

FIG. 37.—Stress-Strain Response for Cyclic Shearing with Progressive Increase in Shearing Stress

there is one situation in which the stress-strain curve can be specified directly. This is the shearing of a material lightly overconsolidated to an initial condition such that no change in volume is observed in drained shear. The shear strain for monotonic loading from a hydrostatic initial effective stress state under conventional triaxial stress conditions is

$$\gamma = \frac{2\kappa}{M(1+e)} \left[ \ln\left(\frac{M}{M-\eta}\right) - \frac{\eta}{M} \right] \quad \ldots \quad (20)$$

in which $\eta$ = the stress ratio = $(\sigma_1' - \sigma_3')/(\sigma_1' + 2\sigma_3')$; $e$ = the void of the material; $M$ = a soil parameter related to the friction angle of the material = $6 \sin \phi'/(3 - \sin \phi')$; and $\kappa$ = a soil parameter related to the swelling index for the material = 2.30 $C_s$. This curve is plotted in Fig. 36(a), and in Fig. 36(b) it is seen that it is very nearly hyperbolic.

When the direction of loading is reversed at some stress ratio, the shear strain after the load reversal is

$$\gamma = \gamma_o + \frac{2\kappa}{M^2(1+e)} \left[ (AM - \eta_o) \ln \left( \frac{AM - \eta_o}{AM - \eta} \right) - (\eta - \eta_o) \right] \quad \ldots \ldots (21)$$

in which $\gamma_o$ = the cumulative strain up to the point where the loading is reversed; $\eta_o$ = the stress ratio at which the reversal takes place; and, $A = +1$ for a compressive stress path and $-1$ for an extension path. An example of the application of this equation to cyclic behavior in which the shearing stress is gradually increased is given in Fig. 37. It is evident that behavior of the type shown in Fig. 3 of the author's paper is modeled well. If the incremental form of the stress-strain equations is used it is possible to calculate soil behavior for any state of overconsolidation and also to arrive at the pore pressure or volume change response.

It is felt that the main virtue of the approach outlined previously is the manner in which it fits into a model for the complete picture of soil behavior, and thus one does not have to depend on a special formulation to handle dynamic problems. Also the values of the same four material properties are sufficient to enable calculation for any state of overconsolidation. The preceding discussion has been in terms of conventional triaxial behavior, an extension to cover plane strain conditions is possible.

One aspect that the model does not handle well is the stiffness at very small strain amplitudes; it predicts a modulus that is rather too large. The reason is that the model was formulated on the assumption that soil has no elastic shear behavior. An important contribution made by Richart and his colleagues to the understanding of soil behavior is the elastic response, presumably the effect of the elasticity of the soil particles, that is observed at very small strain amplitudes. It is a relatively simple matter to incorporate an elastic shear modulus into the treatment given previously. When this is done the dependence of apparent shear modulus and equivalent viscous damping ratio on strain amplitude is modeled very well. The details will be published elsewhere.

APPENDIX.—REFERENCES

85. Pender, M. J., discussion, *Proceedings of the Symposium on Plasticity and Soil Mechanics*, Cambridge University Engineering Department, Cambridge, England, 1973, pp. 140-143.
86. Schofield, A. N., and Wroth, C. P., *Critical State Soil Mechanics*, McGraw-Hill Ltd., London, England, 1968.

# Some Effects of Dynamic Soil Properties on Soil-Structure Interaction[a]

Closure by F. E. Richart, Jr.,[3] F. ASCE

Pender's alternative approach to modeling stress-strain behavior of soils is interesting and apparently is satisfactory for static cases. He notes that the procedure does not handle well the stiffness at very small strain amplitudes, because it was based on the assumption that the soil has no elastic shear behavior.

For evaluating dynamic behavior of soils, it is necessary to consider both small and large amplitude shearing strains, and it is important to keep in mind the orders of magnitude involved. A "small shearing strain" amplitude is about $10^{-5}$ (0.001%) and a "large shearing strain" dynamic amplitude is about $10^{-3}$-$10^{-2}$ (0.10%-1%). In the paper, the writer attempted to summarize the results of dynamic testing of soils during the preceding 15 yr and to describe the influence of the variables noted in Eq. 5. In particular, the writer emphasized the importance of the work by Hardin and Drnevich, which described the shearing stress-strain behavior of the soil ranging from the small strain "elastic" behavior to failure. The development of the hyperbolic stress-strain parameters for dynamic soil behavior forms a basis for studies of dynamic response of soil masses including nonlinear shearing stress-strain relations. The writer also illustrated the application of the Ramberg-Osgood analytical curves to approximate the Hardin-Drnevich empirical data.

The writer believes it is essential to relate laboratory and field results for shear modulus at low shearing strain amplitudes. Then laboratory data permit evaluation of the changes in shear modulus and damping at increased strain levels, or from modifications introduced by changes in the other parameters noted in Eq. 5.

Pender has noted that "it is a relatively simple matter to incorporate an elastic shear modulus into the treatment given previously," and notes that the details will be published elsewhere. When this information becomes available, and a few years' test results to evaluate the influence of parameters comparable to those in Eq. 5 have been obtained, we will be in a better position to evaluate the effectiveness of this alternative approach to modeling stress-strain behavior of soils. The writer thanks Pender for bringing this method to his attention.

---

[a]December, 1975, by F. E. Richart, Jr. (Proc. Paper 11764).
[3]Prof. of Civ. Engrg., The Univ. of Michigan, Ann Arbor, Mich.

# JOURNAL OF THE GEOTECHNICAL ENGINEERING DIVISION

## THE ELEVENTH TERZAGHI LECTURE

Presented at the American Society of Civil Engineers Annual Convention and Exposition, Denver, Colorado

November 6, 1975

GEORGE GEOFFREY MEYERHOF

# Introduction of Eleventh Terzaghi Lecturer

## By George F. Sowers

The Terzaghi Lectures are sponsored by the ASCE Geotechnical Engineering Division in the spirit of Dr. Karl Terzaghi, Hon. M. ASCE and one of the giants in civil engineering evolution of the 20th Century. He was an inspired genius who could unravel the most knotty geotechnical problem so that any one who studied his concepts could comprehend the mechanics and find an answer. At the same time, he enthusiastically preserved the mystical romance of soil mechanics that saw soil as tiny jewels that could be a treasure to those who understood their nature but was treacherous quicksand to the unbeliever.

The Terzaghi lecturers have been selected from those among our profession who have shown the same spirit of inquiry and enthusiasm. They have honored our profession by sharing their ideas and achievements in those lectures. Dr. G. G. Meyerhof, the 11th Lecturer, is certainly one of those who has given his life to our profession in the spirit of Terzaghi. The first Terzaghi lecture was delivered by Ralph Peck in 1963; the successive lecturers have been:

| | |
|---|---|
| 1964, Arthur Casagrande | 1970, T. William Lambe |
| 1966, Laurits Bjerrum | 1971, John Lowe, III |
| 1967, H. Bolton Seed | 1973, Bramlette McClelland |
| 1968, Philip C. Rutledge | 1974, F. E. Richart, Jr. |
| 1969, Stanley D. Wilson | |

Dr. Meyerhof requires little introduction to geotechnical engineers. His publications on foundation engineering and his lectures at technical conferences have made his name familiar to those who have not had the privilege of knowing him personally.

For the nongeotechnical engineer, his kinship to the profession as a whole has been expressed by his earlier years as a structural engineer and his present position as Head of Civil Engineering at the Nova Scotia Technical College.

He graduated from London University, England, in 1938 and immediately engaged in structural engineering design. His earliest publications deal with reinforced concrete, foundation construction, and road design. In 1946 he joined the Staff of the British Building Research Station where his research on the bearing capacity of sand was a technical milestone. This work was recognized by a DSc degree from London University in 1954, although he already had earned the PhD.

In 1953 he emigrated to Canada and worked for two years as an engineer for the Foundation of Canada Engineering Corp. In 1955 he went to Nova Scotia Technical College in Halifax where he has continued both research and consulting work despite administrative duties as Head of Civil Engineering and for several years, Dean of Engineering.

His contributions have been recognized worldwide. He is Fellow of the Royal Society of Canada and a recipient of the Centennial Award of Canada. The Canadian Geotechnical Society, of which he was the first president, gave him their R. F. Legget Award. He has received honorary doctorates from the Technical University of Aachen, West Germany, and the University of Ghent, Belgium.

## Bearing Capacity and Settlement of Pile Foundations

### By George Geoffrey Meyerhof,[1] F. ASCE

**INTRODUCTION**

I should like to thank the American Society of Civil Engineers for the great honor of inviting me to give the Terzaghi Lecture. Exactly 50 years have passed since Karl Terzaghi in 1925 published his classic book on *Erdbaumechanik (Soil Mechanics)*, which created an entirely new branch of civil engineering. A chapter of this book is devoted to the bearing capacity of driven piles and treats in detail the soil deformation and pore-water pressure near piles, the static and dynamic bearing capacities, and the settlement of piles. It seems, therefore, appropriate at this time to review some of the recent concepts in estimating the bearing capacity and settlement of pile foundations in the light of Terzaghi's original contributions. Although these two subjects include some of the most difficult problems in foundation engineering, much progress has been made in our understanding of pile behavior, as was shown in a previous Terzaghi Lecture on the design of deep penetration piles for ocean structures (53).

Since both bearing capacity and settlement of pile foundations depend not only on the nature of the soil and the pile dimensions and layout but also on the method of installing the piles and other factors, each type of pile requires its own approach in estimating the behavior of pile foundations under load. Accordingly, only general principles of the action of some common types of pile foundations under static axial load will be presented in the absence of structural failure of the pile material. It will also be assumed that piles are selected as the most appropriate foundation type on the basis of structural and loading considerations, subsurface conditions, estimated performance, and economy of construction.

The methods of soil exploration used to ascertain the character of natural deposits and the tests to determine the physical and mechanical soil properties required for estimates of the bearing capacity and settlement of piles are generally similar to those of other types of foundations. However, since the behavior of piles depends to a considerable extent on local subsoil conditions, the design and construction of pile foundations is facilitated by related field tests using static and dynamic penetrometers, especially in cohesionless soil, and vane shear

---

Note.—Discussion open until August 1, 1976. To extend the closing date one month, a written request must be filed with the Editor of Technical Publications, ASCE. This paper is part of the copyrighted Journal of the Geotechnical Engineering Division, Proceedings of the American Society of Civil Engineers, Vol. 102, No. GT3, March, 1976. Manuscript was submitted for review for possible publication on June 3, 1975.
[1] Prof. and Head, Dept. of Civ. Engrg., Nova Scotia Technical Coll., Halifax, N.S., Canada.

and piezometer tests in fine-grained cohesive soil. Moreover, for large piled structures this information may be amplified by carefully conducted pressuremeter and plate load tests in large diameter boreholes to determine the initial in-situ stresses, strength, and deformation characteristics of the soil. Additionally, load tests should be made on preliminary piles as a check of estimates of the action of single piles under load, and some construction control of piles driven into cohesionless soil may be obtained by correlating the results of such tests with pile driving formulas and methods of dynamic analysis.

Since installation of piles changes the initial stresses, strength, and deformation properties of the soil mass near the piles, the soil properties governing the bearing capacity and settlement of pile foundations may differ considerably from the original conditions. On account of the complex interaction between the soil and piles during and after construction of the foundation, the behavior of single piles and groups under load can only roughly be estimated from soil tests and semi-empirical methods of analysis based on the results of pile load tests. The empirical data obtained mainly from field observations and some theoretical considerations will be examined in the next sections for driven and bored piles in sand and clay including nonuniform soils.

### BEARING CAPACITY OF PILES IN SAND

When piles are driven into sand, the soil near the piles is compacted to a distance of a few pile diameters. Field investigations of the change of density of sand near single driven piles have been analyzed to obtain some information of the degree of compaction and prestressing of the soil by pile driving and their influence on the point resistance and skin friction of such piles (59). Based on simplifying assumptions approximate estimates have also been made of the deformation of sand near driven piles and the corresponding soil pressures on the point and shaft of a pile (44,87,97).

These and other attempts to predict the ultimate bearing capacity of such piles on a semi-empirical basis indicate that in homogeneous sand both point resistance and average skin friction of a pile would increase with greater depth of penetration. However, large-scale experiments and field observations have shown (39,40,86,95) that the theoretical relationships hold only when the pile point is above a certain critical depth. Below this depth, the point resistance and average skin friction remain practically constant in a homogeneous sand deposit due to effects of soil compressibility, crushing, arching, and other factors. Since no satisfactory method of analysis of pile behavior below the critical depth is available, an empirical approach is necessary at present, as indicated in the following sections.

**Point Resistance.**—The ultimate bearing capacity, $Q_u$, of a pile in homogeneous soil may be expressed by the sum of point resistance $Q_p$ and skin resistance $Q_s$, or

$$Q_u = Q_p + Q_s = q_p A_p + f_s A_s \quad \ldots \ldots \ldots \ldots \ldots \ldots \ldots \ldots \ldots (1)$$

in which $q_p$ = the unit bearing capacity of pile point of area $A_p$; and $f_s$ = the average unit skin friction on shaft of area $A_s$. Further, the ultimate unit point resistance in homogeneous sand may be represented by

$$q_p = p_o N_q \leq q_l \quad \quad \quad \quad \quad \quad \quad \quad \quad \quad \quad \quad \quad \quad \quad \quad (2)$$

in which $p_o$ = the effective overburden pressure at pile point; $N_q$ = the bearing capacity factor with respect to overburden pressure; and $q_l$ = the limiting value of unit point resistance for $D/B \geq D_c/B$ where $B$ = width of pile, $D$ = depth, and $D_c$ = critical depth of penetration of pile.

The semi-empirical relationship between the bearing capacity factor, $N_q$, for driven circular or square piles with various depth ratios $D_b/B$ in the bearing stratum and the angle of internal friction, $\phi$, of the soil before pile driving

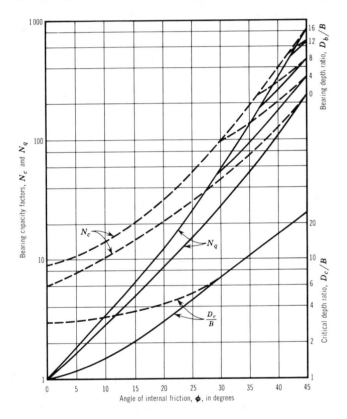

**FIG. 1.—Bearing Capacity Factors and Critical Depth Ratios for Driven Piles**

(16,60) is shown in Fig. 1. It is found that the factor $N_q$ increases roughly linearly with $D_b/B$ and reaches its maximum value at a depth ratio of roughly one-half of the critical depth ratio $D_c/B$ beyond which conventional bearing capacity theory no longer applies (40). The approximate depth ratio $D_c/B$ (23) is also shown in Fig. 1, and for full-sized piles it depends mainly on the friction angle and compressibility of the soil and the ground-water conditions.

The empirical values of the limiting unit point resistance, $q_l$, derived from the limiting static cone resistance, $q_c$, in homogeneous sand (61) are shown in Fig. 2. The relationship between $q_l$ and $\phi$ may be represented approximately by

$$q_l = 0.5 N_q \tan \phi \quad \ldots \quad (3)$$

in which $N_q$ = the bearing capacity factor for short pile and $q_l$ is in tons per square foot (100 kN/m$^2$). The value of $q_l$ would correspond to a limiting effective vertical stress near the pile point at failure varying from about 0.25 tsf (25 kN/m$^2$) for loose sand to 0.5 tsf (50 kN/m$^2$) for dense sand, and this stress remains practically independent of the effective overburden pressure and ground-water conditions beyond the critical depth.

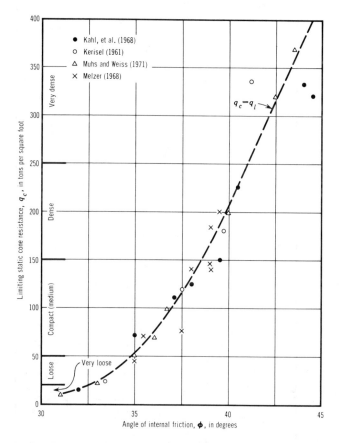

FIG. 2.—Approximate Relation between Limiting Static Cone Resistance and Friction Angle of Sand (1 tsf = 95.8 kN/m$^2$)

Although the values of $N_q$ and $q_l$ depend mainly on the friction angle, $\phi$, they are also influenced by the compressibility of the soil, the method of pile installation, and other factors. Thus, for a given initial $\phi$, bored piles have a unit point resistance of only about one-third to one-half of that of driven piles (22,39,95), and bulbous piles driven with great impact energy have up to about twice the unit point resistance of driven piles of constant section (59). Analysis of the few data on driven tapered piles indicates that the values of $N_q$ and $q_l$ are not significantly affected by pile taper (36,74).

If piles are driven into homogeneous soil to more than the critical depth or if they penetrate through compressible material into a thick bearing stratum which is located below the critical depth of that stratum, the unit point resistance cannot be estimated by conventional bearing capacity theory in terms of $N_q$. The corresponding value of $q_p$ becomes practically independent of the overburden

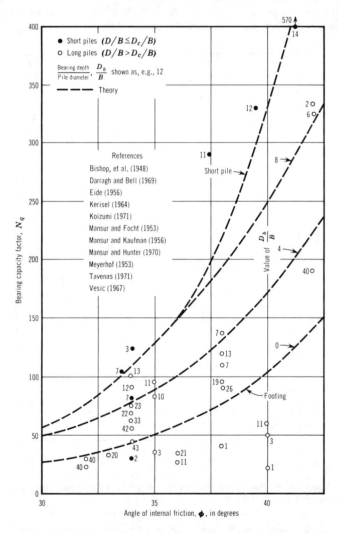

FIG. 3.—Bearing Capacity Factor for Driven Piles in Sand

pressure at the pile point and it depends on the value of $q_l$. This is shown by an analysis of the results of pile load tests, which roughly support the semi-empirical bearing capacity factors, $N_q$, for short piles in sand above the critical depth but not for piles longer than about 15 to 20 pile diameters (Fig. 3).

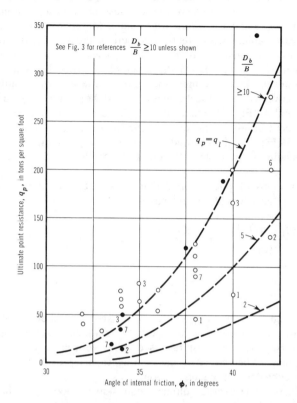

**FIG. 4.—Ultimate Point Resistance of Driven Piles in Sand (1 tsf = 95.8 kN/m²)**

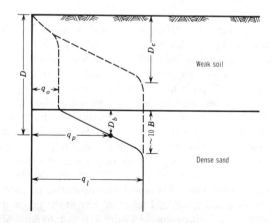

**FIG. 5.—Relation between Ultimate Point Resistance of Pile and Depth in Sand Stratum beneath Weak Soil Layer**

When long piles are driven into sand or through weak strata to a firm sand deposit, the values of $N_q$ deduced from pile load tests for a given initial friction angle, $\phi$, of the sand vary within wide limits. Moreover, the deduced values of $N_q$ decrease rapidly with greater overburden pressure at the tip and they are sometimes even considerably smaller than those for a shallow circular footing, which is impossible (Fig. 3). On the other hand, when the same test results on long piles are analyzed in terms of the limiting value of $q_l$, reasonable agreement is obtained between the empirical values of $q_p$ and the aforementioned relationship between $q_l$ and $\phi$ (Fig. 4), provided the pile points penetrate to a depth, $D_b$, of at least about 10 pile diameters into the sand bearing stratum, as suggested previously (56,58). For loose sand the value of $q_p$ is somewhat greater than the corresponding static cone value due to the greater soil compaction by pile driving.

For penetrations of piles shorter than about $10B$ into the bearing stratum the value of $q_p$ may roughly be estimated by

$$q_p = q_o + \frac{(q_l - q_o) D_b}{10B} \leq q_l \qquad (4)$$

or, conservatively $\quad q_p = \dfrac{q_l D_b}{10B} \leq q_l \qquad \qquad (5)$

in which $q_o$ and $q_l$ are limiting unit point resistance in upper weak and lower firm stratum, respectively (Fig. 5). The latter relationship is also shown in Fig. 4 for comparison with the test results of piles with small embedment in the bearing stratum. Other cases of the bearing capacity of piles in nonuniform soils will be considered in a later section.

**Skin Friction.**—The average ultimate unit skin friction, $f_s$, in homogeneous sand (Eq. 1) may be expressed by

$$f_s = K_s \bar{p}_o \tan \delta \leq f_l \qquad (6)$$

in which $K_s$ = the average coefficient of earth pressure on pile shaft; $\bar{p}_o$ = the average effective overburden pressure along shaft; $\delta$ = the angle of skin friction; and $f_l$ = the limiting value of average unit skin friction for $D/B \geq D_c/B$, which roughly approximates that for the unit point resistance. An estimate of the skin friction and, particularly, of the earth pressure coefficient, $K_s$, on the basis of the friction angle of the sand and the method of pile installation is even more difficult than for the point resistance. Thus, the skin friction depends not only on the aforementioned factors, but also on the compressibility of the soil, the original horizontal stress in the ground, as represented by the earth pressure coefficient at rest $K_o$, and on the pile size and shape (87). Although a rough estimate of the limiting unit skin friction, $f_l$, can be obtained from the results of penetration tests mentioned subsequently, reliable values of $K_s$ and $f_l$ can only be deduced from load tests on piles at the given site.

Analysis of load tests on instrumented piles driven into sand shows (96) that the local ultimate unit skin friction, $f_z$, increases with depth only along the upper portion of a pile to a maximum and then decreases to a minimum at the pile point, the average value of $f_z$ being denoted by $f_s$. Accordingly, the corresponding local coefficient of earth pressure, $K_z$, on the shaft decreases with depth along the pile from a maximum near the top where $K_z$ may approach the passive earth pressure coefficient to a minimum near the pile point where $K_z$ may be less than the initial earth pressure coefficient, $K_o$ (42), the average ultimate value of $K_z$ being denoted by $K_s$. Analysis of the few available results of load tests on short piles above the critical depth in generally homogeneous normally consolidated sand shows that the value of $K_s$ for a given initial friction angle, $\phi$, can scatter considerably from a lower limit of roughly $K_o$ for bored piles or piles jacked into loose sand to about four times this value or more for piles driven into dense sand due to dilatancy effects and other factors (Fig. 6).

As indicated previously, conventional shaft capacity theory in terms of $K_s$ cannot be used for piles longer than about 15 to 20 pile diameters because

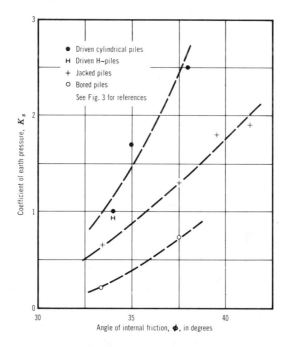

**FIG. 6.—Coefficient of Earth Pressure on Shaft of Piles above Critical Depth in Sand**

the corresponding value of $f_s$ becomes practically independent of the average overburden pressure along the shaft and it is given by $f_l$. This has been shown by analyzing the results of load tests on long piles driven into sand or through weak soil to a firm sand deposit, when the deduced values of $K_s$ for sand were found to decrease rapidly with greater overburden pressure to less than $K_o$, which is impossible. However, even the empirical relationship between the limiting value of $f_s = f_l$ and the friction angle $\phi$ of sand indicates a wide variation from a lower limit for bored piles and piles jacked into normally consolidated sand to an upper limit for piles driven into overconsolidated sand (Fig. 7). Partial downward drag of upper soil into lower strata during pile installation (91) may also explain some of the scatter of the unit skin friction deduced from the results of pile load tests in nonuniform soils.

As would be expected, the ultimate unit skin friction increases with the volume of displaced soil and therefore bored piles or driven piles with a small soil displacement, such as H-piles, have a smaller average skin friction than large displacement piles. Moreover, driven tapered piles have a greater average skin friction than cylindrical piles of the same volume displacement, and for tapers exceeding about 1% of the embedded pile length the value of $f_s$ is roughly 1.5 times that of corresponding cylindrical piles (36,67,71,74).

**Penetrometer Estimates.**—Small variations of the friction angle, $\phi$, of sand considerably influence the values of $K_s$ and $N_q$ for short piles above the critical

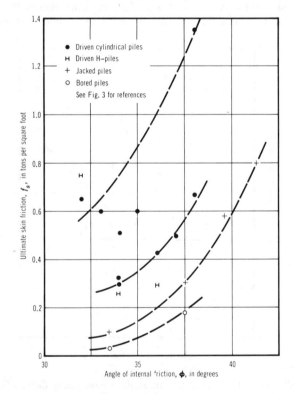

FIG. 7.—Ultimate Skin Friction of Piles in Sand (1 tsf = 95.8 kN/m²)

depth and similarly influence the corresponding values of $f_l$ and $q_l$ for long piles. It is therefore generally preferable to use the results of subsurface soundings by means of penetration tests directly for preliminary estimates of the point resistance and skin friction of piles using semi-empirical and conservative methods which have been checked by field observations.

The static cone penetrometer is a model pile and, when pushed into homogeneous cohesionless soil, the penetration resistance can be correlated with a similarly installed full-sized pile. When a pile is driven into stratified cohesionless soil, however, an estimate of the ultimate point resistance and skin friction can only be made by semi-empirical methods from the static cone and friction resistances (3,21,37). Moreover, when the pile point is above the critical depth in the bearing stratum, the unit point resistance has to be reduced from the limiting static cone resistance, $q_c$, in proportion to the embedment ratio, $D_b/B$, in this stratum. In this case it was suggested (58) that, approximately

$$q_p = \frac{q_c D_b}{10B} \leq q_l \quad \ldots \ldots \ldots \ldots \ldots \ldots \ldots \ldots \ldots \ldots \ldots \ldots (7)$$

which is supported by an analysis of load tests on piles driven to a thick firm cohesionless stratum not underlain by a weak deposit (Fig. 4).

The ultimate skin friction of a driven cylindrical pile is, approximately, given by the unit resistance, $f_c$, of the local friction sleeve of static penetrometers. It has been found that in sand the local friction ratio, $f_c/q_c$, varies roughly between 1/2% and 2% (4), while the corresponding ratio of $f_s/q_p$ for driven cylindrical piles is roughly 1/2% to 1% (Figs. 4 and 7). For bored piles roughly one-third to one-half of the static cone and friction resistances applicable to driven piles may be used for preliminary estimates in cohesionless soil (22,39,59).

By empirical correlations of the resistance of static and dynamic penetration tests for cohesionless soil with given grain size characteristics and analyses of pile load tests, the dynamic cone and standard penetration resistances can also be used in these soils for preliminary estimates of the ultimate point resistance and skin friction of driven piles with an allowance for the embedment ratio of the point and other factors as before (58). Thus, it was suggested that for piles driven to depth $D_b$ into a sand stratum the ultimate unit point resistance, in tons per square foot (100 kN/m²), may be taken approximately as

$$q_p = \frac{0.4 \, N D_b}{B} \leq 4N \quad \ldots \ldots \ldots \ldots \ldots \ldots \ldots \ldots \ldots \ldots \ldots (8)$$

in which $N$ = the average standard penetration resistance, in blows per foot (blows per 0.3 m), near the pile point. It was also suggested that the average ultimate skin friction of driven displacement piles, in tons per square foot (100 kN/m²), is roughly given by

$$f_s = \frac{\bar{N}}{50} \quad \ldots \ldots \ldots \ldots \ldots \ldots \ldots \ldots \ldots \ldots \ldots \ldots \ldots \ldots \ldots \ldots (9)$$

in which $\bar{N}$ = the average standard penetration resistance, in blows per foot (blows per 0.3 m), within embedded length of pile, and that one-half of this value may conservatively be used for piles with small soil displacement, such as H-piles. The upper limit of $q_p$ in Eq. 8 and the value of $f_s$ in Eq. 9 represent the limiting values $q_l$ and $f_l$, respectively.

As a check of these proposed relationships between the ultimate point resistance and average skin friction of piles driven into sand and the recorded standard penetration resistance, the results of pile load tests have been analyzed. It is found that the observed unit point resistance is generally in fair agreement with Eq. 8, except at greater overburden pressures at the point when the ratio of $q_p/N$ decreases (Fig. 8). If the $N$-value corresponding to an effective overburden pressure of 1 tsf (100 kN/m²) is taken as standard and other values near the pile point are empirically corrected as suggested for shallow foundations

(72), the estimates of the point resistance will be somewhat improved, and such corrected $N$-values should therefore be used. The few pile load tests in coarse sand and gravel also support Eq. 8, while for piles driven into nonplastic silt better agreement is obtained by using an upper limit of approximately

$$q_p = 3 N \quad \ldots \ldots \ldots \ldots \ldots \ldots \ldots \ldots \ldots \ldots \ldots \ldots \ldots \ldots \quad (10)$$

in tons per square foot (100 kN/m²), which with the aforementioned overburden pressure correction approaches the relationship suggested before (89).

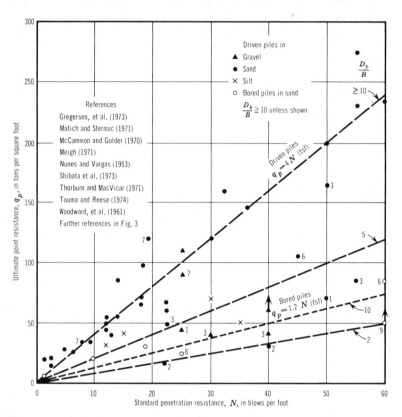

**FIG. 8.—Empirical Relation between Ultimate Point Resistance of Piles and Standard Penetration Resistance in Cohesionless Soil (1 tsf = 95.8 kN/m²; 1 blow/ft = 1 blow/0.3 m)**

The observed average ultimate unit skin friction of piles driven into cohesionless soil generally exceeds that given by Eq. 9, which also provides reasonable estimates for H-piles so that no further reduction needs to be made for such piles (Fig. 9). Some of the scatter of the observed unit skin friction for a given average $N$-value may be attributed to partial downdrag of upper soil as mentioned before and different stress histories of the sand, which are only partly reflected by the measured standard penetration resistance (61). On the other hand, for a given average $N$-value there does not appear to be a significant influence of overburden pressure and grain size characteristics of the soil on the value of $f_s$ (Fig. 9). However, for driven piles with a taper exceeding about 1% about 1.5 times the unit skin friction given by Eq. 9 can be used, as mentioned before.

The ultimate unit point resistance and average skin friction of bored piles in cohesionless soil are less than those of driven piles, and roughly one-third

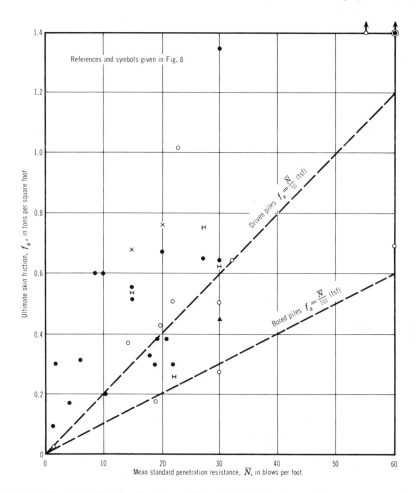

**FIG. 9.—Empirical Relation between Ultimate Skin Friction of Piles and Standard Penetration Resistance in Cohesionless Soil (1 tsf = 95.8 kN/m²; 1 blow/ft = 1 blow/0.3 m)**

of the corresponding values given by Eqs. 8 and 10 for the point resistance and about one-half of the value given by Eq. 9 for the skin friction may be used for preliminary estimates (Figs. 8 and 9). For bulbous piles driven with great impact energy up to about twice the value given by Eqs. 8 and 10 can be used for the ultimate point resistance (59).

**Nonuniform Soils.**—When piles longer than about 15 to 20 pile diameters,

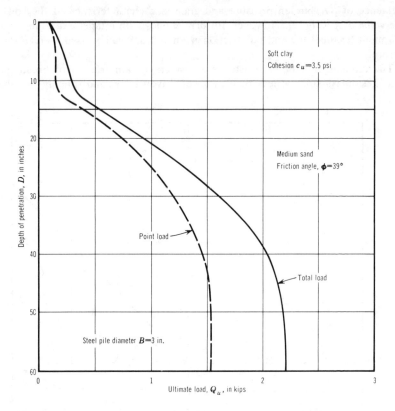

**FIG. 10.—Bearing Capacity of Model Pile Penetrating through Clay Layer into Sand (1 kip = 4.45 kN; 1 in. = 25.4 mm)**

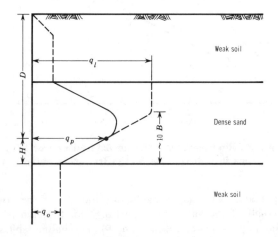

**FIG. 11.—Relation between Ultimate Point Resistance of Pile and Depth in Thin Sand Layer Overlying Weak Soil**

as mentioned previously, penetrate through a weak stratum into a thick firm deposit of cohesionless soil, the ultimate unit point resistance in the bearing stratum increases roughly linearly with the embedment ratio, $D_b/B$, in this stratum, as indicated by Eqs. 4 and 5. Further, below a critical depth ratio of about 10 in the bearing stratum the point resistance remains practically constant at the limiting value, $q_l$, for this stratum (Fig. 5). If the properties of the cohesionless bearing stratum vary near the pile tip, the average values in the corresponding local failure zone between about 4 pile diameters above the tip and 1 pile diameter below the tip should be used in estimating the point resistance

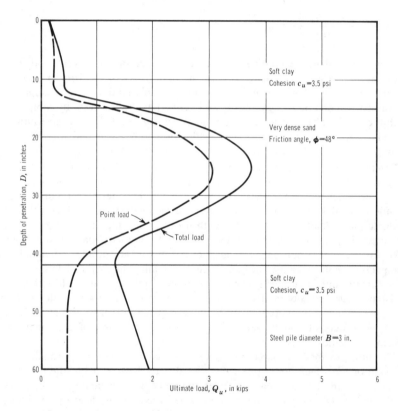

FIG. 12.—Bearing Capacity of Model Pile Penetrating into Thin Sand layer Overlying Clay (1 kip = 4.45 kN; 1 in. = 25.4 mm)

(59). Large-scale experiments with piles penetrating through soft clay into sand (76) or through loose sand into dense sand (93,95) support the aforementioned suggestions (Fig. 10). It was also found that the average ultimate unit skin friction in each cohesionless stratum can roughly be estimated directly from the limiting values of $f_l$ using average soil properties for each layer.

If pile points rest in a relatively thin firm stratum underlain by a weak deposit, the ultimate unit point resistance in the bearing stratum may be governed by the resistance to punching of the piles into the underlying weak soil (Fig. 11). Theory and field observations of piles in a sand layer overlying clay indicate

(35,62) that the rate of decrease of the point resistance with smaller thickness of sand below the tip is similar to that for smaller embedment ratios $D_b/B$ in the sand. This is confirmed by large-scale experiments and model tests on piles penetrating to various depths in a dense sand layer overlying a soft clay or a loose sand deposit (76,93) (Fig. 12). Consequently, approximate estimates of the ultimate unit point resistance of piles in the lower part of a firm cohesionless stratum with a thickness $H$ to the weak deposit below the pile points and smaller than the critical thickness of about $10B$ can be obtained by an expression similar to Eq. 4, i.e.

$$q_p = q_o + \frac{(q_l - q_o)H}{10B} \leq q_l \quad \quad \quad \quad \quad \quad \quad \quad \quad \quad \quad \quad (11)$$

in which $q_o$ and $q_l$ are limiting unit point resistance in lower weak and upper firm stratum, respectively (Fig. 11). In the upper part of the firm stratum the ultimate point resistance would follow Eq. 4. This method is also supported by the corresponding evaluation of static cone penetration tests in comparison with the results of pile load tests in nonuniform soils (23). The average ultimate unit skin friction in each cohesionless soil layer can be estimated approximately from the limiting values, $f_l$, in each layer, as mentioned previously.

**Group Capacity.**—After the ultimate load of single piles has been estimated by one of the aforementioned methods, the bearing capacity of the pile group has to be evaluated. The few available full-scale load tests to failure on pile groups in sand and the results of extensive model tests have shown that the ultimate group load of driven piles with a center spacing of about 2 to 4 pile diameters is greater than that of the sum of the ultimate load of single piles (59). The greater group capacity is due to the overlap of the individual soil compaction zones near the piles and increases mainly the skin resistance, which may exhibit equivalent pier shear failure at small pile spacings, while the point resistance is practically unaffected by group action even at small pile spacings (95). Accordingly, the corresponding ultimate group load in sand not underlain by a weak deposit should generally be taken as the sum of the single pile bearing capacities.

The ultimate load of bored pile groups in sand, however, is smaller than the sum of the single bored pile capacities due to the overlap of the individual shear zones in the soil near the piles without soil compaction. Interference of the individual pile point shear zones would theoretically lead to a reduction of the individual ultimate point loads by roughly one-half for a pile spacing of about 3 pile diameters, and some reduction of the skin resistance must also be expected. The decreased bearing capacity has been confirmed by some load tests on bored pile groups (73). It may, therefore, be suggested that the ultimate group capacity of bored pile groups in sand not underlain by a weak deposit should be taken as about two-thirds of the sum of the single pile capacities at customary pile spacings.

If a pile group in a firm bearing stratum of limited thickness in underlain by a weak deposit, the ultimate group load is given by the smaller value of either the sum of the aforementioned single pile capacities or by block failure of an equivalent pier consisting of the pile group and enclosed soil mass punching through the firm stratum into the underlying weak soil (20,62).

BEARING CAPACITY OF PILES IN CLAY

When piles are driven into saturated clay, the soil near the piles is displaced and it is remolded to a distance of up to roughly 1 pile diameter. Within the highly disturbed zone of the clay at the pile the pore-water pressure induced by the large installation stress dissipates fairly rapidly. After consolidation the clay adjacent to driven piles may have a greater shear strength and smaller water content than at some distance from the piles or before pile driving. In very sensitive clay or in stiff overconsolidated clay, however, the final shear strength or adhesion at the pile may be less than that of the undisturbed soil. The relationship between the increase of bearing capacity and time for a pile driven into saturated clay depends on the nature of the soil, dimensions of the pile and other factors (87).

Near bored concrete piles the clay is usually somewhat remolded and may be considerably softened to a distance of about 1 in. (25 mm) from the pile by the installation and concreting of the shaft. Load tests on such piles have shown (64) that there is no significant increase in the bearing capacity with time on account of the very slow consolidation of the softened clay adjacent to the shaft under the relatively small installation stress. The behavior of piles in unsaturated clay is more complex than for saturated soil. Experience on a regional basis indicates that bearing capacity analysis using shear strength parameters of cohesive-frictional material at the given degree of saturation and stress level gives reasonable agreement with the results of some pile load tests (25,68).

**Point Resistance.**—On the same assumptions, as mentioned for sand, and using Eq. 1 for the ultimate bearing capacity of a pile, the ultimate unit point resistance in homogeneous cohesive soil may be expressed by

$$q_p = c N_c + p_o N_q \leq q_m \quad \ldots \ldots \ldots \ldots \ldots \ldots \ldots \ldots \ldots \ldots \ldots \ldots \ldots (12)$$

in which $c$ = the average unit cohesion of soil near pile point, $N_c$ = the bearing capacity factor with respect to cohesion; $q_m$ = the limiting value of unit point resistance below critical depth; and other symbols are as defined before. The semi-empirical bearing capacity factor, $N_c$, for driven circular or square piles with various depth ratios $D_b/B$ (16,60) and the corresponding approximate critical depth ratio $D_c/B$ are shown in Fig. 1. For a given $\phi$ the value of $N_c$ depends on the same factors as mentioned for $N_q$, but empirical support for the net value of $q_m$ is mainly limited so far to saturated clay.

In saturated homogeneous clay under undrained conditions, theory and observation have shown that the value of $N_c$ below the critical depth varies with the sensitivity and deformation characteristics of the clay from about 5 for very sensitive brittle normally consolidated clay (45,75) to about 10 for insensitive stiff overconsolidated clay (56,81), although a value of 9 is frequently used for bearing capacity estimates of driven and bored piles. Moreover, these values of $N_c$ are based on the initial undrained cohesion of the soil mass near the pile point, using carefully performed undrained triaxial compression tests on large samples of the clay. Furthermore, any disturbance of the clay by pile installation mainly affects the initial point resistance, and subsequent consolidation of the clay will normally lead to a bearing capacity exceeding the undrained value at the end of the construction of the foundation.

**Skin Friction.**—The average ultimate unit skin friction, $f_s$, in Eq. 1 or the equivalent ultimate shaft adhesion, $c_a$, in homogeneous saturated clay is usually expressed by

$$c_a = \alpha\, c_u \quad\quad\quad\quad\quad\quad\quad\quad\quad\quad\quad\quad\quad\quad\quad\quad\quad\quad (13)$$

in which $\alpha$ = the empirical adhesion coefficient for reduction of average undrained shear strength $c_u$ of undisturbed clay within embedded length of pile. The coefficient $\alpha$ depends on the nature and strength of clay, dimensions and method of installation of pile, time effects, and other factors. The values of $\alpha$ vary within wide limits and they decrease rapidly with increasing shear strength. For driven piles the values of $\alpha$ range on the average roughly from unity for soft clay to one-half or less for stiff clay, while for bored piles in stiff clay $\alpha$ is roughly one-half (82,90,91,103). These values of $\alpha$, which represent a maximum adhesion, $c_a$, of roughly 1 tsf (100 kN/m²), indicate that the drained shear strength of the clay would usually govern shaft adhesion.

Thus, immediately after pile driving the shaft adhesion is closely given by the undrained shear strength of remolded clay (27,29,56). However, at later stages and particularly at the end of the foundation construction, the shaft resistance of piles will be governed by the effective drained shear strength parameters, $c$ and $\phi$, of remolded clay failing close to the shaft. The corresponding effective unit skin friction, $f_s$, in homogeneous clay may then be taken as

$$f_s = c + K_s \bar{p}_o \tan \phi \leq c_u \quad\quad\quad\quad\quad\quad\quad\quad\quad\quad\quad\quad (14)$$

in which the symbols are as defined for Eq. 6. Eq. 14 assumes that the excess pore-water pressure sometime after installation and loading of the pile is negligible compared with the effective overburden pressure. Analysis of field observations shows that excess pore pressures at the shaft of single piles driven into saturated soft clay may initially amount up to roughly $5c_u$ to $7c_u$, which approaches the initial horizontal installation stress on the shaft (43). After about one or more months when load tests on piles in clay are usually made, these pore pressures have essentially dissipated (18). Moreover, the initial excess pore pressure induced by pile loading to failure generally appears to be in the range of only about $0.2c_u$ to $0.5c_u$ at the shaft.

It may, therefore, be concluded that the ultimate skin friction of piles in saturated clay can approximately be estimated from the drained shear strength of remolded soil for which the cohesion may usually be taken as zero. On this basis Eq. 14 may be written

$$f_s = \beta \bar{p}_o \leq c_u \quad\quad\quad\quad\quad\quad\quad\quad\quad\quad\quad\quad\quad\quad\quad\quad (15)$$

in which $\beta$ = the skin friction factor and is

$$\beta = K_s \tan \phi \quad\quad\quad\quad\quad\quad\quad\quad\quad\quad\quad\quad\quad\quad\quad\quad\quad (16)$$

Support for this approach is obtained from an analysis of load tests on instrumented piles driven into clay (96) when the local ultimate unit skin friction, $f_z$, is found to increase roughly in direct proportion to depth or the effective overburden pressure along most of the shaft.

For bored piles and for piles driven into saturated soft clay the ultimate coefficient, $K_s$, of earth pressure on the shaft (Eq. 16) may be expected to be close to that of the earth pressure at rest $K_o$ (14,17,27,87,105), as was found

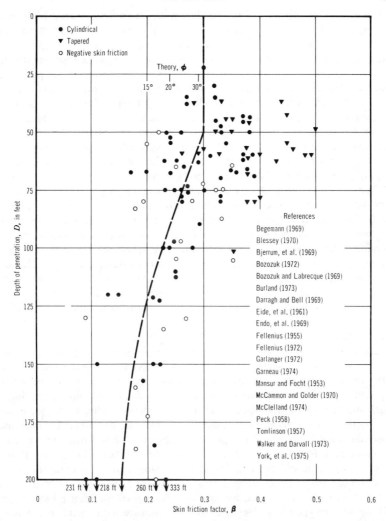

**FIG. 13.—Positive and Negative Skin Friction Factors of Driven Piles in Soft and Medium Clays (1 ft = 0.305 m)**

previously for loose sand. For this condition and homogeneous normally consolidated clay when $K_o = 1 - \sin \phi$, approximately, the skin friction factor, $\beta$, may be represented by

$$\beta = (1 - \sin \phi) \tan \phi \quad \dots \quad (17)$$

which would vary theoretically from about 0.2 to 0.3 for a typical range of $\phi$ for clay. Analysis of the skin friction of piles driven into soft and medium clays shows that the factor $\beta$ decreases with the length of piles from a range of about 0.25 to 0.5 for short piles to a range of about 0.1 to 0.25 for very long piles (Fig. 13). This decrease of $\beta$ with greater pile length may be explained by progressive mobilization of the maximum skin friction due to the compression of long piles for which the effective friction angle of the clay at the shaft may approach the residual value of roughly one-half of the peak angle, $\phi$.

The empirical values of $\beta$ also indicate that $K_s$ is somewhat greater than $K_o$ for driven piles, especially when they are tapered. Accordingly, for preliminary estimates the skin friction factor, $\beta$, of piles driven into soft and medium clays may be taken as about 0.3 (14), provided the depth of penetration is not greater than about 50 ft (15 m). The value of $\beta$ should be reduced for longer piles to about 0.15 for a depth of penetration exceeding about 200 ft (60 m), as indicated in Fig. 13. Since the shaft adhesion is generally not significantly affected by pile taper, the same value of $\beta$ as for cylindrical or H-piles may be used for preliminary estimates in soft and medium clays.

While reconsolidation of clay after pile driving causes some downward drag or negative skin friction, it is usually of little importance for friction piles but may become significant for point-bearing piles. If piles pass through consolidating soft clay or silt the negative skin friction can roughly be estimated from the undrained shear strength of the soil (Eq. 13), when the value of $\alpha$ is found to range from about one-half to over unity for uncoated piles (98). However, since the long-term negative friction is governed by the drained shear strength of the clay, the ultimate negative skin friction on the pile shaft should be evaluated on the basis of Eqs. 15-17. Measurements on instrumented piles driven through uniform clay have confirmed that the value of the local negative unit skin friction, $f_z$, increases roughly linearly with depth (9,12,28,30).

Accordingly, Eq. 15 can be used to determine from pile load tests the relationship between the ultimate negative skin friction factor, $\beta$, along the upper portion of the shaft of uncoated piles and the embedded pile length in soft and medium clays (Fig. 13). It is found that this factor also decreases with increasing pile length and that it is sometimes smaller than the corresponding factor for positive skin friction. While some of this difference may be explained by incomplete consolidation of the soft soil after pile installation and occasionally by insufficient soil movement for mobilization of ultimate skin friction, the main difference is due to the time effect on $f_s$ by the very slow rate of loading (8) and the reduction of vertical stress along the upper portion of piles as a result of negative skin friction (106). This reduction of the effective overburden pressure would lead to a corresponding decrease of the ultimate unit point resistance in cohesionless soil above the critical depth, but it does not affect the limiting value, $q_l$, below that depth. For preliminary estimates the negative skin friction factor, $\beta$, of uncoated piles penetrating to a depth of about 50 ft (15 m) into soft clay and silt may be taken as about 0.3 as for positive skin friction, and this factor may be reduced with greater depth to about 0.2 for piles longer than about 200 ft (60 m).

The ultimate skin friction of piles installed in stiff saturated clay can also be estimated from Eqs. 15 and 16 on the basis of the drained shear strength of remolded soil, provided that the coefficient, $K_s$, is known from previous pile load tests or the value of $K_o$ can be estimated from field or laboratory tests, from which roughly

$$K_o = (1 - \sin \phi) \sqrt{R_o} \qquad (18)$$

in which $R_o$ = the overconsolidation ratio of clay. Thus, for stiff fissured overconsolidated London clay with a value of $K_o$ ranging from about 3 at shallow depth to unity at great depth, analysis of pile load tests shows that for driven piles the average value of $K_s$ varies from roughly $K_o$ to more than $2K_o$ corresponding to a range of $\beta$ from roughly unity to over 2, while for bored piles $K_s$ varies from about $0.7K_o$ to $1.2K_o$ which corresponds to a range of $\beta$ from roughly 0.7 to 1.4 (14,17). These analyses indicate that an average value of $K_s$ equal to $K_o$ tends to underestimate the skin friction of driven piles and

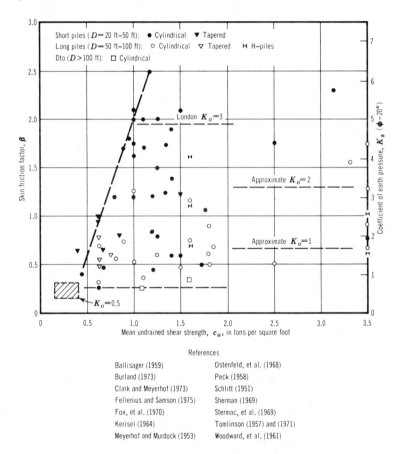

References

| | |
|---|---|
| Ballisager (1959) | Ostenfeld, et al. (1968) |
| Burland (1973) | Peck (1958) |
| Clark and Meyerhof (1973) | Schlitt (1951) |
| Fellenius and Samson (1975) | Sherman (1969) |
| Fox, et al. (1970) | Stermac, et al. (1969) |
| Kerisel (1964) | Tomlinson (1957) and (1971) |
| Meyerhof and Murdock (1953) | Woodward, et al. (1961) |

**FIG. 14.—Skin Friction Factor of Driven Piles in Stiff Clay (1 tsf = 95.8 kN/m²)**

to overestimate somewhat the skin friction of bored piles in London clay due to the corresponding change of the initial horizontal stress in the ground near the shaft by pile installation (Figs. 14 and 15).

On the other hand, if the value of $K_o$ of stiff clay is not known, the corresponding skin friction factor can only be estimated within wide limits. Thus, analysis of the results of load tests on piles driven into stiff clay shows that the skin friction factor, $\beta$, increases with the average undrained shear strength, $c_u$, of the clay (Fig. 14). However, for a given average value of $c_u$ the empirical factor $\beta$ can scatter considerably from a lower limit of about 0.5 for long piles in lightly overconsolidated clay to an upper limit of about 2.5 for short piles in heavily overconsolidated clay, except for very long piles when the residual friction angle reduces the value of β. Fig. 14 shows the approximate values of $K_o$, which have been interpolated between the limits of about 0.5 governing the skin friction of soft normally consolidated clay and about 3 for short piles in London clay. On this basis the coefficient $K_s$ for driven piles in stiff clay is roughly 1.5 times $K_o$. The figure also shows that tapered piles have a greater and H-piles have a somewhat smaller value of β than corresponding cylindrical piles in stiff clay.

A similar analysis of the skin friction of bored piles in stiff clay shows that the factor β increases with the average value of $c_u$ in a similar way but at a smaller rate compared with that of driven piles (Fig. 15). The factor β for bored piles generally ranges from about 0.5 to 1.5 and increases with the degree

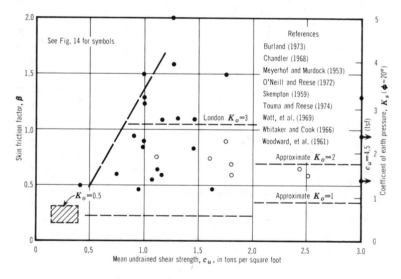

**FIG. 15.—Skin Friction Factor of Bored Piles in Stiff Clay (1 tsf = 95.8 kN/m²)**

of overconsolidation of the clay, as indicated by the approximate interpolated values of $K_o$ shown in the figure. The values of $K_s$ are somewhat smaller than $K_o$ for bored piles in stiff clay and they are roughly one-half of the corresponding values of $K_s$ for driven piles.

The effective horizontal pressure on the shaft of piles driven into clay can also be estimated from semi-empirical relationships based on either the average undrained shear strength and effective overburden pressure along the shaft (53), or more simply from the undrained shear strength, $c_u$, only (18). Thus, the latter approach and a correlation between the plasticity index and the values of $c_u/p_o$ and φ of normally consolidated clay (10) indicates that $K_o$ of such clay varies roughly between 1.5 $c_u/p_o$ for highly plastic clay to over 2 $c_u/p_o$ for clay of low plasticity. Taking the lower value, the unit skin friction of piles driven into saturated clay may then conservatively be expressed as

$$f_s = 1.5\, c_u \tan \phi \quad \dots \dots \dots \dots \dots \dots \dots \dots \dots \dots \dots \dots (19)$$

in which $\phi$ would be expected to decrease from the peak value for short piles to the residual value for very long piles, as mentioned previously. Eq. 19 indicates that the ultimate effective horizontal stress on the shaft of a driven pile in clay may be considerably smaller than the initial horizontal stress after installation mentioned previously. For bored piles the value of $f_s$ may approximately be taken as

$$f_s = c_u \tan \phi \quad\quad\quad\quad\quad\quad\quad\quad\quad\quad\quad\quad\quad\quad\quad (20)$$

These expressions, which are related to Eq. 13, are found to be in reasonable agreement with some load tests on piles in stiff clay (18), and they may be used when the value of $K_o$ of overconsolidated clay is not known.

**Group Capacity.**—In the absence of negative skin friction, the ultimate load of pile groups in clay can be estimated from the smaller value of either block failure of an equivalent pier consisting of the piles and enclosed soil mass or from the aforementioned ultimate bearing capacities of the individual piles. Full-scale load tests on groups of instrumented piles driven into clay (43,77,104) have confirmed that the drained shear strength governs the skin resistance of the individual pile capacities in the groups. Although the initial excsss pore-water pressures between the piles and near the perimeter of the group are of the same order as those observed at the shaft of single piles, the pore pressures in these locations have only partly dissipated at the end of the foundation construction period, and corresponding shear failure would occur under undrained conditions. Thus, the ultimate group load for block failure of an equivalent pier of driven or bored piles in saturated clay should be based on the initial undrained shear strength of the clay for both side and base resistances, as in conventional analysis (88) which is supported by experiments on pile groups with caps resting on the clay (43,101).

However, if the pile caps are not resting on the ground, the group capacity will usually be governed by the sum of the ultimate loads of the single piles with some reduction due to the overlapping zones of shearing deformation in the surrounding soil. This has been confirmed by full-scale load tests on such free-standing groups of friction piles driven into soft and medium clays (1,50,77) and similar model tests (100), which have shown that at customary pile spacings of 3 to 4 pile diameters the ultimate group capacity may be only about two-thirds of the sum of the single pile capacities using the drained remolded shear strength of the clay for the skin friction. The same reduction factor should also be used for preliminary estimates of the ultimate skin resistance of point-bearing pile groups driven or bored through clay not subjected to negative skin friction, when the pile tips rest in a firm deposit. The corresponding ultimate point resistance of such a group is given for driven piles by the sum, or for bored piles by about two-thirds of the sum, of the individual point resistances in a cohesionless bearing stratum or by about two-thirds of the sum of the single point capacities in clay, as mentioned before.

Similarly, the total negative friction on a point-bearing pile group passing through consolidating soft clay is the smaller value of either about two-thirds of the sum of the negative skin friction forces on the individual piles using the drained remolded shear strength of the clay, or the weight of soil within the pile group plus the undrained shear strength of the clay on the perimeter surface area of the equivalent pier. Estimates of the ultimate base resistance of the group with pile points resting above the critical depth in a cohesionless bearing stratum should be based on the effective overburden pressure reduced by negative skin friction on the piles (106) while below the critical depth the base resistance can be estimated directly from the limiting value, $q_l$, without reduction for negative friction.

## Settlement of Pile Foundations

Although installation of piles changes the deformation and compressibility characteristics of the soil mass governing the behavior of single piles under load, this influence extends only to a few pile diameters below the points. Accordingly, the total settlement of a group of driven or bored piles under the safe design load not exceeding one-third to one half of the ultimate group capacity can generally be estimated roughly as for an equivalent pier foundation (88), which is conservative for small groups of driven piles in cohesionless soil. In this method the load of a group of friction piles is usually assumed to be acting on the soil at an effective depth of two-thirds of the pile embedment in the bearing stratum, while for a group of point-bearing piles the equivalent loaded area is taken at the elevation of the pile points.

The deformation and compressibility properties of the soil beneath the pile group can be determined from laboratory tests on undisturbed samples of cohesive soil or they can be estimated from empirical correlations with the results of field tests, such as penetration tests, pressuremeter tests, and plate load tests at the bottom of boreholes. The estimates should also be checked by pile load tests for possible extrapolation to group behavior on the basis of local experience. For pile foundations on sand not underlain by more compressible soil a maximum settlement of 1 in. (25 mm) is frequently allowed for buildings on small pile groups and 2 in. (50 mm) for piled rafts, while about twice these values are usually acceptable for pile foundations on clay (83).

**Groups in Sand.**—Using the concept of an equivalent pier foundation preliminary estimates of the settlement of a pile group in a homogeneous sand deposit not underlain by more compressible soil at greater depth can be made directly from the results of penetration tests as for spread foundations (58). Thus, from empirical correlations between the standard penetration resistance and settlement observations on structures on spread foundations and a recent analysis of corresponding field data (78), the following conservative expression, in inches (25 mm), was derived for the total settlement, $S$, of shallow foundations on saturated sand and gravel (61)

$$S = \frac{2p\sqrt{B}}{N} \quad \quad \quad (21)$$

in which $B$ = the width of pile group, in feet (0.3 m); $p$ = the net foundation pressure, in tons per square foot (100 kN/m$^2$); and $N$ = the average corrected standard penetration resistance, in blows per foot (blows per 0.3 m), within seat of settlement (depth roughly once pile group width in homogeneous soil), while for silty sand twice the right-hand side of Eq. 21 should be used.

It was also suggested that Eq. 21 can be used with a 50% reduction for settlement estimates of deep spread foundations, which may be taken at an effective depth in the bearing stratum of more than about four times the width of the pile group. In other cases the estimated settlement can be interpolated roughly in direct proportion to the ratio of the effective depth to width $D'/B$ of the pile group. Accordingly Eq. 21 can be rewritten, in inches (25 mm) as

$$S = \frac{2p\sqrt{B}\,I}{N} \quad \quad \quad \quad \quad \quad \quad \quad \quad \quad \quad \quad \quad \quad \quad \quad (22)$$

in which $I$ is the influence factor of effective group embedment and is approximately given by

$$I = 1 - \frac{D'}{8B} \geq 0.5 \quad \quad \quad \quad \quad \quad \quad \quad \quad \quad \quad \quad \quad \quad (23)$$

A comparison of this method with observations on foundations supported by driven and bored piles (Fig. 16) shows that the observed maximum settlements

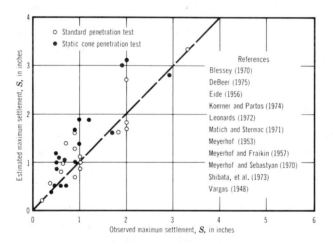

**FIG. 16.—Comparison of Estimated and Observed Settlements of Pile Foundations in Cohesionless Soil (1 in. = 25.4 mm)**

are generally somewhat smaller than estimated, as had been found for spread foundations. This figure also gives a similar comparison between observed and estimated maximum settlements using the results of static cone penetration tests in the relationship (61)

$$S = \frac{pBI}{2q_c} \quad \quad \quad \quad \quad \quad \quad \quad \quad \quad \quad \quad \quad \quad \quad \quad \quad \quad (24)$$

for pile foundations in saturated cohesionless soil, in which $q_c$ = the average static cone resistance within seat of settlement; and all symbols are in consistent units. Both methods appear to give roughly equally reasonable estimates for practical purposes.

The settlement estimates based on the penetration resistance should be modified for any subsequent change in the ground elevation or water table. Moreover, if the thickness of the sand stratum below the effective foundation depth is less than the foundation width, the estimated settlement can be reduced roughly linearly with the corresponding stratum thickness. However, the greatest influence is that of any overconsolidation of the sand in the past, when the evaluation of the penetration resistance leads to a considerable overestimate of settlement unless corrected on the basis of local experience. For the same reason the empirical extrapolation of the total settlement of a pile group in sand from the settlement observed on a test pile at the same load per pile (59,84) can vary within fairly wide limits, even for similar pile dimensions and spacing in the groups (46).

When point-bearing pile groups in sand are subjected to negative skin friction from an upper consolidating clay or silt stratum, the corresponding downdrag force per unit area of the pile group has to be included in the net foundation pressure for settlement estimates. Moreover, for groups of piles passing through clay into sand the effective depth of the pile group in the sand stratum would govern any reduction of the estimated settlement due to the depth of group embedment.

**Groups in Clay.**—The settlement of a pile group in clay or above a clay stratum is conveniently estimated from the initial deformation and consolidation properties of the clay and treating the foundation as an equivalent pier with allowance for the effective foundation embedment and compressible stratum thickness, as mentioned for pile groups in sand. In normally consolidated and particularly in sensitive clay the net foundation pressure, including any negative skin friction, should be kept substantially below the difference between quasi-preconsolidation and effective overburden pressures, if large settlements of the foundation are to be avoided (7). Such settlements can be further decreased if compensated friction pile foundations are used for which piles are installed before basement excavation to reduce heave and recompression of the clay (105).

In overconsolidated and especially in fissured clay the in-situ deformation and compressibility properties deduced from the undrained shear strength of large samples have led to good agreement between observed and estimated settlements of pile foundations (15). In such areas preliminary estimates of the settlement of pile foundations have also been promising by extrapolation from the results of long-term load tests on representative single piles.

While predictions of total settlement can be considerably improved by experience obtained from observations on a regional basis, the rate of settlement of pile foundations in clay can usually only be estimated within wide limits because local drainage conditions are difficult to establish. A better estimate of the coefficient of consolidation of clay and silt strata is frequently obtained by combining in-situ measurements of the coefficient of permeability with laboratory determinations of the modulus of compressibility of large samples. Additionally, these coefficients should be correlated with values deduced from settlement observations on similar piled structures and subsoil conditions on a local basis.

CONCLUSIONS

The behavior of pile foundations under load can only roughly be estimated from the subsurface conditions, type and length of piles, method of installation, size and loading of the foundation, and other factors. For piles in sand and nonplastic silt conventional bearing capacity theory is limited to short piles of less than about 15 to 20 pile diameters. The point resistance of longer piles in cohesionless soil can be estimated from the limiting value of this resistance, using either the friction angle of the soil or preferably the results of static and standard penetration tests directly with an allowance for pile embedment in the bearing stratum and method of pile installation. On the other hand, the skin friction of piles in cohesionless soil varies within wide limits, because it depends additionally on the stress history of the soil, the shape and roughness of the piles, and other factors. Accordingly, penetrometer estimates can furnish only a general guide to the skin friction and should be checked by representative pile load tests.

Preliminary estimates of the bearing capacity of piles in nonuniform soils can be made by simple modifications of the proposed methods of analysis for homogeneous soils. The ultimate load of a group of driven piles in sand can be taken as the sum, and for bored piles as two-thirds of the sum, of the corresponding single pile capacities at customary pile spacings with a check for possible punching failure into an underlying weak deposit.

For piles in saturated clay and plastic silt, conventional bearing capacity theory using the undrained shear strength of undisturbed soil represents mainly the failure condition at the pile points with some allowance for the effect of sensitivity of the soil. On the other hand, several months after installing the piles the shaft resistance is governed by the drained shear strength of remolded soil. The skin friction of driven and bored piles can be estimated from skin friction factors with respect to the average effective overburden pressure, provided the earth pressure coefficient at rest of the deposit is known. In soft and medium normally consolidated clays and silts the positive and negative skin friction factors of driven piles have a similar value and they decrease with pile length due to progressive soil failure at the pile shaft.

For stiff overconsolidated clay the skin friction factor can vary within wide limits with the degree of overconsolidation of the soil, pile shape, method of pile installation, and other factors. If the value of the earth pressure coefficient at rest of the clay is known, the estimated skin friction will provide a lower limit for driven piles and an upper limit for bored piles. In other cases the skin friction factor can only be very roughly estimated from empirical correlations with the average undrained shear strength of the clay for driven and bored piles of various embedded lengths.

The group capacity of driven or bored friction piles with pile caps resting on saturated clay can readily be estimated in the conventional manner from block failure of an equivalent pier using the undrained shear strength of the clay. However, if the pile caps do not rest on the ground, the ultimate load of the group at customary pile spacings is roughly given by two-thirds of the sum of the single pile capacities using the drained remolded shear strength of the clay for the skin friction. The negative skin friction on a point-bearing pile group passing through consolidating soft clay or silt can be similarly estimated.

The settlement of pile foundations under the design load can generally roughly be estimated as for an equivalent pier. Conservative empirical expressions have been presented for preliminary estimates of the total settlement of pile foundations in cohesionless soil using the results of static and standard penetration tests. Conventional methods of estimating the settlement of pile foundations in clay or above a clay stratum are generally satisfactory for the total settlement but the rate of movement is frequently difficult to predict, except on the basis of local experience.

**ACKNOWLEDGMENTS**

During the preparation of this lecture the writer profited from stimulating discussions with many friends in various countries. Some of them provided him with unpublished case records, as mentioned in the references, and are very gratefully acknowledged.

**APPENDIX.—REFERENCES**

1. American Railway Engineering Association Committee, "Steel and Timber Pile Tests—West Atchafalaya Floodway," *AREA Bulletin 489*, 1950, pp. 149-202.
2. Ballisager, C. C., "Bearing Capacity of Piles in Aarhus Septuarian Clay," *Bulletin 7*, Danish Geotechnical Institute, Copenhagen, Denmark, 1959, pp. 14-20.
3. Begemann, H. K., "The Use of the Static Soil Penetrometer in Holland," *New Zealand Engineering*, No. 2, 1963, pp. 41-49.
4. Begemann, H. K., "The Friction Jacket Cone as an Aid in Determining the Soil Profile," *Proceedings of the Sixth International Conference on Soil Mechanics*, Montreal, Canada, Vol. 1, 1965, pp. 17-20.
5. Begemann, H. K., "The Dutch Static Penetration Test with the Adhesion Jacket Cone," *Publication 13*, L.G.M. Laboratory of Grondmechanics, Delft, Netherlands, 1969, pp. 1-86.
6. Bishop, A. W., Collingridge, V. H., and O'Sullivan, T. P., "Driving and Loading Tests on Six Precast Concrete Piles in Gravel," *Geotechnique*, London, England, Vol. 1, 1948, pp. 49-58.
7. Bjerrum, L., "Engineering Geology of Norwegian Normally-Consolidated Marine Clays as Related to Settlements of Buildings," *Geotechnique*, London, England, Vol. 17, 1967, pp. 81-118.
8. Bjerrum, L., "Problems of Soil Mechanics and Construction on Soft Clays," *Proceedings of the Eighth Conference on Soil Mechanics*, Moscow, U.S.S.R., Vol. 3, 1973, pp. 111-159.
9. Bjerrum, L., Johannessen, I. J., and Eide, O., "Reduction of Negative Skin Friction on Steel Piles to Rock," *Proceedings of the Seventh International Conference on Soil Mechanics*, Mexico, Vol. 2, 1969, pp. 27-34.
10. Bjerrum, L., and Simons, N. E., "Comparison of Shear Strength Characteristics of Normally Consolidated Clays," *Proceedings of the Research Conference on Shear Strength Cohesive Soils*, ASCE, 1960, pp. 711-726.
11. Blessey, W. E., "Allowable Pile Capacity, Mississippi Deltaic Plain," *Proceedings of Design Installation of Pile Foundations*, Lehigh University, Bethlehem, Pa., 1970, pp. 87-106.
12. Bozozuk, M., "Downdrag Measurements on a 160-ft. Floating Pipe Test pile in Marine Clay," *Canadian Geotechnical Journal*, Vol. 9, 1972, pp. 127-136.
13. Bozozuk, M., and Labrecque, A., "Downdrag Measurements on 270-ft. Composite Piles," *ASTM Special Technical Publication 444*, American Society for Testing and Materials, 1969, pp. 15-40.
14. Burland, J. B., "Shaft Friction of Piles in Clay—A Simple Fundamental Approach," *Ground Engineering*, Vol. 6, 1973, pp. 30-42.
15. Butler, F. G., "Settlement on Heavily Overconsolidated Clays," *Proceedings of the Conference Settlement of Structures*, Cambridge, England, 1974, pp. 531-578.
16. Caquot, A., and Kerisel, J., *Traité de Mécanique des Sols*, Gauthier-Villars, Paris, France, 1966.
17. Chandler, R. J., "The Shaft Friction of Piles in Cohesive Soils in Terms of Effective Stress," *Civil Engineering Public Works Review*, Vol. 63, 1968, pp. 48-51.
18. Clark, J. C., and Meyerhof, G. G., "The Behaviour of Piles Driven in Clay—Inves-

tigation of the Bearing Capacity Using Total and Effective Strength Parameters," *Canadian Geotechnical Journal*, Vol. 10, 1973, pp. 86-102.
19. Darragh, R. D., and Bell, R. A., "Load Tests on Long Bearing Piles," *ASTM Special Technical Publication 444*, American Society for Testing and Materials, 1969, pp. 41-67.
20. Davisson, M. T., "Settlement Histories of Two Pile Supported Grain Silos," *Proceedings of the Conference on Performance Earth-Supported Structures*, Purdue University, Lafayette, Ind., Vol. 1, Part 2, 1972, pp. 1155-1167.
21. De Beer, E. E., "Etude des Fondations sur Pilotis et des Fondations Directes," *Annales des Travaux Publics des Belgique*, Vol. 46, 1945, pp. 229-307.
22. De Beer, E. E., "Some Considerations Concerning the Point Bearing Capacity of Bored Piles," *Proceedings of the Symposium on Bearing Capacity of Piles*, Roorkee, India, 1964, pp. 178-204.
23. De Beer, E. E., "Méthodes de Déduction de la Capacité Portante d'un Pieu à Partir des Résultats des Essais de Pénétration," *Annales des Travaux Publics des Belgique*, Vol. 72, Nos. 4-6, 1971, pp. 1-142.
24. De Beer, E. E., Unpublished Records, 1975.
25. Dos Santos, M. P., and Gomes, N. A., "Experiences with Piled Foundations of Maritime Structures in Portuguese East Africa," *Proceedings of the Fourth International Conference on Soil Mechanics*, London, England, Vol. 2, 1957, pp. 27-34.
26. Eide, O., "Baereevne av Peler i Sand," *Publication 18*, Norwegian Geotechnical Institute, Oslo, Norway, 1956, pp. 1-15.
27. Eide, O., Hutchinson, J. N., and Landva, A., "Short and Long-Term Test Loading of a Friction Pile in Clay," *Proceedings of the Fifth International Conference on Soil Mechanics*, Paris, France, Vol. 2, 1961, pp. 45-53.
28. Endo, M., et al., "Negative Skin Friction Acting on Steel Pipe-Piles in Clay," *Proceedings of the Seventh International Conference on Soil Mechanics*, Mexico, Vol. 2, 1969, pp. 85-92.
29. Fellenius, B., "Results of Tests on Piles at Gothenburg Railway Station," *Bulletin 5*, Swedish State Railways, 1955.
30. Fellenius, B. H., "Down-Drag on Piles in Clay due to Negative Skin Friction," *Canadian Geotechnical Journal*, Vol. 9, 1972, pp. 323-337.
31. Fellenius, B. H., and Samson, L., Unpublished Records, 1975.
32. Fox, D. A., Parker, G. F., and Sutton, V. J. R., "Pile Driving in the North Sea Boulder Clays," *Proceedings of the Second Conference on Offshore Technology*, Houston, Tex., Vol. 1, 1970, pp. 535-546.
33. Garlanger, J. E., "Prediction of the Downdrag Load at Cutler Circle Bridge," *Conference, Massachusetts Institute of Technology*, Cambridge, Mass., 1972.
34. Garneau, R., "Capacité de Pieux Flottants dans une Argile Molle," *Proceedings Conference of the Quebec Good Roads Association*, Quebec, Canada, 1974, pp. 42-64.
35. Geuze, E. C., "Résultats d'Essais de Pénétration en Profondeur et de Mise en Charge de Pieux," *Annales de l'Institute Technique du Batiment et des Travaux Publics*, Paris, France, Vol. 6, 1953, pp. 313-319.
36. Gregersen, O. S., Aas, G., and Dibiagio, E., "Load Tests on Friction Piles in Loose Sand," *Proceedings of the Eighth International Conference on Soil Mechanics*, Moscow, U.S.S.R., Vol. 2.1, 1973, pp. 109-117.
37. Huizinga, T. K., "Application of Results of Deep Penetration Tests to Foundation Piles," *Proceedings of the Building Research Congress*, London, England, Vol. 1, 1951, pp. 173-179.
38. Kahl, H., Muhs, H., and Meyer, W., "Ermittlung der Grösse und des Verlaufs des Spitzendrucks bei Drucksondierungen in Ungleichförmigen Sand-Kies-gemischen und in Kies," *Mitteilungen DEGEBO*, Berlin, West Germany, No. 21, 1968, pp. 1-36.
39. Kerisel, J., "Fondations Profondes en Milieu Sableux," *Proceedings of the Fifth International Conference Soil Mechanics*, Paris, France, Vol. 2, 1961, pp. 73-83.
40. Kerisel, J., "Deep Foundations Basic Experimental Facts," *Proceedings Conference on Deep Foundations*, Mexico, Vol. 1, 1964, pp. 5-44.
41. Koerner, R. M., and Partos, A., "Settlement of Building on Pile Foundation in Sand," *Journal of the Geotechnical Engineering Division*, ASCE, Vol. 100, No. GT3, Proc. Paper 10405, Mar., 1974, pp. 265-278.

42. Koizumi, Y., "Field Tests on Piles in Sand," *Soils and Foundations*, Japan, Vol. 11, 1971, pp. 29-49.
43. Koizumi, Y., and Ito, K., "Field Tests with Regard to Pile Driving and Bearing Capacity of Piled Foundations," *Soils and Foundations*, Japan, Vol. 7, 1967, pp. 30-53.
44. Ladanyi, B., "Etude Thëorique et Expërimentale de l'Expansion dans un Sol Pulvérulent d'une Cavité Présantant une Symétrie Spherique ou Cylindrique," *Annales des Travaux Publics de Belgique*, Vol. 62, 1961, pp. 105-148, 365-406.
45. Ladanyi, B., "Bearing Capacity of Deep Footings in Sensitive Clays," *Proceedings of the Eighth International Conference on Soil Mechanics*, Moscow, U.S.S.R., Vol. 2.1, 1973, pp. 159-166.
46. Leonards, G. A., "Settlement of Pile Foundations in Granular Soil," *Proceedings, Earth and Earth-Supported Structures*, Purdue University, Lafayette, Ind., Vol. 1, Part 2, 1972, pp. 1169-1184.
47. Mansur, C. I., and Focht, J. A., "Pile Loading Tests, Morganza Floodway Control Structure," *Proceedings*, ASCE, Vol. 79, Paper No. 324, 1953, pp. 324-1-31.
48. Mansur, C. I., and Hunter, A. H., "Pile Tests—Arkansas River Project," *Journal of the Soil Mechanics and Foundations Division*, ASCE, Vol. 96, No. SM5, Proc. Paper 7509, Sept., 1970, pp. 1545-1582.
49. Mansur, C. I., and Kaufman, R. I., "Pile Tests, Low Sill Structure, Old River, La.," *Journal of the Soil Mechanics and Foundations Division*, ASCE, Vol. 82, No. SM4, Proc. Paper 1079, pp. 1079-1-33.
50. Masters, F. M., "Timber Friction Pile Foundations," *Transactions*, ASCE, Vol. 108, Separate No. 2174, 1943, pp. 115-140.
51. Matich, M. A. J., and Stermac, A. G., "Settlement Performance of the Burlington Bay Skyway," *Canadian Geotechnical Journal*, Vol. 8, 1971, pp. 252-271.
52. McCammon, N. R., and Golder, H. Q., "Some Loading Tests on Long Pipe Piles," *Geotechnique*, London, England, Vol. 20, 1970, pp. 171-184.
53. McClelland, B., "Design of Deep Penetration Piles for Ocean Structures," *Journal of the Geotechnical Engineering Journal*, ASCE, Vol. 100, No. GT7, Proc. Paper 10665, July, 1974, pp. 705-747.
54. Meigh, A. C., "Some Driving and Loading Tests on Piles in Gravel and Chalk," *Proceedings of the Conference on Behaviour of Piles*, London, England, 1971, pp. 9-16.
55. Melzer, K. J., "Sondenuntersuchungen in Sand," *Mitteilungen der Vereinigung der Grosskesselbetreiber*, Aachen, Germany, No. 43, 1968, pp. 1-345.
56. Meyerhof, G. G., "The Ultimate Bearing Capacity of Foundations," *Geotechnique*, London, England, Vol. 2, 1951, pp. 301-332.
57. Meyerhof, G. G., "An Investigation for the Foundations of a Bridge on Dense Sand," *Proceedings of the Third International Conference on Soil Mechanics*, Zurich, Switzerland, Vol. 2, 1953, pp. 66-70.
58. Meyerhof, G. G., "Penetration Tests and Bearing Capacity of Cohesionless Soils," *Journal of the Soil Mechanics and Foundations Division*, ASCE, Vol. 82, No. SM1, Proc. Paper 866, Jan. 1956, pp. 866-1-19.
59. Meyerhof, G. G., "Compaction of Sands and Bearing Capacity of Piles," *Journal of the Soil Mechanics and Foundations Division*, ASCE, Vol. 85, No. SM6, Proc. Paper 2292, Dec. 1959, pp. 1-30.
60. Meyerhof, G. G., "Some Recent Research on the Bearing Capacity of Foundations," *Canadian Geotechnical Journal*, Vol. 1, 1963, pp. 16-26.
61. Meyerhof, G. G., "Penetration Testing in Countries Outside Europe," *Proceedings of the European Symposium on Penetration Testing*, Stockholm, Sweden, Vol. 2.1, 1974, pp. 40-48.
62. Meyerhof, G. G., "Ultimate Bearing Capacity of Footings on Sand Layer Overlying Clay," *Canadian Geotechnical Journal*, Vol. 11, 1974, pp. 223-229.
63. Meyerhof, G. G., and Fraikin, L. A., "The Design of Sorel Stadium with Special Reference to its Foundations," *EIC Journal*, Vol. 40, 1957, pp. 270-275.
64. Meyerhof, G. G., and Murdock, L. J., "An Investigation of the Bearing Capacity of Some Bored and Driven Piles in London Clay," *Geotechnique*, London, England, Vol. 3, 1953, pp. 267-282.
65. Meyerhof, G. G., and Sebastyan, G. Y., "Settlement Studies on Air Terminal Building

and Apron, Vancouver International Airport, British Columbia," *Canadian Geotechnical Journal*, Vol. 7, 1970, pp. 433-456.
66. Muhs, H., and Weiss, K., "Untersuchung von Grenztragfähigkeit und Setzungsverhalten Flachgegründeter Einzelfundamente in Ungleichförmigen Nichtibindigen Boden," *Mitteilungen DEGEBO*, Berlin, West Germany, No. 26, 1971, pp. 1-39.
67. Nordlund, R. L., "Bearing Capacity of Piles in Cohesionless Soils," *Journal of the Soil Mechanics and Foundations Division*, ASCE, Vol. 89, No. SM3, Proc. Paper 3506, May, 1963, pp. 1-35.
68. Nunes, A. J. C., and Vargas, M., "Computed Bearing Capacity of Piles in Residual Soil Compared with Laboratory and Load Tests," *Proceedings of the Third International Conference on Soil Mechanics*, Zurich, Switzerland, Vol. 2, 1953, pp. 75-79.
69. O'Neill, M. W., and Reese, L. C., "Behavior of Bored Piles in Beaumont Clay," *Journal of the Soil Mechanics and Foundations Division*, ASCE, Vol. 98, No. SM2, Proc. Paper 8471, Feb., 1972, pp. 195-213.
70. Ostenfeld, A. C., Frandsen, A. G., and Vefling, G., "Pile Driving and Pile Load Tests in Test Areas and Under Caissons, Motorway Bridge across Lillebaelt," Bygn, Medd., Copenhagen, Vol. 41, 1968, pp. 33-54.
71. Peck, R. B., "A Study of the Comparative Behavior of Friction Piles," *Special Report 36*, Highway Research Board, Washington, D.C., 1958, pp. 1-72.
72. Peck, R. B., Hanson, W. E., and Thornburn, T. H., *Foundation Engineering*, John Wiley and Sons, Inc., New York, N.Y., 1974.
73. Press, H., "Die Tragfähigkeit von Pfahlgruppen in Beziehung zu der des Einzelpfahles," *Bautechnik*, Vol. 11, 1933, pp. 625-627.
74. Robinsky, E. I., and Morrison, C. F., "Sand Displacement and Compaction Around Model Friction Piles," *Canadian Geotechnical Journal*, Vol. 1, 1964, pp. 81-93.
75. Roy, M., Michaud, D., and Tavenas, F. A., "The Interpretation of Static Cone Penetration Tests in Sensitive Clays," *Proceedings of the European Symposium on Penetration Testing*, Stockholm, Sweden, Vol. 2.2, 1974, pp. 323-330.
76. Sastry, V. V., "Bearing Capacity of Steel Piles Penetrating through Clay into Sand," thesis presented to the Nova Scotia Technical College at Halifax, Nova Scotia, in 1975, in partial fulfillment of the requirements for the degree of Doctor of Philosophy.
77. Schlitt, H. G., "Group Pile Loads in Plastic Soils," *Proceedings*, Highway Research Board, Washington, D.C., Vol. 31, 1951, pp. 62-81.
78. Schultze, E., and Sherif, G., "Prediction of Settlement from Evaluation of Settlement Observations for Sand," *Proceedings of the Eighth International Conference on Soil Mechanics*, Moscow, U.S.S.R., Vol. 1.3, 1973, pp. 225-230.
79. Sherman, W. C., "Instrumented Pile Tests in a Stiff Clay," *Proceedings of the Seventh International Conference on Soil Mechanics*, Mexico, Vol. 2, 1969, pp. 227-232.
80. Shibata, T., Hijikuro, K., and Tominaga, M., "Settlement of a Blast Furnace Foundation," *Proceedings of the Eighth International Conference on Soil Mechanics*, Moscow, U.S.S.R., Vol. 1.3, 1973, pp. 239-242.
81. Skempton, A. W., "The Bearing Capacity of Clays," *Proceedings of the Building Research Congress*, London, England, Vol. 1, 1951, pp. 180-189.
82. Skempton, A. W., "Cast In-situ Bored Piles in London Clay," *Geotechnique*, London, England, Vol. 9, 1959, pp. 153-173.
83. Skempton, A. W., and MacDonald, D. H., "The Allowable Settlement of Buildings," *Proceedings of the Institution of Civil Engineers*, London, England, Vol. 5, 1956, pp. 727-784.
84. Skempton, A. W., Yassin, A. A., and Gibson, R. E., "Théorie de la Force Portante des Pieux dans le Sable," *Annales de l'Institut Technique du Batiment et des Travaux Publics*, Paris, France, Vol. 6, 1953, pp. 285-290.
85. Stermac, A. G., Selby, K. G., and Devata, M., "Behaviour of Various Piles in Stiff Clay," *Proceedings of the Seventh International Conference on Soil Mechanics*, Mexico, Vol. 2, 1969, pp. 239-245.
86. Tavenas, F. A., "Load Tests Results on Friction Piles in Sand," *Canadian Geotechnical Journal*, Vol. 8, 1971, pp. 7-22.
87. Terzaghi, K., *Erdbaumechanik auf Bodenphysikalischer Grundlage*, Deuticke, Vienna, Austria, 1925.
88. Terzaghi, K., and Peck, R. B., *Soil Mechanics in Engineering Practice*, John Wiley and Sons, Inc., New York, N.Y., 1967.

89. Thorburn, S., and MacVicar, R. S., "Pile Load Tests to Failure in the Clyde Alluvium," *Proceedings of the Conference on Behaviour of Piles*, London, England, 1971, pp. 1-7.
90. Tomlinson, M. J., "The Adhesion of Piles Driven in Clay Soils," *Proceedings of the Fourth International Conference on Soil Mechanics*, London, England, Vol. 2, 1957, pp. 66-71.
91. Tomlinson, M. J., "Some Effects of Pile Driving on Skin Friction," *Proceedings of the Conference on Behaviour of Piles*, London, England, 1971, pp. 107-114.
92. Touma, F. T., and Reese, L. C., "Behavior of Bored Piles in Sand," *Journal of the Geotechnical Engineering Division*, ASCE, Vol. 100, No. GT7, Proc. Paper 10651, July, 1974, pp. 749-761.
93. Valsangkar, A. J., Unpublished Records, 1975.
94. Vargas, M., "Building Settlement Observations in Sao Paulo," *Proceedings of the Second International Conference on Soil Mechanics*, Rotterdam, the Netherlands, Vol. 4, 1948, pp. 13-21.
95. Vesic, A. S., "A Study of Bearing Capacity of Deep Foundations," *Report B-189*, Georgia Institute of Technology, Atlanta, Ga., 1967.
96. Vesic, A. S., "Load Transfer in Pile-Soil Systems," *Proceedings of the Conference on Design Installation of Pile Foundations*, Lehigh University, Bethlehem, Pa., 1970, pp. 47-73.
97. Vesic, A. S., "Expansion of Cavities in Infinite Soil Mass," *Journal of the Soil Mechanics and Foundations Division*, ASCE, Vol. 98, No. SM3, Proc. Paper 8790, Mar., 1972, pp. 265-290.
98. Walker, L. K., and Darvall, P., "Dragdown on Coated and Uncoated Piles," *Proceedings of the Eighth International Conference on Soil Mechanics*, Moscow, U.S.S.R., Vol. 2.1, 1973, pp. 257-262.
99. Watt, W. G., Kurfurst, P. J., and Zeman, Z. P., "Comparison of Pile Load-Test-Skin-Friction Values and Laboratory Strength Tests," *Canadian Geotechnical Journal*, Vol. 6, 1969, pp. 339-352.
100. Whitaker, T., "Experiments with Model piles in Groups," *Geotechnique*, London, England, Vol. 7, 1957, pp. 147-167.
101. Whitaker, T., "Some Experiments on Model Piled Foundations in Clay," *Proceedings of the Symposium on Pile Foundations*, Stockholm, Sweden, 1960, pp. 124-139.
102. Whitaker, T., and Cooke, R. W., "An Investigation of the Shaft and Base Resistance of Large Bored Piles in London Clay," *Proceedings of the Symposium on Large Bored Piles*, London, England, 1966, pp. 7-49.
103. Woodward, R. J., Lundgren, R., and Boitano, J. D., "Pile Loading Tests in Stiff Clays," *Proceedings of the Fifth International Conference on Soil Mechanics*, Paris, France, Vol. 2, 1961, pp. 177-184.
104. York, D. L., Miller, V. G., and Ismael, N. F., "Long-Term Load Transfer in End Bearing Pipe Piles," *Canadian Geotechnical Journal*, Vol. 12, 1975, (in press).
105. Zeevaert, L., "Compensated Friction—Pile Foundation to Reduce the Settlement of Buildings on Highly Compressible Volcanic Clay of Mexico City," *Proceedings of the Fourth International Conference on Soil Mechanics*, London, England, Vol. 2, 1957, pp. 81-86.
106. Zeevaert, L., "Reduction of Point Bearing Capacity Because of Negative Friction," *Proceedings of the First Panamanian Conference on Soil Mechanics*, Mexico, Vol. 3, 1959, pp. 1145-1152.

**APPENDIX II.—NOTATION**

*The following symbols are used in this paper:*

$A_p$ = area of pile point;
$A_s$ = area of pile shaft;
$B$ = width;
$c$ = unit cohesion;
$c_a$ = unit adhesion;

$c_u$ = undrained shear strength;
$D$ = depth;
$D_b$ = depth in bearing stratum;
$D_c$ = critical depth;
$f_c$ = unit resistance of local friction sleeve;
$f_l$ = limiting unit skin friction;
$f_s$ = average unit skin friction;
$H$ = thickness of bearing stratum below pile point;
$I$ = influence factor of embedment of pile group;
$K_o$ = coefficient of earth pressure at rest;
$K_s$ = coefficient of earth pressure on shaft;
$N$ = standard penetration resistance;
$N_c, N_q$ = bearing capacity factors;
$p$ = net foundation pressure;
$p_o$ = effective overburden pressure;
$Q_p$ = point resistance;
$Q_s$ = skin resistance;
$Q_u$ = ultimate bearing capacity of pile;
$q_c$ = limiting cone resistance;
$q_l$ = limiting unit point resistance;
$q_o$ = unit point resistance in weak stratum;
$q_p$ = unit bearing capacity of pile point;
$R_o$ = overconsolidation ratio;
$S$ = total settlement;
$\alpha$ = adhesion coefficient;
$\beta$ = skin friction factor;
$\delta$ = angle of skin friction; and
$\phi$ = angle of internal friction.

## BEARING CAPACITY AND SETTLEMENT OF PILE FOUNDATIONS[a]

### Discussion by Theodore K. Chaplin,[2] F. ASCE

The Eleventh Terzaghi Lecture reflects a great amount of work, and the author should be congratulated on a stimulating and practical contribution, especially for the emphasis on critical depth. As exact solutions are essentially restricted to elasticity theory (which cannot directly include dilatancy), empirical methods will continue to be widely used. Most of the following points refer to piles in sand.

Compaction usually seems to involve some grain crushing with relatively large porosity reductions near the point of the pile or penetrometer. The surrounding zone has a moderate increase in stress but much less contraction, or even some expansion (39). Farther away, relatively little volume change seems likely, despite the stress increase that tends to accumulate in a horizontal direction. Creep of sand has been known to cause considerable reductions in pile penetration resistance, over several hours or days, which may sometimes be very important. If the grain material is not fairly clean quartz (as presumed for all the author's tests), the crushing resistance measured at zero lateral strain could be greatly reduced. For highly angular quartz sand at porosities over about 47%, the penetration resistance might be very small even if the porosity is near the minimum for that sand. The author's comments on these aspects would be welcomed.

More information about the types of test, stress levels, and methods of interpretation which gave the $\phi$-values in Figs. 1-7 would be particularly useful. How do the upper and lower curves in Fig. 1 differ? The number of points in Fig. 4 above the deep-embedment curve at lower $\phi$-values suggests that the shape of point in elevation, e.g., a leading spike to cause volume reduction, might have much more effect in loose than in dense sand.

Expressing the limiting point resistance, $q_1$, from Fig. 2 in terms of $N_q$, Eq. 3, seems cumbersome. For (medium) quartz sand it would be preferable to use empirical expressions such as

$$q_1 \simeq 0.41 \, (\phi - 28°)^{2.5} \text{ tsf} \quad \quad \quad \quad \quad \quad \quad \quad \quad \quad (25a)$$

$$q_1 \simeq 44 \, (\phi - 28°)^{2.5} \text{ kN/m}^2 \quad \quad \quad \quad \quad \quad \quad \quad \quad (25b)$$

As coarse quartz sand grains are far weaker than fine grains, and the shock waves from a dynamic penetration test can split grains, one would expect the ratio, $q_c/N$, (Eq. 8), to depend on grain size and mineral type. From many references for normal (essentially quartz) sands the writer's colleague John Billam (personal communication, 1976) has found

$$\frac{q_c}{N} \simeq 6.5 d^{0.25} \text{ tsf} \quad \quad \quad \quad \quad \quad \quad \quad \quad \quad \quad \quad (26a)$$

---

[a] March, 1976, by George Geoffrey Meyerhof (Proc. Paper 11962).
[2] Sr. Lect., Dept. of Civ. Engrg., Univ. of Birmingham, Birmingham, England.

$$\frac{q_c}{N} \simeq 700 d^{0.25} \, \text{kN}/\text{m}^2 \qquad (26b)$$

in which $d$ is in millimeters.

For model piles fully penetrating through a sand layer, Fig. 12, a comparison curve for a much thicker sand layer of the same porosity would show better where the effect of the underlying soft clay really begins.

In Fig. 13, the high values of the skin friction factor, $\beta$, for driven taper piles in clay are very striking. A simple explanation is that a taper pile in effect expands laterally during penetration and so tends to close up the axial "scores" or circumferential gaps previously made in the clay by lower parts of the pile, thus increasing the area of adhesive contact.

A clear distinction between continuous and step-taper piles would be helpful throughout the paper. Is there any evidence that continuous-taper piles in sand do give more skin friction than step-taper or straight-sided piles under the same conditions? The author told the writer in 1971 (while visiting Birmingham) that a continuous taper pile in sand had a greater bearing capacity than a straight-sided pile of breadth equal to the maximum breadth of the taper pile; could the author give more information?

Tovey (107) reported that in London Clay some jacked piles gave 40% lower ultimate load than corresponding driven piles, because the jacking allowed more reorientation of the particles. Might some analogous effect account for the author's remarkably low skin friction values for jacked piles in sand? Is it likely that the even lower skin friction values for bored piles in sand are due to the use of drilling mud, or to disturbance of sand structure by drilling?

The author's comments are requested on the physical conditions controlling critical depth, and particularly the ratio between limiting point resistance and the odometer "turnover" critical pressure for thin samples at the same initial porosity.

Finally, one feels the value of the paper would have been much enhanced (like so many papers) by as much information as possible about particle shape, measured limiting porosities, and grading for all sands. Stress-strain curves for clay (or $E_c/c_u$ ratios) would similarly help readers to make their own interpretations.

## Appendix.—Reference

107. Tovey, N. K., "Some Applications of Soil Microscopy to Soil Engineering," *Soil Microscopy*, G. K. Rutherford, ed., The Limestone Press, Kingston, Ontario, Canada, 1974, pp. 119-142.

# BEARING CAPACITY AND SETTLEMENT OF PILE FOUNDATIONS[a]

## Discussion by Jean Biarez[3] and Pierre Foray[4]

The writers appreciate the very interesting and complete synthesis about pile foundations presented by the author and wish to discuss further point bearing capacity of piles in sand.

**Limit Point Resistance.**—As in the examples given by the author, the experiments of the writers have shown that beyond some critical depth, $D_c$, point resistance, $q_p$, remains constantly equal to $q_l$ in homogeneous medium. The author gives an approximate relation between $q_l$ and $\phi$, the angle of friction of the sand, noting that $q_l$ is also influenced by the compressibility of the soil.

The writers think that, if $\phi$ is the parameter that can describe correctly the part in $p_o N_q$, it cannot alone explain the part in $q_l$, which essentially depends upon compressibility phenomena.

Experiments on various sands (110) have shown that the relationship between $q_l$ and $\phi$ is valid only for sands, the physical and mechanical properties of which are not very different [Fig. 17(a)].

A representation of the $q_l$ function of relative density of the sand, $D_R$, gives a good grouping of the experiments [Fig. 17(b)]. It seems that relative density $D_R$ takes the soil compactibility into better account.

The transition between the phase of lateral extension of the soil towards the surface, represented by $N_q$, and the phase of compression of the soil in the vicinity of pile point, represented by $q_l$, has to be related with the evolution of the mechanical behavior of sand under the mean applied stress. This evolution may be characterized by the notion of "critical pressure $p_c$" (108), mean stress applied in a triaxial test, in which corresponding critical density is the initial density of the sand.

At low depth, the mean stress under the pile-point is lower than $p_c$, and the soil dilates itself (lateral extension). At great depth, the mean stress tends to become higher than $p_c$, the soil compresses itself, and $q_c$ becomes equal to $q_l$. The writers have shown (108–110) that $q_l$ can be expressed explicity as a function of $p_c$:

$$q_l = K_c p_c \qquad (27)$$

in which $K_c$ is a function of $\phi$ or $D_R$; and $p_c$ is expressed by the "critical diagram" of the sand (108).

**Overlayered Stratum.**—As stated by the author, experiments (110) show that $q_l$ has the same value for a pile driven in an homogeneous medium to a depth

---

[a] March, 1976, by George Geoffrey Meyerhof (Proc. Paper 11962).
[3] Prof., Central School of Paris, Paris, France.
[4] Master-Asst., Soil Mechanics Lab., Mechanics Inst., Grenoble Scientific and Medical Univ., Grenoble-Cedex, France.

$D \geq D_c$, or in a bearing stratum overlayered by a compressible material (Fig. 18).

**Critical Depth.**—For a homogeneous medium, the writers propose a relation between $D_c$ and $p_c$, or more simply $q_l$. In homogeneous medium this relationship is not very different from that presented by the author. For overlayered stratum,

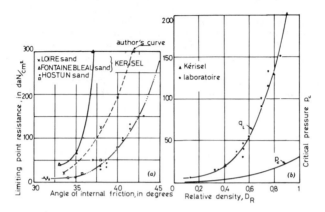

**FIG. 17.—Correlation Between Limiting Point Resistance $q_l$ and: (a) Angle of Internal Friction $\phi$; (b) Relative Density $D_R$**

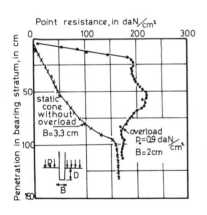

FIG. 18.—Comparison Between Penetration in Medium with and without Overload

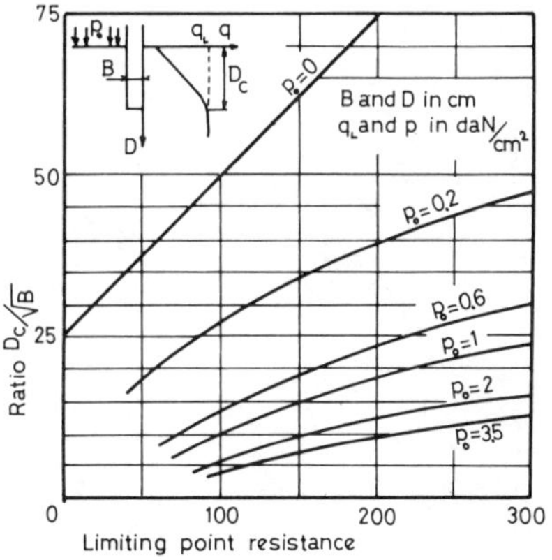

FIG. 19.—Experimental Determination of Critical Depth in Overloaded Medium

experimental results [Figs. 17(b) and 19] have shown that the value of $D_c$ decreases rapidly with the value of the overload applied to the bearing stratum (111). For overloads corresponding to about 1 kN/m², the critical depth in the bearing stratum is lower than 10 diameters, i.e., about 3 or 5 diameters. It shows that for long piles, the current rule of 3–5 diameters of penetration in the bearing stratum allows a mobilization of the full point resistance of the pile.

APPENDIX.—REFERENCES

108. Biarez, J., and Gresillon, J. M., "Essais et suggestions pour le calcul de la force portante des pieux en miliou pulvérulent," *Geotechnique*, London, England, Vol. 22, No. 3, 1972.
109. Foray, P., and Puech, A., "Influence de la compressibilité sur la force portante limite des pieux en milieu pulvérulent," *Annales de l'Institut Technique du Bâtiment et des Travaux Publics*, Paris, France, May, 1976.
110. Puech, A., "De l'influence de la compressibilité sur la force portante limite des fondations profondes," thesis presented to the University of Grenoble, at Grenoble, France, in 1975, in partial fulfillment of the requirements for the degree of Doctorat, Third Order.
111. Puech, A., et al., "Contribution to the Study of Static and Dynamic Penetrometers," *European Symposium on Penetration Testing*, Stockholm, Sweden, June, 1974.

## Bearing Capacity and Settlement of Pile Foundations[a]
### Closure by George Geoffrey Meyerhof,[5] F. ASCE

In reply to the discussers the writer's approximate empirical relation between the limiting static cone penetration resistance and angle of internal friction of homogeneous cohesionless soil below the critical depth (Fig. 2) was obtained for fairly clean air-dry quartz sand and gravel using penetrometers with a Dutch or similar type 60° cone with sleeve and a 1.4-in. (36-mm) diam (61). The friction angle of the soil was determined by conventional triaxial or direct shear tests at effective pressures between about 0.5 tsf and 5 tsf (50 $kN/m^2$ and 500 $kN/m^2$). On the other hand, Biarez and Foray carried out tests on penetrometers with a flat plate attached to the end of cased rods, which allowed some sand to be displayed into the void above the plate and thereby led to a much smaller limiting point resistance than found for the cones. Moreover, results of cone penetration tests in various cohesionless soils show that the friction angle of the soil is a better index parameter for the limiting cone resistance than the relative density which, for soils of various grain-size characteristics, gives considerably different cone resistances for the same relative density (38,55). The same would also apply to the ultimate unit point resistance of piles in cohesionless soil.

As mentioned in the paper, the effects of soil compressibility, crushing, arching, and other factors, such as stress history, influence the previously mentioned relationship for different cohesionless soils, especially in the dense and very dense states (61). Since the soil properties vary with the ground-water conditions and due to pore pressure effects, submerged sand and gravel have smaller values of the limiting static cone resistance than similar dry soils (Fig. 20). The percentage difference between the cone resistance of dry and submerged soils decreases as the friction angle of the soil increases, and it is roughly equivalent to a reduction of 1° of the friction angle due to submergence of the soil. It is of interest to note that the ultimate point resistance of piles longer than about 15 pile diameters–20 pile diameters driven at least about 10 pile diameters into a deep submerged sand bearing stratum is approximately the same as the limiting static cone resistance in similar dry sand (Fig. 4) because the bearing capacity of driven piles in sand may be roughly 50% greater than that of similar jacked piles or cones.

The approximate critical depth ratios for piles in homogeneous submerged soil shown in Fig. 1 are greater than those for dry soil. This also applies to the aforementioned critical bearing depth ratio of roughly 10 for full-sized piles in a deep cohesionless bearing stratum (Fig. 5), which was the average value obtained from an analysis of load tests on long piles driven through a weak

---

[a] March, 1976, by George Geoffrey Meyerhof (Proc. Paper 11962).
[5] Prof. and Head, Dept. of Civ. Engrg., Nova Scotia Technical Coll., Halifax, N.S., Canada.

deposit into submerged sand. It was found that the critical bearing depth ratio has a great scatter and varies on the average from about 8 for loose sand to about 12 for dense sand with a total range from about 5-15 in practice, even for overburden pressures exceeding 2 tsf (200 kN/m$^2$) on the bearing stratum (Fig. 21). This field data and additional research on the bearing capacity of long piles in layered soils show (112) that the critical bearing depth ratio, $D_{bc}/B$, is smaller than the corresponding critical depth ratio, $D_c/B$, for short piles in homogeneous soil and depends on the thickness and friction angle or penetration resistance of the bearing stratum and on the ratio of the penetration resistance of the weak deposit to that of the bearing stratum.

As shown by Biarez and Foray, the critical bearing depth ratio depends also to some extent on the effective overburden pressure on the bearing stratum. However, their model pile tests in sand loaded by an air cushion surcharge

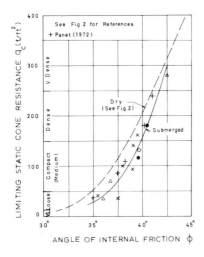

FIG. 20.—Approximate Relation between Limiting Static Cone Resistance and Friction Angle of Sand (1 tsf = 95.8 kN/m$^2$)

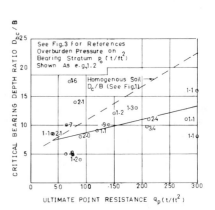

FIG. 21.—Critical Bearing Depth Ratio for Driven Piles in Sand Stratum beneath Weak Soil Layer (1 tsf = 95.8 kN/m$^2$)

to simulate an overlying weak deposit gave much smaller values of this critical depth ratio than found for full-sized piles in practice. The difference may be explained by the dry state of the sand and the type of surcharge loading used in the model tests, which not only increased the ultimate point resistance with greater surcharge above the limiting value (Fig. 18, 108) but also considerably reduced the critical depth ratio by decreasing the soil displacements, arching, and compressibility effects of the overburden near the model piles compared with those occurring near piles in two-layered soils under field conditions.

Chaplin has discussed some of the factors influencing the point resistance of piles and penetrometers, and he suggests further empirical relations based on the friction angle and grain-size characteristics of sand. The shear strength parameters in the figures by the writer were those given in the various references to which the reader is directed for further details due to space limitations of the paper. Additional information on the bearing capacity of piles in layered soils has confirmed (112) that for full-sized piles driven into a thick submerged cohesionless bearing stratum underlain by a weak deposit the critical depth ratio above the weak deposit is roughly 10 pile diameters (Fig. 11). This critical depth ratio has about the same magnitude and depends on the same factors as previously mentioned for the critical bearing depth ratio, $D_{bc}/B$, and for smaller diameters of piles and cone penetrometers the critical depth ratios increase due to scale effects (112).

Analyses of the behavior of tapered piles in the paper referred to piles with a continuous taper, which may also be used as a reasonable approximation to similar step-taper piles (67). Further, it was mentioned that for average tapers exceeding about 1% of the embedded pile length the ultimate unit skin friction of piles driven into cohesionless soil is roughly 1.5 times that of corresponding cylindrical piles. Accordingly, the total skin friction of tapered piles in practice frequently exceeds that of a cylindrical pile with the diameter equal to the maximum embedded diameter of a tapered pile. The ultimate unit point resistance in sand is, however, not affected by pile taper. The ultimate bearing capacity of piles jacked into cohesionless soil is roughly midway between those of corresponding bored and driven piles. The difference in pile behavior would seem to be mainly due to the effect of the method of installation on the changes of the original stresses, strength, and deformation properties of the soil mass near the piles. These changes also influence the critical depth ratios to some extent, and the main factors controlling these ratios in cohesionless soil were summarized previously.

The writer is grateful to the discussers for their helpful comments and information on a subject of great practical interest.

## APPENDIX.—REFERENCES

112. Meyerhof, G. G., and Valsangkar, A. J., "Bearing Capacity of Piles in Layered Soils," *Proceedings*, Ninth International Conference on Soil Mechanics, Vol. 2, Tokyo, Japan, 1977.
113. Panet, M., "Etude des Fondations sur Sable en Modèle Réduit," *Annales de l'Institut Technique du Batiment et Travaux Publics*, Paris, France, Vol. 25, 1972, pp. 62-84.

# JOURNAL OF THE GEOTECHNICAL ENGINEERING DIVISION

## THE TWELFTH TERZAGHI LECTURE

Presented at the American Society of Civil Engineers Annual Convention and Exposition, Philadelphia, Pennsylvania

LYMON C. REESE

# Introduction of Twelfth Terzaghi Lecturer

## By Kenneth L. Lee

In this bicentennial year in the history of our nation and in this city of Philadelphia where the founding constitution was written 200 years ago, it is altogether fitting and proper that as geotechnical engineers we pause at this time for a look back into our own professional history. Let us recall that it was also exactly 200 years ago that the French engineer, C. A. Coulomb, published his classical memoir on statics, which included a treatise on the subject of lateral earth pressures against retaining walls and on the strength of soils. These 200-year old principles still form the basis of most of the calculations involving soil strength and earth pressure problems made by modern engineers at this time.

It was just over 50 years ago that Karl Terzaghi culminated some of his early research in publishing his book, *Erdbaumechanik*, thus laying the foundations for modern soil mechanics as we know it today. Responding to the growing professional area described in Terzaghi's book and papers, the Soil Mechanics and Foundations Division of ASCE was first organized in 1936, exactly 40 years ago. Karl Terzaghi was as much interested in geology as he was in soils. Thus, it is significant to note in this brief bicentennial review that exactly 25 years ago this month marks the birth of a new joint technical committee on Engineering Geology. The sponsoring parent organizations were the Geological Society of America and the Soil Mechanics and Foundations Division of ASCE. The first chairman was Karl Terzaghi.

This evening we have assembled not only to honor the memory of Karl Terzaghi, the founding father of our modern profession, but also to honor one of a later generation who has contributed widely and generously to the field of knowledge in geotechnical engineering, and from which we all have reaped benefits. In October 1960, the Soil Mechanics and Foundations Division established the Karl Terzaghi Lecture and at about yearly intervals a distinguished engineer is invited to present the Terzaghi Lecture at an appropriate meeting of the American Society of Civil Engineers.

Many previous Terzaghi Lecturers are in the audience this evening. The text of their lectures have all been published in the Journal of the Division, and the first nine lectures have been republished in a special memorial volume. The more recent lectures will also be published in a similar memorial volume in the near future.

To the list of the 11 distinguished engineers who have presented the Terzaghi Lectures in previous years, we are pleased this evening to hear the Twelfth Terzaghi Lecture in this series presented by Professor Lymon C. Reese. The first Terzaghi Lecture was delivered by Ralph Peck in 1963; the successive lectures have been:

| | |
|---|---|
| 1964, Arthur Casagrande | 1970, T. William Lambe |
| 1966, Laurits Bjerrum | 1971, John Lowe, III |
| 1967, H. Bolton Seed | 1973, Bramlette McClelland |
| 1968, Philip C. Rutledge | 1974, F. E. Richart, Jr. |
| 1969, Stanley D. Wilson | 1975, George Geoffrey Meyerhof |

Lymon Reese is a long-time resident of Texas. His early years were spent in west Texas. After a period of active service in the Navy during World War II in the Aleutian Islands and Okinawa, he returned to Texas where he attended the University of Austin, and received his Bachelor of Science and Master of Science degrees in Civil Engineering in 1949 and 1950. After a brief period of teaching at Mississippi State College and a period of additional graduate study at the University of California, Berkeley, where he received his PhD degree, he returned to a teaching and research career at the University of Texas in 1955, where he still resides.

At the University of Texas, Professor Reese has held many academic ranks, including all grades of professor and Chairman of the Civil Engineering Department, and is currently the Associate Dean for Program Planning. He is also currently the T.U. Taylor Professor of Civil Engineering.

In addition to his university work, Lymon Reese has held a number of full-time and part-time career positions, including various field engineering and construction assignments, and has served as consultant to a large number of organizations on foundation engineering problems.

Lymon Reese is a master of progressive innovative thinking, stimulation, and inspiring leadership. He is quick to distill the essential details from a complex technical problem and then proceed in a deliberate and thoughtful manner to develop an engineering solution. He has the leadership capacity to inspire both students and colleagues with realistic challenges so that difficult problems quickly crumble under the enthusiastic attacks of his research teams.

Through a combination of professional experiences, academic inquiry, and skillful leadership, Lymon Reese and his associates have significantly advanced the theory and the practice in the field of deep pile foundations and made them readily available to the entire profession through his many practical yet scholarly publications. He has pioneered in performing field studies of instrumented piles and has developed analytical methods now widely used in the design of pile foundations for major structures. His name is internationally known in the subject area of piles and deep foundations, and he is a popular speaker at many meetings and symposia dealing with these important topics. He had been honored for his work by receiving the Middlebrooks Award from our Society and in 1975 he was given the prestigious honor of election to membership in the National Academy of Engineering.

Lymon Reese is as generous with his time as he is with his talents. He serves as a consultant to many companies and government agencies. He has spent two summers teaching in India for the United States Agency for International Development. As Associate Dean for Program Planning, the University of Texas has benefitted greatly from his progressive leadership.

He has had a long and active career of service in the American Society of Civil Engineers, both in the Texas Section and at the National level. He holds the rank of Fellow in the Society. He was President of the Texas Section in 1968–1969 and he continues with committee assignments. At the National level, he has served as Chairman of the Computer Applications Committee and was Chairman of the highly successful Specialty Conference on Analysis and Design in Geotechnical Engineering, held at the University of Texas in 1974.

He is a member of the American Society of Engineering Education, the National

Society of Professional Engineers, several honor societies and fraternities, and is a registered civil engineer in the State of Texas.

Lymon Reese is happily married and the father of three children.

Ladies and gentlemen, the Twelfth Terzaghi Lecture will now be presented by Lymon C. Reese, on the subject, "The Design and Construction of Drilled Shafts."

# DESIGN AND CONSTRUCTION OF DRILLED SHAFTS

By Lymon C. Reese,[1] F. ASCE

**FOREWORD**

I am, of course, deeply honored to be asked to present this year's Terzaghi Lecture, principally because of the stature of the lecture as established by those distinguished engineers who preceded me, but also because the lecture bears the name of the man who is rightly credited with establishing the discipline that occupies our attention.

Other Terzaghi Lecturers have indicated how their work was influenced by Dr. Terzaghi. I will be no exception. I think that it is not well known that Terzaghi had a close association with the University of Texas and with Professor Raymond F. Dawson. Terzaghi presented papers at six of the eight Texas Conferences on Soil Mechanics and Foundation Engineering that were organized by Dawson. As a matter of fact, the second and third of those conferences were entitled, "The Terzaghi Lectures." He presented a total of 14 papers at those six conferences beginning in 1939 and concluded with the notable paper, "Submarine Slope Failures," at the eighth and last conference in 1956. Professor Hudson Matlock and I were honored to be on the program with Terzaghi when we collaborated on our first paper on laterally loaded piles. Terzaghi visited a test site near Austin in 1956 where a program of lateral load tests were in progress.

Terzaghi accepted an appointment as Distinguished Professor at the University of Texas for the spring semester in 1941. I have talked with Professor Dawson on many occasions about Terzaghi's life in Texas. Great curiosity, boundless energy ("Let's climb just one more mountain before we call it a day, Raymond"), the ever-present camera, notebook, and sketch book, attention to detail, and hours spent in preparing notes in the evening on the day's observations marked Terzaghi's time in Texas as elsewhere.

Not long after Terzaghi's professorship in Texas, I began my research under Professor Daswon and started my work on deep foundations. Later, when in California, I was influenced by a statement of Terzaghi's in the *Proceedings of the British Building Research Congress* (30). Terzaghi wrote, "Among important topics for further research is the effect of pile penetration on the consistency of clays; the investigation can be performed only in the field at appropriately selected sites." I worked on that topic at Berkeley and have continued to do

---

Note.—Discussion open until June 1, 1978. To extend the closing date one month, a written request must be filed with the Editor of Technical Publications, ASCE. This paper is part of the copyrighted Journal of the Geotechnical Engineering Division, Proceedings of the American Society of Civil Engineers, Vol. 104, No. GT1, January, 1978. Manuscript was submitted for review for possible publication on February 24, 1977.

[1] T. U. Taylor Prof. of Civ. Engrg. and Assoc. Dean, Coll. of Engrg., Univ. of Texas at Austin, Austin, Tex.

research on deep foundations ever since. I believe that I was influenced into that area of study by Terzaghi's writings and by his lasting influence on the Texas program.

## INTRODUCTION

The drilled shaft is constructed by drilling a cylindrical hole, or a cylindrical and underreamed hole, into the soil and subsequently filling that hole with concrete. The drilled shaft may also be termed a drilled pier, a caisson, an underreamed foundation, a bored pile, or a bored and cast-in-place pile. While the most common application of the drilled shaft is to sustain large axial loads, it may also be used to sustain horizontal loads as in a retaining structure.

The excavation for drilled shafts by machine appeared in the United States in the 1920's (15). After World War II, a large number of small drilling contractors appeared in the United States and Great Britain. They operated principally in the United States in areas of Texas, California, Michigan, Illinois, and in the London area of England. At most of these locations there were cohesive soils that permitted the excavation of free-standing holes (22). At the present time, the drilled shaft industry is operating virtually around the world. The Association of Drilled Shaft Contractors is an international organization of contractor members and of associate members who manufacture equipment.

This paper presents information on construction procedures, deficiencies in drilled shafts related to design and construction, results from an extensive field research program, and recommendations for the design of drilled shafts to sustain axial load.

## DESCRIPTION OF CONSTRUCTION METHODS

Three different construction methods will be described; however, there can be many variations in the methods. The descriptions present a number of important points concerning current construction techniques. Methods of construction are important because there is a critical relationship between construction and design (3,23,28).

**Dry Method.**—The dry method is applicable to soils above the water table that will not cave or slump when the hole is drilled to its full depth. A soil that meets this condition is a homogeneous, stiff clay.

Fig. 1(a) shows the excavation being carried to the full depth by use of a crane attachment; underreaming was not employed. Concreting is shown in Fig. 1(b) with some free fall of the concrete. Unless a special mix is employed with characteristics that have been proved by field experiments, it is not acceptable to allow concrete placed by the free fall method to bounce from the sides of the excavation or to fall through a rebar cage. Fig. 1(c) shows the placement of a rebar cage in the upper portion of the drilled shaft; the final portion of the concrete is poured with a tremie without free fall. The completed shaft is shown in Fig. 1(d).

**Casing Method.**—The casing method is applicable to sites where soil conditions are such that caving or excessive deformation will occur when a hole is excavated. A stratum must exist below the caving soil that has low permeability and sufficient strength so that a drilled hole will not collapse.

If it is assumed that some dry soil of sufficient stiffness to prevent caving exists near the ground surface, as shown in Fig. 2(a), the construction procedure can be initiated as with the dry method. When the caving soil is encountered, a slurry is introduced in the hole and the excavation proceeds, as shown in Fig. 2(b). Bentonite or some other admixture can be used. Depending on the condition of surface soil, the elevation of the top of the slurry column may be just above the caving soil, or it may be brought near the ground surface, as shown in the figure. An alternate method of construction that is frequently used is to "mud" the hole down without using bentonite. Thus, the slurry is mixed in place from the natural soil and water.

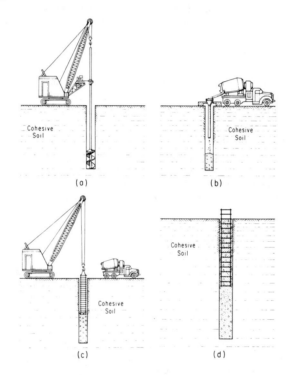

**FIG. 1. Dry Method of Construction: (a) Initiating Drilling; (b) Starting Concrete Pour; (c) Placing Rebar Cage; (d) Completed Shaft**

Drilling is continued until the stratum of caving soil is pierced and a stratum of impermeable soil is encountered. As shown in Fig. 2(c), a casing is introduced at this point, a "twister" or "spinner" on the kelly is used to rotate and push the casing into impermeable soil an amount sufficient to effect a seal.

As shown by the figures, the outside of the casing is slightly smaller than the inside of the drilled hole. In connection with the size of the casing, construction costs are minimized, if "OD" pipe is allowed.

A bailing bucket is placed on the kelly and the slurry is removed from the casing, as shown in Fig. 2(d). A smaller drill is introduced into the hole, and the hole is drilled to the projected depth, as shown in Fig. 2(e). A belling tool can be employed, as shown in Fig. 2(f), and the base of the drilled shaft can be enlarged. During this operation, as shown in the figures, slurry is retained in the annular space between the outside of the casing and the inside of the upper drilled hole. Therefore, it is critical that the casing be sealed in the impermeable formation. It is sometimes necessary to place teeth on the bottom of the casing in order to penetrate a sufficient depth into the impermeable formation. As may be understood, the casing method cannot be employed if a seal is impossible to obtain, or if there is no impermeable formation into which the lower portion of the hole can be drilled. Many controversies have arisen where soil borings have failed to reflect properly whether or not a formation of low permeability exists at a reasonable depth into which the base of the drilled shaft can be placed.

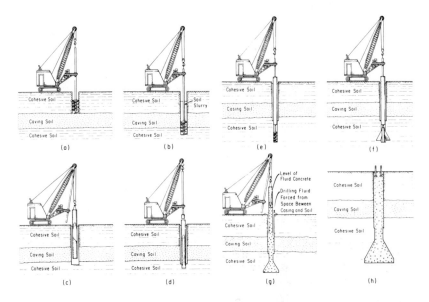

FIG. 2.—Casing Method of Construction: (a) Initiating Drilling; (b) Drilling with Slurry; (c) Introducing Casing; (d) Casing Is Sealed and Slurry Is Being Removed from Interior of Casing; (e) Drilling Below Casing; (f) Using Underreaming Tool; (g) Removing Casing; (h) Completed Shaft

If reinforcing steel is to be used with drilled shafts constructed by the casing method, the rebar cage must extend to the full depth of the excavation. The cage will be designed from the standpoint of meeting two requirements: (1) Structural requirements for bending and for column action imposed by loads from the superstructure; and (2) stability requirements during its placing and during the placing of concrete.

After reinforcing steel has been placed, the hole should be completely filled with fresh concrete with good flow characteristics. The casing may only be pulled when there is sufficient hydrostatic pressure in the column of concrete to force the slurry that has been trapped behind the casing from the hole, as shown in Fig. 2(g). If the hydrostatic pressure of the fluid concrete at the bottom of the casing is too low, the slurry will fall into the excavation and cause a defective foundation. Problems of this sort have been described by Baker (2) and by Peck (23). If the concrete has been placed to a sufficient height but if it has taken a partial set or has too low a slump, the friction between the concrete and the inside of the casing can be sufficient to lift the column of concrete when the casing is lifted, causing a failure (9).

As may be understood from an examination of Fig. 2(g), the upper portion of the column of concrete must move downward with respect to any rebar cage when the casing is pulled, causing a downward force to be exerted on the rebar cage. The magnitude of the downward force will depend on the shearing

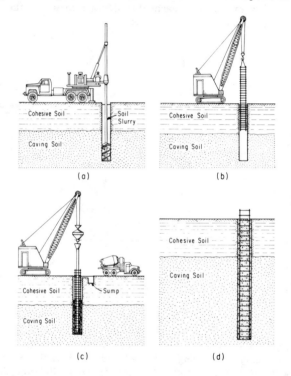

FIG. 3.—Slurry Method of Construction: (a) Drilling to Full Depth with Slurry; (b) Placing Rebar Cage; (c) Placing Concrete; (d) Completed Shaft

resistance of the fresh concrete at the velocity of flow that exists and on the area of the elements of the rebar cage. The rebar cage may fail by torsional buckling, by slipping at joints, and possibly by single-bar buckling. The completed shaft is shown in Fig. 2(h).

There are instances in which the soil profile is such that only a thin stratum of caving soil exists at the site. In such a case, it is possible to eliminate the use of slurry and introduce the casing when the caving formation is encountered. The casing is pushed and twisted through the thin stratum into impermeable soil below. Excavation can then proceed, along with the other steps in the construction process, as described in the preceding.

**Slurry Displacement Method.**—The soil conditions where the slurry displacement method is applicable could be any of the conditions described for the casing method. Perhaps the most beneficial use of the slurry displacement method would be at sites where it is impossible to seal a casing except by special techniques, such as freezing the soil.

The dry method is employed until a caving formation is encountered. At this point, slurry is introduced into the hole as for the casing method, and drilling is continued. Alternately, the hole is "mudded" down if the formation will stand without caving without the use of bentonite. Fig. 3($a$) shows the excavation being carried to the full depth by use of a truck-mounted rig, with the slurry in place. The character of the slurry is sufficient that particles of granular soil are put in suspension. If the excavation is to be carried through a stratum of clay, the excavated clay will of course be lifted through the slurry and brought to the ground surface. A drilling tool should be employed so that the column of slurry can flow through the tool in order to prevent the development of negative pressure beneath the drilling tool and the possible collapse of the hole.

If reinforcing steel is to be used, the rebar cage is placed in the slurry as shown in Fig. 3($b$). After the rebar cage has been placed, the concrete is poured with a tremie. The end of the tremie should be closed with an appropriate valve (a plywood plate can be used) until the tip of the tremie reaches the bottom of the hole. The placement of concrete in the tremie will then open the valve. Concreting can proceed with precautions being taken to ensure that the tip of the tremie is always below the column of fresh concrete (25). As shown in Fig. 3($c$), the column of concrete will rise in the hole and displace the column of slurry which is of lower density. The completed foundation is shown in Fig. 3($d$).

## CONSTRUCTION PROBLEMS

Drilled shafts of good quality can be built using each of the construction procedures described previously and using numerous variations of those procedures; however, problems can arise in construction that will lead to a drilled shaft of inferior quality. Some of these problems are listed and covered briefly as a means of more fully describing the drilled shaft foundation:

1. Selection of Incorrect Construction Procedure. The dry method of construction cannot be used if the drilled hole will not stand open or if water flows into the hole. The casing method cannot be employed if the casing cannot be sealed at its base in an impermeable formation. A hole cannot be drilled economically through a layer of boulders if the size of the boulders is larger than about a third of the diameter of the hole. Because the performance of a drilled shaft foundation is dependent on the construction procedure that is used, it is obviously important that the subsurface investigations be performed in such a way as to allow the appropriate construction procedure to be employed.

2. Improper Excavation. The hole may be drilled at the wrong location and with an accidental batter and underreams may be omitted or improperly sized. The hole may be allowed to collapse during drilling or the base of the hole may be improperly cleaned.

3. Poor Concrete. As noted later, the design of the concrete mix is important in drilled shaft construction. However, a mix may be properly designed but the quality of the concrete in the drilled shaft may be poor because of excessive time in making the pour. Also, excessive water is sometimes added at the job site.

**TABLE 1.—Axial Load Tests of Instrumented Drilled Shafts**

| Location (1) | Identifying symbol (2) | Date tested (3) | Dimensions Below Groundline, in feet | | | | Method of construction (8) | Failure load,[a] in kips (9) | Reference number (10) |
|---|---|---|---|---|---|---|---|---|---|
| | | | Stem diameter (4) | Stem length (5) | Underream diameter (6) | Underream height (7) | | | |
| Austin | MTO | 2-67 | 2.0 | 12. | — | — | dry | 320 | 24 |
| San Antonio | SA | 5-68 | 2.5 | 26.8 | — | — | dry | 1,880 | 27,34 |
| Houston | HB and T | 6-69 | 2.79 | 60. | — | — | slurry and casing to 54 ft; bottom 6 ft dry | 1,660 | 5 |
| Houston | S1 | 8-68 | 2.5 | 23.1 | — | — | dry | 280 | 22,27 |
| Houston | S2 | 3-69 | 2.5 | 18.5 | 7.5 | 4.5 | dry | 1,070 | 22,27 |
| Houston | S3 | 10-69 | 2.5 | 23. | — | — | dry | 200 | 22,27 |
| Houston | S4 | 12-69 | 2.5 | 45. | — | — | slurry and casing to 40 ft; bottom 5 ft dry | 640 | 22,27 |
| George West | US59 | 9-70 | 2.5 | 25.4 | — | — | dry | 840 | 27,32 |
| George West | HH | 10-70 | 2.0 | 19.8 | — | — | dry | 720 | 27,32 |
| Houston | G1 | 10-71 | 3.0 | 54.8 | — | — | slurry | 900 | 27,32 |
| Houston | G2 | 10-71 | 2.5 | 73.5 | — | — | slurry | 1,340 | 27,32 |
| Houston | BB | 10-71 | 2.5 | 45.0 | — | — | slurry | 1,200 | 27,32 |
| Bryan | Bry | 3-73 | 2.5 | 42.0 | — | — | dry | 850 | 11 |
| Puerto Rico | PR2 | 7-73 | 3.0 | 52.0 | — | — | dry | 1,890 | 11 |
| Puerto Rico | PR3 | 8-73 | 3.0 | 87.0 | — | — | dry | 1,856 | 11 |
| Austin | MT1 | 1-74 | 2.5 | 23.8 | — | — | slurry and casing to 19.3 ft; bottom 4.5 ft dry | 1,200 | 1 |
| Austin | MT2 | 1-75 | 2.5 | 24.0 | — | — | slurry | 1,220 | 1 |
| Austin | MT3 | 1-75 | 2.5 | 24.0 | — | — | dry | 1,140 | 1 |
| Dallas | DT1 | 2-75 | 3.2 | 23.0 | — | — | casing | 900 | 1 |

[a] Maximum load shaft could sustain if base was in clay or load at a settlement equal to 5% of base diameter if base was in sand.
Note: 1 ft = 0.305 m; 1 kip = 4.45 kN.

4. Improper Placement of Concrete. Some of the procedures associated with improper placement of concrete are: (a) Pouring concrete through water; (b) segregation of concrete during placement; (c) failure to complete pour with submerged tremie, which leaves section of weak concrete in shaft; and (d) vibration of concrete, which leads to collapse of weak surface soils. Vibration should be avoided but some careful rodding at the top of the shaft may be used.

5. Improper Pulling of Casing. The casing may be pulled with insufficient concrete inside, allowing slurry to run into the hole or allowing the hole to

collapse. If a rebar cage is being employed, the casing can be pulled too rapidly and cause buckling of the rebar cage.

## DESCRIPTION OF RESEARCH PROGRAM

During the last decade, a research program has been underway that has involved field, laboratory, and analytical studies. The most striking aspect of the research program is that axial load tests in the field have been performed on 19 full-sized drilled shafts that have been instrumented for the determination of axial load

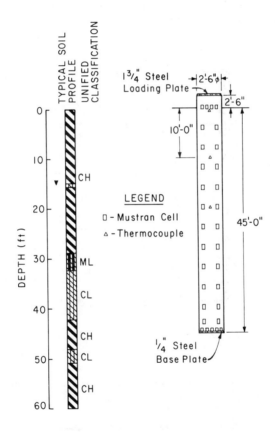

FIG. 4.—Test Shaft S4 (1 in. = 25.4 mm; 1 ft = 0.305 m)

as a function of depth. A brief description of the tests that have been performed is presented in Table 1. Each of the field tests involve the following: (1) Soil borings, laboratory testing, and in-situ soil studies to determine significant engineering properties of the soils; (2) preparation of instrumentation in the laboratory and the installation of the instrumentation on a rebar cage; (3) test shaft and the reaction shafts; (4) load testing and data collecting; and (5) analysis of data. The quick-load-test method (13) was used in the testing program, a method similar to the constant-rate-of-penetration method (12).

Fig. 4 presents an example of a shaft tested at a Houston site (22). The family of curves that was obtained for the test, showing the distribution of load as a function of depth, is presented in Fig. 5. The plotted points on the figure indicate the loads as determined from the output of the load cells. While the load-cell readings are not precise, the curves that are developed from the data are believed to represent a reasonably accurate indication of the behavior of the drilled shaft. It is possible to gain an understanding of a number of the aspects of the interaction between the foundation and the supporting soil from such data. For example, the load-settlement curves for the base and for the sides of the shaft shown in Fig. 6 were obtained from analyses of the curve showing total load versus settlement and the curves in Fig. 5 (8). Such

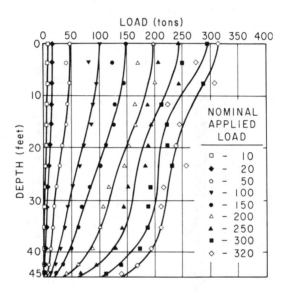

**FIG. 5.—Load Distribution Curves for Test Shaft S4 (1 ft = 0.305 m; 1 ton = 8.90 kN)**

analytical techniques were employed in developing curves summarizing results of the load-test program, shown in the following sections.

## SUMMARY OF SIGNIFICANT RESULTS FROM TESTS IN STIFF CLAYS

Prior to presenting results from the field testing of instrumented shafts, some general comments are necessary about the determination of the shear strength. Early in the test program it was thought that the effective stress approach might be used in clays but unsuccessful attempts were made to measure the earth stresses at the interface of the concrete and soil (7,24). Also, it was impossible to obtain good quality soil samples for laboratory testing at some of the test sites (1,11,22). Therefore, the usual approach in foundation engineering (35) was adopted. The undrained shear strength of clay is used in developing correlations between drilled shaft behavior and properties of clay. Where possible, the strength of the clay was obtained by use of triaxial testing but in-situ methods were employed in some cases. In most instances, the relative density of sand was obtained by use of data from standard penetration testing. Additional details concerning soil testing are presented in the reference cited in Table 1.

The curve showing the distribution of load on the shaft at failure can be differentiated to obtain the maximum unit load transfer as a function of depth. The undrained shear strength of the soil at a particular depth can be divided into the maximum load transfer to obtain the dimensionless shear strength reduction factor, $\alpha$. The plot of $\alpha$ as a function of relative depth, in which

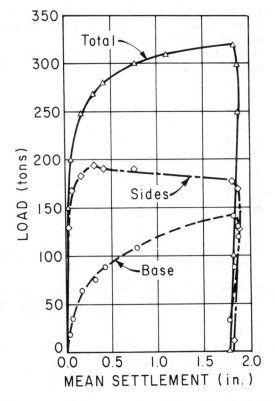

FIG. 6.—Load-Settlement Curves for Test Shaft S4 (1 in. = 25.4 mm; 1 ton = 8.90 kN)

relative depth is defined as the depth below ground surface divided by the penetration of the cylindrical portion of the shaft, is shown in Fig. 7. There is a considerable amount of scatter in the curves but the trend of the curves is well defined. The value of $\alpha$ is low near the ground surface, has a maximum value near the midheight of the cylindrical portion of the shaft, and it is low again near the tip of the shaft. The average value of $\alpha$ is about 0.6 and the average of the maximum value ranges from about 0.7–0.8. The low value of $\alpha$ near the ground surface is probably due to the low normal pressure at the interface of the concrete and the soil because of the initial low hydrostatic pressure in the fresh concrete at the ground surface. The low value of $\alpha$ near the base of the shaft is thought to be related to the mechanics of load transfer (10).

The magnitude of the average value of $\alpha$ for drilled shafts is nearly the same as that obtained by other investigators (31,37) for driven piles, but the mechanisms related to load transfer for drilled shafts and for driven piles are different. With regard to drilled shafts, some mechanisms causing load transfer in side resistance to be less than the in-situ shear strength of the clay were examined by Meyerhof and Murdock (21). Two of the factors mentioned were the opening of cracks in fissured clay due to soil movement toward the drilled hole and movement of water from fresh concrete to partially saturated clay. Data from experiments on moisture movement have been presented (7,21,22).

Bearing capacity factors were computed for tests in clay and are shown in Fig. 8. These factors were computed by dividing the ultimate stress at the base of the foundation by the undrained shear strength, with the strength of

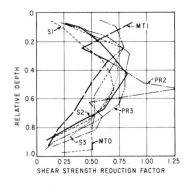 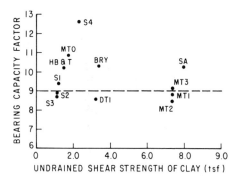

FIG. 7.—Relative Maximum Load Transfer in Clay as Function of Relative Depth

FIG. 8.—Bearing Capacity Factor for Clay as Function of Undrained Shear Strength (1 tsf = 95.8 kN/m²)

the clay being taken as the average of that for two diameters below the base of the shaft. Five of the values fall between 8.5–9.0, and there are seven values above 9.0. These data do not indicate that the bearing capacity factor is a function of shear strength.

Fig. 9 is a dimensionless plot showing the downward movement that is necessary to develop the maximum load transfer in skin friction. The abscissa for the plot is the load actually carried in skin friction divided by the theoretical maximum load, where the theoretical maximum load is computed by multiplying the average undrained shear strength by the circumferential area of the shaft. Thus, the maximum abscissa for a curve is the average value of $\alpha$ that was developed for the test indicated. The ordinate for the plot is the downward movement of the midheight of the shaft divided by the diameter, expressed as a percentage. As would be expected, there is a considerable amount of scatter in the curves, with the maximum skin friction being developed at an average downward movement ranging from about 0.5%–2% of the shaft diameter. The average of six load tests reported by Whitaker and Cooke (35) are also plotted for comparison. Some of the tests, e.g., US 59; are for sites where there was some sand along the sides of a shaft but the results were plotted in Fig. 9 because most of the soil along the sides was clay. The large value of $\alpha$ for the SA test is believed to be due to underestimating the magnitude of the shear strength.

A dimensionless plot showing downward movement of the shaft required to develop end bearing is shown in Fig. 10. The abscissa of the plot is the unit end bearing divided by nine times the undrained shear strength at the base of the shaft. Thus, the maximum abscissa of each of the curves would be unity if the measured bearing capacity factor had been nine. The ordinate is the settlement of the base of the shaft divided by the term $2B\epsilon_{50}$, in which $B$ = the diameter of the base and $\epsilon_{50}$ = the strain at a stress of one-half

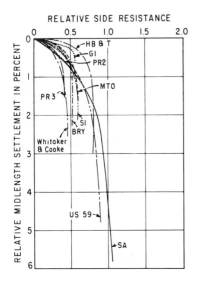 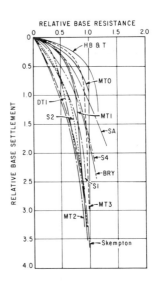

FIG. 9.—Relative Side Resistance for Clay Versus Relative Midlength Settlement

FIG. 10.—Relative Base Resistance versus Relative Base Settlement

of the compressive strength of the soil. The term $2B\epsilon_{50}$ was selected because of a suggestion by Skempton (29). Skempton's curve, which agrees with some experimental data available to him, is shown in the figure. As can be seen, there is a considerable amount of scatter in the experimental load-settlement curves and most of the curves are stiffer than Skempton's suggestion. Much of the scatter results from the difficulty of obtaining stress-strain curves of good quality. The larger stiffness of the experimental curves for all but three of the tests may result from assuming too large a value of $\epsilon_{50}$ (0.01 was selected in making the analyses when no stress-strain curves were available) or because laboratory stress-strain curves prove to be less stiff than curves for in-situ soil. The settlements in Figs. 9 and 10 are "short-term" and do not include, of course, the settlement due to consolidation.

## SUMMARY OF SIGNIFICANT RESULTS FROM TESTS IN SAND

Only a few of the instrumented drilled shafts that were tested were installed at sites where there was sand. In no instance did the entire soil profile consist of sand. In several of the tests there were strata of sand along the sides of

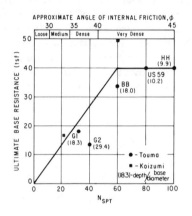

FIG. 11.—Ultimate Base Resistance in Sand Versus $N_{SPT}$ (1 tsf = 95.8 kN/m²)

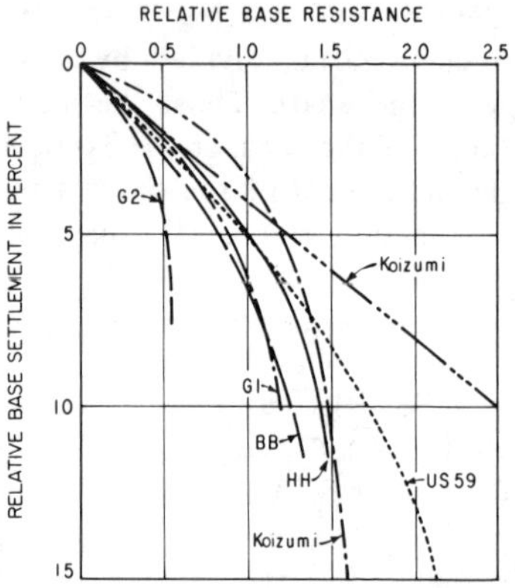

FIG. 12.—Relative Base Resistance in Sand Versus Relative Base Settlement

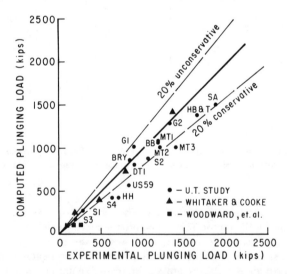

FIG. 13.—Comparison of Experimental and Computed Plunging Loads (1 kip = 4.45 kN)

the shafts and load was transferred in skin friction. However, data on load transfer in skin friction were meager and will not be summarized here. At five of the test locations, the drilled shafts had their tips bearing in sand and data concerning base resistance will be summarized.

The ultimate base resistance in sand as a function of blow count from the standard penetration test is shown in Fig. 11. Results from two additional tests reported by Koizumi (18) are also indicated in the figure. The ultimate base resistance is assumed to be the resistance that exists at a settlement equal to 5% of the base diameter. The numbers in parentheses near the plotted points are the ratios of the depth to base diameter. In most instances the base of the foundation is thought to be below the critical depth (19,33). The values for unit base resistance are probably somewhat less than values obtained for driven piles (33) because relative density will be increased by pile driving (20). The solid line in the figure indicates a design curve that might be employed, using an appropriate factor of safety.

Experimental curves in dimensionless form that present relative base resistance as a function of relative base settlement are shown in Fig. 12. The abscissa shows the experimental bearing stress divided by the ultimate bearing stress at failure, as taken from Fig. 11. The ordinate shows the experimental settlement divided by the diameter of the base of the shaft, as a percentage. As can be seen from the figures, bearing stress continues to increase with increasing relative settlement, except for one of the tests. That the bearing stress continues to increase with increasing settlement is in general agreement with results obtained by other investigators (17,18).

## DESIGN PROCEDURES FOR DRILLED SHAFTS

Procedures are outlined in this section for the design of drilled shafts in clay and in sand under compressive load. The design procedures consist of two steps: (1) The computation of the ultimate load; and (2) the computation of settlement under working load (6,14,36). As has been indicated, the design must also take into account the construction procedures that are employed. Considering the number of variables that are involved in the design, the data that are available on the behavior of drilled shafts under axial load are meager; therefore, the design recommendations cannot be used indiscriminately, but should be useful for three situations:

1. For sizing of test shafts. The design recommendations will allow the selection of a shaft of appropriate size for a test-loading program. Shafts of various diameters and penetrations can be studied in order to optimize the design.
2. For adjusting the size of the shafts after results from test loading are available. The results from the test-loading program can be analyzed in terms of the design recommendations and a clear direction can be indicated for adjusting the size, if necessary, of the production shafts to sustain working loads.
3. For other uses. On some occasions, conditions may be such that the design recommendation can be used for final design of drilled shafts. Of course, appropriate factors of safety must be employed considering all of the features of the design problem.

**Design Procedures for Stiff Clays.**—The equations that are recommended for the design of drilled shafts in clays are as follows:

$$(Q_T)_{ult} = (Q_S)_{ult} + (Q_B)_{ult} \quad \dots \dots \dots \dots \dots \dots \dots \dots \dots \dots \dots (1)$$

$$(Q_S)_{\text{ult}} = \int_{H_1}^{H_2} \alpha_c s_u \, dA_s \quad \ldots \ldots \ldots \ldots \ldots \ldots \ldots \ldots \ldots \ldots \ldots \ldots \ldots (2)$$

$$(Q_B)_{\text{ult}} = N_c s_u A_B \quad \ldots \ldots \ldots \ldots \ldots \ldots \ldots \ldots \ldots \ldots \ldots \ldots \ldots \ldots \ldots (3)$$

in which $(Q_T)_{\text{ult}}$ = ultimate axial load; $(Q_S)_{\text{ult}}$ = ultimate load is side resistance; $(Q_B)_{\text{ult}}$ = ultimate in end bearing; $H_1$ and $H_2$ = dimensions selected so that a portion of the soil near the top of the shaft and a portion of the soil near the base of the shaft are ignored in computing side resistance; $\alpha_c$ = shear strength reduction factor for clays; $s_u$ = undrained shear strength of the clay; $dA_s$ = differential circumferential area of the drilled shaft; $N_c$ = bearing capacity factor; and $A_B$ = area of base of shaft. An appropriate factor of safety is selected to compute the working load. The details of the computation of the ultimate load have been covered elsewhere (27).

A check must be made to see whether or not the estimated settlement is less than the allowable. The estimated settlement is obtained by reference to Figs. 9 and 10. The settlement in Figs. 9 and 10 is the immediate settlement, and any settlement due to consolidation must be added.

**Design Procedures for Sands.**—The equations that are recommended for the design of drilled shafts in sand are as follows:

$$(Q_T)_f = (Q_S)_f + (Q_B)_f \quad \ldots \ldots \ldots \ldots \ldots \ldots \ldots \ldots \ldots \ldots \ldots \ldots \ldots (4)$$

$$(Q_S)_f = \int_0^H \alpha_s p_z \tan \phi \, dA_s \quad \ldots \ldots \ldots \ldots \ldots \ldots \ldots \ldots \ldots \ldots \ldots (5)$$

$$(Q_B)_f = \frac{q_t}{K} A_B \quad \ldots \ldots \ldots \ldots \ldots \ldots \ldots \ldots \ldots \ldots \ldots \ldots \ldots \ldots \ldots (6)$$

in which $(Q_T)_f$ = failure load on shaft for a specified downward movement of the shaft; $(Q_S)_f$ = failure load in side resistance; $(Q_B)_f$ = failure load in end bearing; $H$ = penetration of shaft below ground surface; $\alpha_s$ = reduction factor for sand; $p_z$ = effective overburden stress at depth $z$; $\phi$ = angle of internal friction of sand; $dA_s$ = differential circumferential area of shaft; $q_t$ = maximum bearing stress at a downward movement of 5% of the diameter of the base; and $A_B$ = the area of the base of the shaft.

The factor $K$ is employed to reduce the value of $q_t$ to that for a specified downward movement and may be computed by assuming that the load-settlement curve for the base is a straight line. An appropriate factor of safety must be selected. The details of the computation of the failure load have been covered elsewhere (27).

A check must be made to see whether or not the estimated settlement is less than the allowable. The estimated settlement of the base may be obtained by referring to Fig. 12. Unfortunately, data are limited on side resistance versus settlement; however, available information (30) suggests that the settlement necessary to develop the full side resistance in sand is similar to that described in Fig. 9 pertaining to clay.

**Comparison of Computations with Experiments.**—The procedures outlined previously were used to compute the plunging loads for for those drilled shafts tested by the writer and his associates as well as for a few additional tests reported in the literature (35,37). The results of those computations are plotted versus the loads determined by experiment, as shown in Fig. 13. The design recommendations appear to be unconservative in only three cases and conservative for 18 cases. While detailed arguments cannot be set forth in this paper, it is believed that the major source of error lies in the inability to determine accurately the shear strength of the soils. Nevertheless, the results in Fig. 13 seem to offer reasonable support for the design procedures that have been recommended.

**Additional Comments About Design.**—There are a number of factors in the design of drilled shafts beyond the computation procedures that are outlined. Some of these factors are:

1. Careful determination of axial load (and choice of load factors). In some instances, it is necessary to take into account the fact that surface soils may put a down drag on the drilled shaft and in other instances the surface soils may be expansive and will impose an upward load.

2. Consideration of horizontal loads. Drilled shafts can take significant horizontal loads if properly reinforced. There are well-developed design procedures for dealing with such problems (26).

3. Inadequate soil borings. As indicated earlier, the soil boring must present valid information that will allow the proper construction procedure to be selected. With regard to design, the boring and subsequent laboratory tests should provide information that will allow the engineering properties of the soil to be evaluated accurately. If it is suspected that cavities exist at the construction site, an investigation should be designed that will eliminate the possibility of constructing a drilled shaft just above such a cavity. Also, soil investigation should identify water-deficient clays at a depth that can have access to free water as a result of the construction program.

4. Improper design of concrete, or rebar cage, or both. The concrete should be designed such that it has the appropriate strength, flow characteristics, and setting characteristics. The design should be such that concrete can be placed without difficulty and that it will completely fill the excavated hole. The method of trial mixes or some similar method should be employed to ensure that the time for the initial set is retarded sufficiently to allow the concrete to be placed without damage. In this respect, it is important to consider the availability of concrete and the size of the concrete pour. If a rebar cage is employed, the space between the rebars and the outside of the rebar cage and the inside of the excavated hole should be such as to allow the concrete to flow freely. The maximum size of the course aggregate that is employed is an important consideration in this regard.

## Proof of Design by Load Tests

Many building codes require that a load test be performed if a deep foundation is designed making use of resistance due to skin friction. Such a load test could be designed only to show that the deep foundation can sustain a load of perhaps twice the design load without excessive settlement. The writer is strongly of the opinion that many opportunities exist for significant savings in foundation costs if the load testing program involves the loading of a drilled

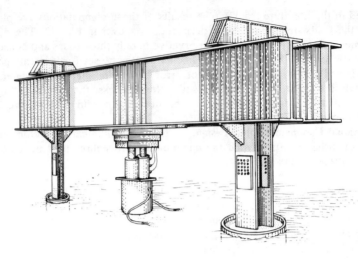

FIG. 14.—Diagram of Arrangement for Performing Load Tests

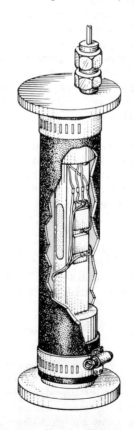

FIG. 15.—Sketch of Mustran Cell

shaft to failure. The ability to make savings is enhanced if internal instrumentation is employed in the drilled shaft to allow the determination of the distribution of axial load as a function of depth. In addition, of course, a valuable contribution can be made to the technical literature from such a test program.

A testing arrangement that has been employed successfully a number of times is shown in Fig. 14. Reaction shafts designed to take tensile load are installed on each side of the test shaft, as shown. The tensile load is transferred to the reaction shaft by a structural shape that is embedded in the reinforced reaction shaft an appropriate distance. An alternate procedure that has been employed with success (1) involves the placing of plastic tubes in the reaction

FIG. 16.—Instrumented Rebar Cage Being Placed

shaft to receive high strength rebars. The continuously threaded rebars are screwed into nuts that bear on a plate at the base of the shaft. After the test is completed, the rebars can easily be retrieved by unscrewing them from the base nuts.

A load is applied to the test shaft by a hydraulic ram or rams that bear against crossbeams that transfer the load to two reaction beams. Some hydraulic rams are specially designed so that friction is minimized, allowing the load to be obtained from a measurement of hydraulic pressure in the rams. However, the use of a calibrated load cell is recommended. An arrangement such as shown in the figure has been utilized to apply loads of up to 1,000 tons.

A number of devices can be employed to allow a determination to be made of the axial load in the drilled shaft as a function of depth. Electrical resistance strain gages may be attached to vertical rebars, or telltales (16) may be employed. A device that has served well is shown in Fig. 15 (4). The mustran cell consists of a 1/2-in. × 1/2-in. (12.7-mm × 12.7-mm) bar on which is affixed electrical resistance strain gages. End caps are attached to the bar and the strain gages are protected by a rubber hose as shown in the figure. The lead wires are brought through the upper end cap by use of special fittings and extend through the concrete to the ground surface. The ends of all the lead wires are placed inside a container pressurized with dry nitrogen. Nitrogen passes down the sheath on the lead wires and pressurizes the interior of each mustran cell and prevents the entrance of moisture. The cells are placed at various levels along the length of a drilled shaft, usually in pairs so as to eliminate any effects from accidental bending, and have proved effective in measuring axial load in a drilled shaft for as long as 1 yr (38). Cells are designed such that their stiffness is approximately equal to that of a concrete that is displaced. They are calibrated prior to installation; however, one or more levels of cells are placed above the ground surface and are used for obtaining readings that are used for interpreting the readings of the cells below the ground surface. Fig. 16 shows a rebar cage, instrumented with mustran cells, that is being placed in an excavation.

## ADVANTAGES AND DISADVANTAGES OF DRILLED SHAFTS

Some of the disadvantages of drilled shafts for a foundation are:

1. An excellent subsurface investigation must be carried out in order that designs are made properly and in order that the appropriate construction procedure is selected.
2. Drilled shafts of good quality are critically dependent on the construction techniques that are employed.
3. The appropriate inspection of the construction requires a considerable amount of knowledge and experience. It is normally not possible to investigate the completed shaft to see whether or not a good construction job has been obtained.
4. The shear strength of the supporting soil in general is reduced by the construction operation.
5. Drilled shafts of small diameter cannot successfully be constructed.
6. Failures of a drilled shaft can be expensive because a single shaft is usually designed to carry a load of large magnitude.

Some of the advantages of drilled shafts for a foundation are:

1. Drilled shafts can be successfully constructed in soils where it might be difficult to install other types of deep foundations.
2. Soil movements during construction due to heave or vibration are minimized.
3. Soil exposed during the construction operation and inspection can reveal whether or not the soil is consistent with that predicted by the subsurface investigation.

4. The size of the drilled shaft can be readily adjusted during the construction operation so that variations in subsurface conditions can be accommodated.

5. Drilled shafts can be built rapidly as compared to some other types of deep foundations.

6. Construction materials are readily available and construction equipment is generally available in all parts of the United States as well as many other parts of the world.

7. Pile caps can be eliminated on many jobs, leading to appreciable economy.

8. Construction noise is tolerable as compared to other types of construction of deep foundations.

## Conclusions

1. A good subsurface investigation is important in order to obtain engineering properties of the soil for design computations, and in order to select the most appropriate construction procedure.

2. The details of construction procedures are important because virtually all of the failures of drilled shaft foundations have been related to improper construction procedures. All of such failures can be avoided.

3. The preliminary research findings reported herein and elsewhere should prove useful. While the research findings do not provide definitive procedures for the design of drilled shafts, the results provide an excellent guide for the design of field load tests and should prove helpful to designers in other ways.

4. The performance of load tests of drilled shafts that are instrumented for measuring the distribution of axial load with depth is economically justified for many jobs. Efficient procedures for the design of load tests and for the design of instrumentation have been developed and the use of these procedures, in addition to leading to significant savings in the cost of a foundation, can constitute a permanent contribution to technical literature.

## Acknowledgments

The writer wishes to acknowledge with gratitude those companies or agencies who have made financial contributions to the research program that has been described or who have sponsored portions of the research activity, as follows: Texas Department of Highways and Public Transportation, Federal Highway Administration, Association of Drilled Shaft Contractors, Farmer Foundation Company, Martin and Martin Construction Company, Brown and Root, Florida Department of Transportation, and The University of Texas at Austin.

The writer also wishes to acknowledge the significant contributions to the research program by his past and present colleagues, research assistants, and graduate students, as follows: J. N. Anagnos, W. R. Barker, J. C. Brown, D. B. Campbell, J. W. Chuang, H. H. Dalrymple, C. J. Ehlers, D. E. Engling, W. R. Hudson, M. W. O'Neill, G. W. Quiros, F. T. Touma, V. J. Vijayvergiya, R. C. Welch, J. A. Wooley, and S. J. Wright.

## Appendix.—References

1. Aurora, R. P., and Reese, L. C., "Behavior of Axially Loaded Drilled Shafts in

Clay-Shales," *Research Report 176-4*, Center for Highway Research, The University of Texas at Austin, Austin, Tex., Mar., 1976.
2. Baker, C. N., Jr., "Drilled Piers, Construction Considerations," Lecture Series on Deep Foundations, Boston Society of Civil Engineers Section, ASCE, Apr., 1975.
3. Baker, C. N., Jr., and Khan, F., "Caisson Construction Problems and Correction in Chicago," *Journal of the Soil Mechanics and Foundations Division*, ASCE, Vol. 97, No. SM2, Proc. Paper 7934, Feb., 1971, pp. 417–440.
4. Barker, W. R., and Reese, L. C., "Instrumentation for Measurement of Axial Load in Drilled Shafts," *Research Report 89-6*, Center for Highway Research, The University of Texas at Austin, Austin, Tex., Nov., 1969.
5. Barker, W. R., and Reese, L. C., "Load Carrying Characteristics of Drilled Shafts Constructed with the Aid of Drilling Fluids," *Research Report 89-9*, Center for Highway Research, The University of Texas at Austin, Austin, Tex., Aug., 1970.
6. Burland, J. C., Butler, F. G., and Dunican, P., "The Behavior and Design of Large Diameter Bored Piles in Stiff Clay," *Proceedings*, Symposium on Large Bored Piles, Institution of Civil Engineers, London, England, 1966, pp. 51–71.
7. Chuang, J. W., and Reese, L. C., "Studies of Shearing Resistance between Cement Mortar and Soil," *Research Report 89-3*, Center for Highway Research, The University of Texas at Austin, Austin, Tex., May, 1969.
8. Coyle, H. M., and Reese, L. C., "Load Transfer for Axially Loaded Piles in Clay," *Journal of the Soil Mechanics and Foundations Division*, ASCE, Vol. 92, No. SM2, Proc. Paper 4702, Mar., 1966, pp. 1–26.
9. D'Appolonia, E., D'Appolonia, D. J., and Ellison, R. D., "Drilled Piers," *Foundation Engineering Handbook*, D. Van Nostrand Reinhold, New York, N.Y., 1975, pp. 601–615.
10. Ellison, R. D., D'Appolonia, E., and Thiers, G. R., "Load-Deformation for Bored Piles," *Journal of the Soil Mechanics and Foundations Division*, ASCE, Vol. 97, No. SM4, Proc. Paper 8052, Apr., 1971, pp. 661–678.
11. Engling, D. E., and Reese, L. C., "Behavior of Three Instrumented Drilled Shafts Under Short Term Axial Loading," *Research Report 176-3*, Center for Highway Research, The University of Texas at Austin, Austin, Tex., May, 1974.
12. Fellenius, B. H., "Test Loading of Piles and New Proof Testing Procedure," *Journal of the Geotechnical Engineering Division*, ASCE, Vol. 101, No. GT9, Proc. Paper 11551, Sept., 1975, pp. 855–869.
13. Fuller, R. M., and Hoy, H. E., "Pile Load Tests Including Quick-Load Test Method, Conventional Methods and Interpretations," *Pile Foundations*, Highway Research Board, No. 333, Washington, D.C., 1970, pp. 74–84.
14. Gardner, W. S., "Considerations in the Design of Drilled Piers," Lecture Series on Deep Foundations, Boston Society of Civil Engineers Section, ASCE, Apr., 1975.
15. Greer, D. M., "Drilled Piers State of the Art, 1969," *W.C.A. Geotechnical Bulletin*, Vol. III, No. 2, Woodward-Clyde and Associates, New York, N.Y., Sept., 1969.
16. Hansen, V., and Kneas, F. N., "Static Load Tests for Bearing Piles," *Civil Engineering*, ASCE, Vol. 12, No. 10, Oct., 1942, pp. 545–547.
17. Kerisel, Jean, "Deep Foundations Basic Experimental Facts," *Proceedings*, Deep Foundations Conference, Sociedad Mexicana de Mecania de Suelos, Mexico City, Mexico, Dec., 1964, pp. 1–31.
18. Koizumi, Y., et al., B.C.P. Committee, "Field Tests on Piles in Sand," *Soils and Foundations*, Vol. 11, No. 2, Tokyo, Japan, June, 1971.
19. Meyerhof, G. G., "Bearing Capacity and Settlement of Pile Foundations," *Journal of the Geotechnical Engineering Division*, ASCE, Vol. 102, No. GT3, Proc. Paper 11962, Mar., 1976, pp. 197–228.
20. Meyerhof, G. G., "Compaction of Sands and Bearing Capacity of Piles," *Journal of the Soil Mechanics and Foundations Division*, ASCE, Vol. 85, No. SM6, Proc. Paper 2292, Dec., 1959, pp. 1–29.
21. Meyerhof, G. G., and Murdock, L. J., "An Investigation of the Bearing Capacity of Some Bored and Driven Piles in London Clay," *Geotechnique*, London, England, Vol. 3, No. 7, 1953, pp. 267–282.
22. O'Neill, M. W., and Reese, L. C., "Behavior of Axially Loaded Drilled Shafts in Beaumont Clay," *Research Report 89-8*, Center for Highway Research, The University of Texas at Austin, Austin, Tex., Dec., 1970, 749 pages.
23. Peck, R. B., "Pile and Pier Foundations," *Journal of the Soil Mechanics and*

Foundations Division, ASCE, Vol. 91, No. SM2, Proc. Paper 4273, Mar., 1965, pp. 33–38.
24. Reese, L. C., and Hudson, W. R., "Field Testing of Drilled Shafts to Develop Design Methods," *Research Report 89-1*, Center for Highway Research, The University of Texas at Austin, Austin, Tex., Apr., 1968.
25. Reese, L. C., O'Neill, M. W., and Touma, F. T., "Bored Piles Installed by Slurry Displacement," *Proceedings*, Eighth International Conference on Soil Mechanics and Foundations Engineering, Moscow, U.S.S.R., Vol. 3, 1973, pp. 203–209.
26. Reese, L. C., "Laterally Loaded Piles," *Proceedings of the Seminar Series*, Design, Construction, and Performance of Deep Foundations; Geotechnical Group and Continuing Education Committee, San Francisco Section, ASCE; and Continuing Education in Engineering, University Extension, and The College of Engineering, University of California, Berkeley, Calif.; Aug., 1975.
27. Reese, L. C., Touma, F. T., and O'Neill, M. W., "Behavior of Drilled Piers Under Axial Loading," *Journal of the Geotechnical Engineering Division*, ASCE, Vol. 102, No. GT5, Proc. Paper 12135, May, 1976, pp. 493–510.
28. Reese, L. C., "Construction of Drilled Shafts," *Proceedings*, Cimientos Profundos Colados en Sitio (Cast-in-place Deep Foundations), Reunion conjunta ADSC-AMMS, Sociedad Mexicana de Mecanica de Suelos, June, 1976, pp. 2–37.
29. Skempton, A. W., "The Bearing Capacity of Clays," *Proceedings*, Building Research Congress, Division 1, Part III, London, England, 1951, pp. 180–189.
30. Terzaghi, K., "The Influence of Modern Soil Studies on the Design and Construction of Foundations," *Proceedings*, Building Research Congress, Division 1, Part III, London, England, 1951, pp. 139–145.
31. Tomlinson, J. J., "Adhesion of Piles in Stiff Clays," *Report 26*, Construction Industry Research and Information Association, London, England, Nov., 1970.
32. Touma, F. T., and Reese, L. C., "The Behavior of Axially Loaded Drilled Shafts in Sand," *Research Report 176-1*, Project 3-5-72-176, Center for Highway Research, The University of Texas at Austin, Austin, Tex., Dec., 1972.
33. Vesic, A. S., "Tests on Instrumented Piles, Ogeechee River Site," *Journal of the Soil Mechanics and Foundations Division*, ASCE, Vol. 96, No. SM2, Proc. Paper 7170, Mar., 1970, pp. 56–584.
34. Vijayvergiya, V. N., Hudson, W. R., and Reese, L. C., "Load Distribution for a Drilled Shaft in Clay Shale," *Summary Report 89-5*, Center for Highway Research, The University of Texas at Austin, Austin, Tex., Mar., 1969.
35. Whitaker, T., and Cooke, R. W., "An Investigation of the Shaft and Base Resistance of Large Bored Piles in London Clay," *Proceedings*, Symposium on Large Bored Piles, Institution of Civil Engineers, London, England, 1966, pp. 7–49.
36. Woodward, R. J., Gardner, W. S., and Greer, D. M., *Drilled Pier Foundations*, McGraw-Hill Book Co., Inc., New York, N.Y., 1972.
37. Woodward, R. J., Lundgren, R., and Boitano, J. D., "Pile Loading Tests in Stiff Clay," *Proceedings*, Fifth International Conference on Soil Mechanics and Foundation Engineering, Paris, France, Vol. 2, 1961, pp. 177–184.
38. Wooley, J. A., and Reese, L. C., "Behavior of an Axially Loaded Drilled Shaft Under Sustained Loading," *Research Report 176-2*, Center for Highway Research, The University of Texas at Austin, Austin, Tex., May, 1974.

## Design and Construction of Drilled Shafts[a]
### Discussion by Ibrahim H. Sulaiman,[2] M. ASCE

The author is to be congratulated for his review of drilled shaft foundations, and for the fruitful years spent in research with his co-workers. The writer read most of the series of publications including the Terzaghi lecture with much interest. The comment on Terzaghi's influence on the author's objective brings to mind the statement made by Terzaghi concerning researchers on pile-driving formula. In 1925 Terzaghi (40) wrote.

> Therefore no pile-driving formula of universal validity can possibly be obtained. This result is not surprising. Far more surprising are the strenuous efforts made to solve the insoluble problem, since an expenditure of $200 for a loading test is sufficient to get an accurate individual solution of a problem whenever needed.

Not that the problem of drilled shafts is insoluble. Many designs have been used safely without load tests. The need for load tests as the author states is still evident for large projects. However, expenditure for the test is currently more like $10,000. Owners of small projects are reluctant to spend on tests of drilled shaft foundations. This problem was circumvent in stiff to hard clay by designing for the entire load to be supported by the underream, and neglecting skin friction on the shaft. The author's recommendation will produce more savings in this area, by allowing for some skin friction in the design.

The author neglects the effective stress approach in computation of skin friction. He states that the low value of shear strength reduction factor $\alpha$ near the base of the shaft is thought to be related to the mechanics of load transfer (10). The writer feels that this item needs clarification. Review of Ref. 10, which is a finite element analysis of drilled shafts, shows that the base load induces a tension crack just above or adjacent to the shaft base. This is not surprising, since Geddes (39) showed that a force at a point in the interior of a semi-infinite solid induces tensile stresses a substantial distance up towards the surface of the solid. The net result is a reduction in the compressive confining stress along the shaft, especially near the base. This explains the mechanics of load transfer. In effect, it seems that the low value of $\alpha$ implies a reduction in skin friction near the base of the shaft due to a reduction in confinement. Therefore, skin friction in clay is also a function of confining pressure. Consequently, it should be evaluated in terms of effective stress.

Finally, little mention is made about the effective that jointing in clay has upon collapse of underreams during construction. The writer has witnessed

---
[a] January, 1978, by Lymon C. Reese (Proc. Paper 13503).
[2] Civ. Engr., Spencer J. Buchanan and Assoc., Inc., Bryan, Tex.

severe underream collapse in plastic to stiff jointed clay, so much so that their use was abandoned.

APPENDIX.—REFERENCES

39. Geddes, J. D., "Stresses in Foundation Soils Due to Vertical Subsurface Loading," *Geotechnique*, London, England, Vol. 16, No. 3, Sept., 1966.
40. Terzaghi, K., "Modern Conceptions Concerning Foundation Engineering," *Journal of the Boston Society of Civil Engineers*, Vol. 12, No. 10, Dec., 1925.

# Design and Construction of Drilled Shafts[a]
## Closure by Lymon C. Reese,[3] F. ASCE

Sulaiman notes that the expense of a load test, in the range of $10,000, is not justified for owners of small projects. A relatively simple computation can be made to ascertain whether or not a load test is justified. The drilled shaft foundation can be designed considering end-bearing only and considering both end-bearing and skin friction, employing as a guide the correlations between load transfer and soil properties that have been published. If the computations indicate that the potential savings by counting on skin friction are more than the cost of the load test, the load test is justified if time allows. The preceding approach is based on the idea that the particular building code requires a load test if skin friction is counted on in design or that the designer requires proof of the magnitude of the skin friction resistance.

In connection with load tests, the argument can be made effectively that it is frequently cost effective in performing a load test on drilled shafts to make use of internal instrumentation to allow the measurement of axial load with depth (41). Without instrumentation, the load test can only prove or disprove that a particular design is satisfactory. With the use of instrumentation, load transfer parameters can be determined, allowing a redesign of the drilled shaft if necessary, and a contribution can be made to the technical literature.

Sulaiman mentions the desirability of using the effective stress method in developing correlations between load transfer relationships and soil properties. The writer subscribes to that concept but, there are formidable difficulties because of problems associated with predicting soil strength parameters and effective stresses. The strength parameters and the effective stresses are related to the in-situ conditions, the foundation geometry, the construction process, and the nature of the loading. The strength parameters and the effective stresses will vary radially and along the length of the foundation. No comprehensive approach has yet been proposed for dealing with the complete range of variables. However, some interesting proposals have been made and results from additional research studies will no doubt be forthcoming.

Sulaiman notes that the collapse of the underreams for drilled shafts is sometimes a problem. In this connection it is interesting that most difficulties that have occurred with the use of drilled shafts in practice have been associated with construction problems. There are acceptable solutions for virtually every one of these problems and such failures are rapidly being eliminated as knowledge of drilled shafts becomes more widespread. With regard to the collapse of an underream, it is a good practice to drill a test excavation if collapse is expected.

---

[a] January, 1978, by Lymon C. Reese (Proc. Paper 13503).
[3] T. U. Taylor Prof. of Civ. Engrg. and Assoc. Dean, Coll. of Engrg., Univ. of Texas at Austin, Austin, Tex.

If collapse occurs, the underream can be replaced with an appropriate length of straight shaft.

Sulaiman's comments are constructive and are appreciated.

**APPENDIX.—REFERENCE**

41. Reese, L. C., "The Design and Evaluation of Load Test on Deep Foundation," presented at the June 25–30, 1978, American Society for Testing and Materials Symposium on Deep Foundations, held at Boston, Mass.

# JOURNAL OF THE GEOTECHNICAL ENGINEERING DIVISION

## THE THIRTEENTH TERZAGHI LECTURE

Presented at the American Society of Civil Engineers Annual Convention and Exposition, San Francisco

October 20, 1977

**ROBERT F. LEGGET**

## INTRODUCTION OF THIRTEENTH TERZAGHI LECTURE
### By Richard E. Gray

The Karl Terzaghi Lectureship was established in memory of Dr. Terzaghi by the Geotechnical Engineering Division in 1960. The first lecture was presented at San Francisco in 1963 by Dr. Ralph Peck. Selection as a Terzaghi Lecturer is among the highest honors a geotechnical engineer may achieve. The lecturer is selected on the basis of his contributions to the technical or professional stature, or both, of geotechnical engineering.

The topic of tonight's lecture, "Geology and Geotechnical Engineering" is appropriate to the occasion. Dr. Terzaghi started out to quantify geologic materials and developed soil mechanics. Dr. Peck, in describing Terzaghi's lecture visits to the University of Illinois, said (2):

> He never gave a lecture in soil mechanics. They were always lectures in geology, geomorphology, and how they related to a problem, to which, incidentally, some of the engineering science of soil mechanics had an application. He was a geologist at heart although he was an engineer's engineer at the same time. But he always regarded soil mechanics as a branch of engineering geology which in turn was a branch of geology.

Dr. Robert F. Legget's philosophy of geotechnical engineering is akin to that of Dr. Terzaghi's. Born in Liverpool, England, of Scottish parents and educated in engineering at the University of Liverpool, Dr. Legget holds 12 honorary degrees from universities in Canada, the United States, and Europe. He was engaged in civil engineering work in England, Scotland, and Canada until 1936. He then started teaching civil engineering at Queens University and moved onto the University of Toronto in 1938.

In 1947 he was invited to Ottawa to establish the Division of Building Research of the National Research Council of Canada and served as its Director until 1969.

Dr. Legget has held many important professional and technical offices and has been the recipient of many awards. Notable are: (1) The first recipient of the William Smith Medal of the Geological Society of London for distinguished achievements in applied geology; (2) the Dumont Medal by the Geological Society of Belgium which is given at 5-yr intervals to honor an international figure in geology—in this case for stimulating cooperation between geologists and engineers; and (3) the First Gold Medal of the Canadian Council of Professional Engineers.

Many here have used his text, *Geology and Engineering*, first published in 1939. Dr. Legget has authored or coauthored seven books and innumerable engineering and geological papers. One of his recent books, *Cities and Geology*, directs our attention forward to future challenges in applied geology and emphasizes the need for planning by interdisciplinary teams if we are to develop a better urban environment.

Dr. Legget has been a leader and catalyst in bridging the gap between geology and geotechnical engineering. His peers have acknowledged his unique interdisciplinary abilities and efforts by electing him President of the Geological Society of America and the American Society for Testing and Materials, plus honorary membership in the Association of Engineering Geologists and ASCE.

I was most fortunate to hear his presidential address to the Geological Society of America in San Francisco entitled, "Soil: Its Geology and Use," (1) in which Dr. Legget described soil mechanics and emphasized the need for, and benefits of, collaboration between geologists and soils engineers. This evening, he will again serve as a bridge between the professions, geology and soil mechanics, by re-emphasizing the value of geology in geotechnical engineering.

We sincerely regret Mrs. Legget is not with us this evening and send her our best wishes.

It is with great personal pleasure that I present the 13th Terzaghi Lecturer, Dr. Robert F. Legget.

## APPENDIX.—REFERENCES

1. Legget, R. F., "Soil: Its Geology and Use," *Bulletin*, Geological Society of America, Vol. 78, No. 12, Dec., 1967.
2. Yatsu, E., Ward, A. J., and Adams, F., *Mass Wasting,* 4th Guelph Symposium on Geomorphology, 1975.

# GEOLOGY AND GEOTECHNICAL ENGINEERING

By Robert F. Legget,[1] Hon. M. ASCE

I came to the United States and hoped to discover the philosopher's stone by accumulating and coordinating geological information in the construction camps of the U.S. Reclamation Service. It took me two years of strenuous work to discover that geological information must be *supplemented* by numerical data which can only be obtained by physical tests carried out in a laboratory. The observations which I made during these years crystallized into a program for physical soil investigations . . . (Terzaghi, 1936).

These words came to me, echoing down the years, as I pondered the Society's invitation to deliver this, the thirteenth Terzaghi Lecture. The words came at the start of the Presidential Address at the First International Conference on Soil Mechanics and Foundation Engineering. I was privileged to be one of the happy company of no more than 200 who, on a day in June, 1936 at Harvard University, listened to Dr. Terzaghi's challenging address at this inauguration of the new discipline of Soil Mechanics. Despite all the years between, I can still recall so vividly with what emphasis he stressed that word "supplemented," warrant indeed for the italics that I have used.

The complementarity of geology and soil mechanics was the continuing theme in the many talks which I was privileged to have with this great friend. His first publication was a translation of Geikie's *Outlines of Field Geology;* at the time of his death he had started work on a book about Engineering Geology. Well do I remember a day in 1944 when we spent an afternoon together [prior to a memorable evening lecture (Terzaghi, 1944)] walking on Mount Royal in Montreal. It was, indeed, almost a charge that he then gave me, in his own inimitable manner, in his review of the engineering soil problems of Canada as he then saw them.

Never did I dream that one day I would be given the opportunity of passing on that charge to others, in the way that the delivery of this lecture makes possible. Accordingly, I am grateful indeed to the society for the invitation to deliver this lecture, as I am also to the Executive Committee of the Geotechnical Engineering Division for their agreement to my reviewing a subject which differs

---

Note.—Discussion open until August 1, 1979. To extend the closing date one month, a written request must be filed with the Editor of Technical Publications, ASCE. This paper is part of the copyrighted Journal of the Geotechnical Engineering Division, Proceedings of the American Society of Civil Engineers, Vol. 105, No. GT3, March, 1979. Manuscript was submitted for review for possible publication on March 31, 1978.

[1] Consultant; formerly Dir.—1947–1969—Div. of Building Research, National Research Council of Canada; Ottawa, Canada.

so markedly from the topics dealt with by my predecessors in this notable series.

Not only do I feel reasonably certain that Dr. Terzaghi would not have disagreed with this choice of subject for a lecture in his honor but, as recently as 1973, one of his closest collaborators had this to say:

> Every earth structure is constructed in or of a medium. . . . Geology should be used to greater advantage. We deal with geological materials, yet geological techniques, geological reasoning, and the implications of geology are rarely utilized to maximum advantage (Peck, 1973).

Is it not strange, indeed disquieting, that Dr. Ralph Peck should have had to make this eloquent plea in the address that he gave as President to the Eighth Soil Mechanics Conference in 1973? In the splendid advance made by the steady progress of soil mechanics, or Geotechnique to use the far more expressive and more accurate term, during the four decades from 1936, is it possible that the fundamental and prime importance of geology in all geotechnical investigations has, all too often, been overlooked? I regret to say that I think it has. And perhaps this is so just because the significance of geology in all civil engineering works is so obvious, so obvious indeed that it can so readily be completely overlooked.

Every structure built by man depends for its safety and stability upon the ground on which or in which it is built; upon geology. One would therefore think that every civil engineer would be given a sound introduction to elementary geological principles in his basic training. Would that this were so. Every time an excavator, be it hand shovel or mammoth dragline, moves "spoil" from one location to another, geology is involved. Since geological conditions are never the same at any two locations, one would imagine that contractors would have at least a general appreciation of the importance of the earth sciences in the successful (and profitable) conduct of their operations. All too often, this simple fact is so "obvious" that they never even think about it until trouble arises.

Such pedestrian observations in a Terzaghi Lecture would be inappropriate were they not unfortunately more than warranted. Let two recent statements published be cited in confirmation. In a paper that is to be found in the April 1977 issue of the *Journal of the Geotechnical Engineering Division* of this Society, there will be found the remarkable inference that engineers should interpret soil conditions between boreholes on contract drawings in order to assist contractors with their work (Thompson and Tannenbaum, 1977). The most elementary acquaintance with Geology makes any such suggestion invalid. There can never be any certainty about geological conditions between adjacent boreholes, even 5 ft apart, until excavation has actually opened up the ground. The idea of any such idea being current today, and in academic circles, is alarming. (See Fig. 1.)

Equally serious is a statement to be found in the reply to the percipient discussion of a fine paper describing the construction of the new suspension bridge linking Europe and Asia across the Bosporus, presented to and published by the Institution of Civil Engineers. This 942-m span bridge is a notable structure, already carrying traffic far exceeding original expectations. Its completion was delayed by 5 months due to difficulties encountered in excavation for one of the anchorage piers. When this was raised in discussion, and an account of the work involved was requested, with a suggestion that possibly the preliminary investigation was not adequate, the eminent authors replied: "In our opinion most site investigations prove to be 'inadequate' but in this case we and the Client's engineers did as much as possible in the time available" (Brown and Parsons, 1975, 1976). They went on to describe the borings that had been put down, the laboratory tests carried out on a sample of rock, and the desirability of sinking test pits. The work "Geology," however, does not appear in their statement.

These two examples, unfortunately, are not isolated cases, carefully selected to support the thesis of this lecture. They are all too typical of the continuing

FIG. 1.—Dr. Terzaghi on Quabbin Dam, Massachusettes, during First International Conference on Soil Mechanics and Foundation Engineering in Summer of 1936: [Figures in foreground are Mr. F. E. Schmitt (Editor, Engineering News-Record) left; and Dr. Rudolph Tillman, (City Engineer of Vienna, Austria) right]

neglect of geology in geotechnical investigations. It is indeed a distressing thing to find two members of a noted firm of British Consulting Engineers, a firm distinguished for its innovations in structural engineering, stating publicly that, in their opinion "most site investigations prove to be inadequate"—and this more than 40 yr after the inauguration of Geotechnique as a recognized discipline in civil engineering. Possibly they are right, although my own experience does not support this suggestion. Right or wrong, however, the publication of such a statement is surely profoundly serious, confirming the thought that possibly a review of the place that geology should occupy in such investigations may still be helpful to some, even though obvious to others, in bringing again to professional attention a somewhat neglected aspect of Geotechnique.

At the outset, it is essential to remove any possibility of misunderstanding about the recognition that geology is only one part of the several services that, together, make up Geotechnique of today. Nobody appreciates more than I, the importance and full significance of the advances that have so happily been made in the testing of earth materials, soil and rock; in the development of instrumentation both simple and complex for assisting with such laboratory work, as also for detailed studies in the field; in the aid which comes from sciences such as mineralogy and geochemistry in the support of geotechnical studies; and above all, perhaps, in the quite remarkable progress made in the development of theories of soil action, and soil interaction with structures, now often involving the use of matrix and finite element techniques, assisted so often by the use of computer services. These laboratory and theoretical studies and methods have indeed made such remarkable progress in the last two decades that there is just a possibility that they may have surpassed our ability to apply them in the field with desirable certainty. All that geological studies can do in ensuring this goal is to check the validity of the *assumptions* upon which all such theoretical approaches must be based.

There is, clearly, a desirable balance between the practical and the theoretical aspects of Geotechnique that is essential, one that must always take into account the fact that no two foundation conditions are ever identical, unlike the situation in other branches of civil engineering. If only as a reminder of this, every soil mechanics laboratory could well have on its walls a framed copy of these early words of Dr. Terzaghi:

> The geological origin of a deposit determines both its pattern of stratification and the physical properties of its constituents. . . . Therefore, the knowledge of the relation between physical properties and geological history is of outstanding practical importance (Terzaghi, 1955).

Correspondingly, in the field, the aid that can come from theories of soil action and from laboratory tests on good samples must always be kept in mind, even though final decisions on foundation and all other geotechnical problems must ultimately depend upon judgment. No computer is ever going to decide when a suitable foundation bed has been reached, or when tunnel supports are necessary. In the final analysis it is human judgment that makes possible the safe uses of the earth. And judgment is based upon sound experience that, whether so recognized or not, includes an instinctive appreciation of the significance of all geological factors. This unique approach to every individual geotechnical problem and design should be recognized by all civil engineers, and especially by structural engineers since it is in such contrast to the certainty of their materials and loads. The vital need for the exercise of such professional judgment is part of the great challenge of civil engineering.

The thesis of this lecture, therefore, will be that geological investigations are the essential first step in all geotechnical studies; that they may not have been fully appreciated and used by all practitioners; that there is now a wealth of information to assist in their proper use; and that, in view of this suggested neglect, desirably "something should be done about it." Inevitably, therefore, this will be a "dissertation on the obvious." My title could well be "Back to Terzaghi." But was it not Herbert Spencer who once said that "it is only by constant and varied iteration that essential truths can be forced upon reluctant minds"?

By way of introduction, let us review briefly some well-known examples of civil engineering works in the design and construction of which geology was, or was not, used to advantage.

**Dams.**—The fact that the Mission Dam of the British Columbia Power Authority is now named the Terzaghi Dam is well known (see Fig. 2), the paper describing the many problems that had to be faced in its design and construction, published posthumously, was Dr. Terzaghi's last written contribution to the profession (Terzaghi and Lacroix, 1964). Not far from this justly famous dam is another dam of the same Authority, which could equally well have carried Terzaghi's name. He was again the consultant; again he studied the problems of the site in his own inimitable way. On the basis of his observations, he posited that a dam could be built at the site selected for topographical reasons, a matter about which there had been grave doubt. As always with difficult jobs, he

FIG. 2.—Plaque, Naming Mission Dam of B.C. Hydro and Power Authority Terzaghi Dam, Unveiled by Mrs. Terzaghi at Sixth International Conference on Soil Mechanics and Foundation Engineering in Montreal, Sept. 8, 1965, in Its Final Location at Dam Amid Coast Mountains of British Columbia

wrote an account of the problems, which became the last publication of his lifetime (Terzaghi, 1960).

This is the Cheakamus Dam, the only major dam known to have been built on top of a massive landslide. Geological studies, notably by W. H. Mathews, had shown that about a century ago a major rock slide, probably involving 20,000,000 cu yd of rock debris, had flowed down a small creek emerging from Lake Garibaldi in the Coast Range of western Canada, filling the adjacent valley of the Cheakamus River for a distance of 2.5 miles up and down stream to a depth of up to 150 ft, burying a forest previously there, and forming Stillwater Lake when natural conditions had readjusted themselves. (See Fig. 3).

Based on a detailed study of the site and aided by earlier geological studies, without which subsurface conditions would have been inexplicable, Terzaghi decided that the dam which was required to raise the water level of the Lake by 65 ft, in order to take advantage of a potential power development, could be built. Completed between 1955 and 1957 as a rockfill structure, using rockfall debris as the main material, the dam is about 1,500 ft long and 88 ft high. The raised water level permits water to be diverted through a 6.5-mile long tunnel to a vertical drop of 1,125 ft, generating 190,000 hp, in a power house

FIG. 3.—Cheakamus Dam of British Columbia Hydro and Power Authority, on Cheakamus River: (Photo from Authority)

FIG. 4.—Reservoir, in Mountains of Mourne, Northern Ireland, Formed by Silent Valley Dam (Seen in Foreground), View Taken Shortly after Dam Was Finished: (Photo: Water Service—Eastern Division, Department of the Environment for Northern Ireland, Belfast)

on the Squamish River. Dr. Terzaghi visited the site no less than 18 times during the course of construction, adjusting the design as excavation revealed new features of the complex geology of the heterogeneous rockslide material. The dam stands today, performing quite satisfactorily, one of the supreme examples of adopting most unfavorable geological conditions for the service of man.

For obvious reasons it will be desirable to take an example of dam construction from the past, rather than from recent years, in illustrating what can happen when the geology of a site is not studied, although some recent examples can readily be called to mind. The Silent Valley Dam in northern Ireland was one such tragic example about which I heard a good deal at the outset of my own work in civil engineering, and which I have never since forgotten. It did not fail with consequent loss of life; it has performed well since its completion. Its construction, however, brought tragedy to some of those involved and incredible worries to its designers, builders and owners. (See Fig. 4).

For creating a new impounding reservoir for the water supply of Belfast, Northern Ireland, a dam was planned on the small Kilkeep River in the early

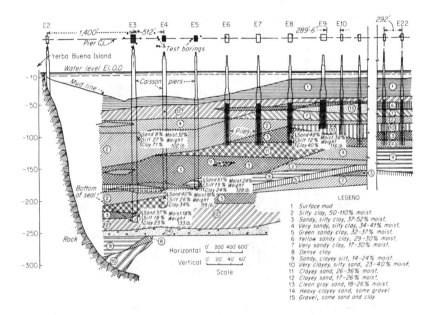

**FIG. 5.—San Francisco-Oakland Bay Bridge:** [Geological section along east channel section, showing results of subsurface exploration and location of piers (Engineering News-Record)]

1930s. A topographically suitable site was chosen; some test drilling had been carried out as early as 1913, the results of which suggested that solid rock would be encountered at depths up to a maximum of 50 ft below ground level. So far as can now be determined, no geological study of the site was undertaken. A contract was awarded in 1924 on the assumption noted, but when excavation reached the anticipated bottom level, large boulders were revealed instead of bedrock. Eventually, and after one of the most remarkable construction operations on record, the necessary cut off wall was founded on bedrock 180 ft below ground level, excavation having to be carried out in water-bearing sand that necessitated the use of the maximum permissible pressure for compressed air operations. The cost, in time and worry, can best be left to the imagination but boulders would almost certainly have been predicted in that location from even a cursory geological study. Some of the boulders were later found, by geologists, to be Scandinavian in origin (McIldowie, 1936).

**Bridges.**—The San Francisco area presents some excellent examples of bridge foundations in the design of which geology played its full part in the first phase of geotechnical investigations. Pride of place must be given to the San Francisco Oakland Bay Bridge, still performing as if it had just been completed, now more than four decades later. Its substructure was described in the third volume of the Proceedings of the First Soil Mechanics Conference (Proctor, 1936). Although the founding on bedrock of the large caisson for the central anchorage of the twin suspension bridge crossing of the west channel naturally attracted most publicity, the founding of the piers for the east channel crossing was the more interesting geologically. Bedrock depths were beyond the capacity

FIG. 6.—Railway Bridge across South Channel of St. Lawrence River near Cornwall-Massena, Shortly after Central Pier Had Failed on Sept. 6, 1898, Dropping 2 Spans into River, with Serious Loss of Life: (Photo: No. PA 108929 from Public Archives of Canada; Dominion Bridge Collection)

of the test drilling equipment (being more than 300 ft) so each pier was carefully founded on an appropriate stratum of soil, adequate to give requisite bearing, on the basis of a detailed subsurface study along the line of the bridge, the complex geological structure being delineated as the extensive test boring program proceeded. After 40 yr of service, it is so easy to forget the outstanding geotechnical engineering responsible for the safe construction and continuing performance of this great structure (See Fig. 5).

In the James Forest Lecture which Dr. Terzaghi gave at the Institution of Civil Engineers in London, England, in 1939, he said that "some of my colleagues established an observation-well at the eastern approach to the San Francisco Oakland Bay Bridge, and named it 'the monument to the theory of consolidation' " (Terzaghi, 1939). Professor H. Bolton Seed has kindly made diligent inquiries about the location of this "monument" but has been unable to trace it, or any record of it. If any reader has information that might lead to its discovery, advice of this would be welcomed.

Turning to the past again for a counter-balancing example, a grim reminder of a bridge failure directly due to geological neglect was provided by the salvage in 1958 of the twisted steel of the superstructure of a bridge across part of the River St. Lawrence, during the construction of the St. Lawrence Seaway and Power project. In 1898 one of the midriver piers constructed to carry a steel cantilever bridge over the United States section of the St. Lawrence, for the New York Central Railroad at Cornwall Ontario, collapsed, dropping two bridge spans into the river and tragically killing 15 men. The pier had been founded on a boulder in a layer of so-called "hardpan" which had not been investigated, apparently, either geologically or by borings. After the accident this stratum was found to be only 2 ft thick. The "hardpan" (a nongeological term that should *never* be used by engineers, its best definition being that it

**FIG. 7.—Portal of Broadway Tunnel near Oakland, Calif. (Photo: Department of Transportation, State of California)**

is material that was harder to excavate than the contractor had expected) was probably the glacial till that was to cause such problems in excavation for the Seaway locks and canal 60 yr later and to which further reference will be made [(Waddell, 1916) see Fig. 6].

**Tunnels.**—The construction of the Broadway Tunnel through the Grizzly Peak Hills on California Highway No. 24, in the outskirts of Oakland, designed to provide improved highway communication to the east, provides an example of the inadequate use of Geology. It was completed in the midthirties but for reasons that will shortly be obvious the problems associated with its excavation were not publicly described until 1950. There were two parallel bores, about 900 m long and 11 m wide. A special highway district was formed to build the facility. The district engaged a prominent geologist; he made a favorable report which was mentioned in the contract documents (and made available to contractors) but not officially endorsed or quoted. A combination of six experienced contractors was awarded the job and it now seems clear that they accepted this preliminary report and made no geological study of their own. Very little in the way of supports or lining was anticipated but the driving of a pilot drift showed how erroneous this impression was. None of the ground penetrated could be left unsupported prior to lining. Two bad cave-ins took place, one taking three lives. The contractors withdrew from their contract, which was completed by other constructors; they sued the highway district, unsuccessfully, for over $3,000,000. Very few who use the tunnel today know of this tangled history but there are few more telling examples of how inadequate geological advice, unrelated to proper geotechnical investigations, can be so misleading [(Page, 1950) see Fig. 7].

California has many notable modern tunnels that could be cited as examples in which geology has played its proper role in preliminary studies and during

FIG. 8.—Construction of First Section of Toronto Subway Proceeding by Cut-and-Cover under Yonge Street in Downtown Toronto, Looking South, with Final Lift of Excavation in Progress: (Photo: Toronto Transit Commission. Canada Pictures photo)

construction. In great contrast, there may be mentioned the building of Canada's first subsurface urban transit system, the Toronto subway. The first 4.6-mile section of this double-track standard-gage electrified line started in the busiest section of downtown Toronto, now a city of more than two million on the north shore of Lake Ontario. Every available geological report on the Toronto area was studied as the start of geotechnical investigations. Records of all known boreholes in the vicinity of the works were carefully searched out, and all information available from recent deep excavations. On the basis of what was thus found, a test boring program was designed and carried out, holes at about 400-ft centers being sunk at selected sidewalk locations. Undisturbed soil sampling was undertaken as were standard tests on soil samples so obtained. Some holes were cased and ground-water level variations were recorded for more than a year before construction started in 1949. All the information so assembled was utilized in the preparation of the contract drawings for the reinforced concrete tunnel structure (here built generally by cut and cover). In addition, all records and samples of all materials encountered were put on display for all tenderers to examine. Bidding was close. The successful tenderer, a consortium, completed the work satisfactorily within the contract time. Such claims as there were, were all readily settled in amicable discussions between engineer and contractor [(Legget and Schriever, 1960) see Fig. 8].

**Buildings.**—One must go to Mexico City for the world's most remarkable examples of building foundations illustrating both the neglect and the full use of geological information. Facing each other across Place la Alameda are the Palace of Fine Arts and the Tower Latino Americana. (See Fig. 9). Beneath both buildings are the complex strata of varied water-bearing soils that underlie

FIG. 9.—Palace of Fine Arts, and the Tower Latino Americana which Face Each Other across Place la Alameda, Mexico City: (Photos: Secretaria de Turismo, Mexico)

the city and yet one building exhibits world-renowned settlements, both total and differential, while the other is performing so well that its adjustable first floor has not had to be moved since the building was first occupied. The massive masonry building that is the Palace of Fine Arts was started in 1904 and completed 30 yr later. It is believed that a study of foundation conditions was made but nothing is known of the use of any such information in design, the architect being reported to have said that "if a structure is pleasant to my eye it is structurally sound." Lamentably, this remarkable attitude is not unique to him. The result is that the whole building has settled more than 3 m, as can be observed by looking at the front steps, with differential settlement so severe that, looking the full length of the building from one of the front corners, one wonders how a masonry structure can withstand such distortion (Thornley, et al., 1955).

Indicative of advances in building design, and in geotechnical engineering, is the Tower Latino Americana just across the Square; even with its 43 stories it weighs less than half the Palace of Fine Arts (24,000 tonnes as compared with 58,500 tonnes). It was completed in 1951, its design based on a most thorough study of the subsurface conditions to a depth of 70 m below ground level, carried out with full appreciation of the urban geology of the site. The foundation consists essentially of end bearing piles driven to sound bearing on a stratum of sand 33.5 m below the surface. So complex was the geology of the site that even the construction of the foundation structure involved engineering design of a high order but all was successfully completed, design and construction constituting together a notable chapter in the history of outstanding foundation engineering (Zeeraert, 1957).

One or more of these examples will probably be familiar to those who have studied the development of geotechnical engineering. They are but eight of a vast number of similar examples that could be cited, from all over the world, illustrating clearly the complete interdependence of geology and geotechnical engineering either through the neglect of geology or, more happily, through its full utilization as an essential part of subsurface investigation. Familiar though they may be, they appear to be worthy of this brief mention since it is well to be reminded occasionally that "he to whom the present is the only thing that is present knows nothing of the age in which he lives."

Fully to appreciate the significance of the geological aspects of subsurface investigation, it is naturally desirable to look at examples in a little more depth than has so far been done. Three more cases have therefore been selected, two showing the dire results to which neglect of geology can lead, and one demonstrating the powerful aid that the science can provide on a major engineering project. These examples have been selected in order to illustrate also problems that have to be faced in terrain that has been glaciated, in the borderland between glaciated and unglaciated country, and in unglaciated areas covered, therefore, with residual soil. Since about three quarters of the land surface of the globe is overlain by soil, these three geological divisions are of considerable significance in all civil engineering work.

**Kinzua Dam.**—Allegheny Reservoir is an important flood control reservoir in the upper reaches of the Allegheny River in northwestern Pennsylvania and southwestern New York, extending about 30 miles up river from about 9 miles east of Warren, Pa. to the downstream end of Salamanca, N.Y. It was created by the construction of the Kinzua Dam by the U.S. Corps of Engineers, the project being essentially complete by 1966. The reservoir when full covers 21,180 acres, maximum flood storage level being 1,365 ft above sea level. Flooding of the reservoir necessitated the relocation of 83 miles of highways and 37 miles of railways. The dam is a combined concrete gravity and earthfill structure 1,879 ft long, its crest 179 ft above stream bed level. Total national investment in the project was $109,000,000, the dam structure accounting for $22,600,000. It is estimated that, excluding the major benefits contributed by the project to the minimizing of damage from Hurricane Agnes in 1972, the direct benefits attributable to the Kinzua Dam amounted to 87% of the total cost of the project within its first 10 yr of operation. And the assistance given by geology in determining the final site for the dam ensured a saving of at least $4,000,000.

In the vicinity of the dam, the Allegheny River flows generally southward from near Salamanca, N.Y., swinging to a westerly direction at Big Bend near the village of Kinzua. In this pleasant hill and valley country, there was an "obvious" location for the dam just downstream from the confluence of Kinzua Creek and the main river, at the upper end of an almost gorge-like section of the valley into which the river flows from a quiet stretch in the unglaciated Kinzua Valley. When the project was first given serious consideration in 1936, this site was that naturally selected, three lines of NX (3-in.) borings being put down on parallel axes. These revealed sound bedrock but at depths up to 75 ft below stream bed near the center of the narrow valley. Estimates of cost, based on preliminary designs, were prepared on the assumption that the dam would be a solid concrete gravity structure. (See Fig. 10.)

**FIG. 10.—General View of Kinzua Dam on Alleghany River, Pennsylvania: (Photo: U.S. Corps of Engineers, Pittsburgh District Office Copyright Photo by Gordon Mahan)**

Then occurred a 20-yr delay in any further consideration of the project, detailed studies being resumed in 1955. During this period, geologists of the Pittsburgh office of the U.S. Corps of Engineers, notably Shailer Philbrick to whose writings this summary is indebted, were able to study the complex geology of the Allegheny Valley upstream and downstream of the location proposed for the dam at Big Bend. Present-day geomorphology is the final result of a complex succession of events associated with successive glaciations. Studies were not confined to the field. Every available earlier geological report on the area was examined, valuable information being found in several, notably in a report by J. F. Carll (1828–1904) of the Second Pennsylvania Geological Survey. In studies of gas and oil wells in this area he had noticed that the bedrock surface slopes downward to the north, the opposite of what would be expected from the topography and present direction of river flow. He deduced from this that the preglacial drainage of this area was to the north rather than to the south, as at present.

Philbrick and his fellow workers were able to confirm this deduction. Geophysical work and extensive test boring and sampling in 1955-56 confirmed that in the vicinity of Big Bend there was a bedrock shelf 450 ft wide at an elevation 37 ft higher than the bedrock at the 1936 (the "obvious") site. Wisconsin outwash materials filled the deeper preglacial valley on the north side of the present river valley at the Big Bend location but this would permit of the use of an earthfill dam for the major part of the river crossing. A concrete gravity structure founded on the bedrock shelf on the south side of the valley would be more than adequate for the necessary spillway structures. A combined structure

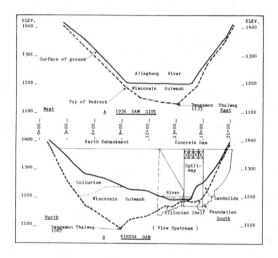

FIG. 11.—Cross Sections of Alleghany River at Site Originally Chosen for Kinzua Dam, and at Site Finally Used after Detailed Geological Studies, as Explained in Text: [Sections courtesy of S. S. Philbrick, reproduced by permission from *Geomorphology and Engineering*, p. 184 (Philbrick, 1976)]

was then designed on the basis of these geological findings, showing a saving in capital cost (as compared with the 1936 design) of over $4,000,000, in 1957 dollars. Suitable material was available in the vicinity for the earth fill. A slurry-trench concrete wall (1,066 ft long and 2 ft 6 in. wide) was installed as a cut-off in the earthfill, embedded 2 ft into bedrock in view of the pervious nature of the fill used and of the outwash material in the valley bed. Outwash gravel might have been used for concrete aggregate, as is customary in the Pittsburgh district, but the contractor for the dam chose to use crushed limestone brought by rail (at an unusually favorable rate) from a quarry 165 miles away [(Philbrick, 1976) see Fig. 11].

It may be thought that it was the 20-yr delay in the construction of the Kinzua Dam that enabled this application of geological studies to be so constructively utilized. The delay did permit of more thorough investigations than usual to be made, but it is to be noted that the original deduction of reversed drainage in preglacial times was first propounded by Carll as early as 1880. In his paper describing this work, Philbrick lists four other earlier papers on the geology of the region which assisted in the elucidation of the geomorphology of the valley. It is also of significance that the test drilling in the vicinity of the site chosen in 1936, on the basis of perfectly acceptable topographic interpretation, although well done, was not of itself sufficient to indicate that the original site could be improved upon if the underlying bedrock geology was fully understood. Even the best of test boring programs may prove to be less than fully effective unless carried out in full conformity with the geology of the region and not just of the site under study.

The example of the Kinzua Dam might be thought to be an unique case since it involved the complete reversal in the direction of flow of a major river. It is unusual but far from unique. Geological and geophysical investigations on Whitewask Creek, near Kingra, Tasmania, Australia, revealed the existence of a buried Tertiary drainage system, quite unrelated to the regime of the existing Creek, with buried porous soils of such extent that two possible dam sites were abandoned (Stevenson and Moore, 1976). And then there is the now classical case of the Vrynwy Dam in North Wales for the water supply of Liverpool, England, located before the end of the last century, on the basis of carefully deduced glacial geology, so accurately that excavation revealed that a shift of the axis of only a few hundred yards would have increased the cost by well over $1,000,000 (Deacon, 1896).

**Asia Insurance Building, Singapore.**—Considerations of the neglect of geology involve the recounting of failures. This is never a pleasant task and yet it is only by the constructive study of failures that some advances in technology can be made. As has been said so often, if one does not consider mistakes constructively and find out why they happened, the same thing may happen again with even worse consequences. In contrast, therefore, with the utilization of geological studies for the Kinzua Dam, consider next a case in which geology was apparently completely neglected. It was in the construction of a notable commercial building that is today an integral part of the background to Raffles Quay in Singapore, known to all who have visited this part of the Far East. Its location indicated that soils encountered would be residual soils, with the probability that boulders of unweathered rock might be encountered. The famous quay had been built on reclaimed land so that other subsurface problems might well have been anticipated.

The Asia Insurance Building is a steel-framed 18-story building, founded on a site reclaimed from the sea in the 19th century. The beach formation beneath the site varies from 10 ft–14 ft in thickness and was thought to be underlain by fill and decomposed shale and sandstone. Erection of the buildings was to be an early post-war project. Four borings were put down in 1949 and came to refusal from 30 ft–42 ft below the surface. No diamond drilling equipment was then available in Singapore and so a chopping bit was used to obtain fragments of the sandstone encountered at the bottom of the holes. On the basis of this investigation, a foundation structure of cylindrical piers bearing on rock was designed and work was started (see Fig. 12).

It was soon found that what had been assumed to be bedrock was merely the upper surface of boulders, as encountered by pure chance in all four boreholes.

The boulders were of weathered Triassic shale and sandstone, as elementary acquaintance with the geology of Singapore might have suggested. Further borings were put down after proper equipment had been brought in and a halt called to construction work. The new borings encountered tough clay and boulders down to depths of over 100 ft. Some of the boulders were of granite, brought to the port as ballast and dumped overboard at an earlier quay. Samples of the clay were obtained and tested; on the basis of the results so obtained a new foundation structure was designed. Belled-out drilled piers were utilized to give an appropriate bearing pressure (3.5 tons/sq ft instead of 10 tons/sq ft). The cylinders already in place had to be grouted up and underpinned, a

FIG. 12.—Asia Insurance Building, Raffles Quay, Singapore: (Photo: Public Utilities Board, Singapore)

delicate and difficult operation successfully completed but (naturally) at great cost. As an example of superlatively executed foundation construction this job was outstanding but the exercise of all the special skills involved should not have been necessary (Nowson, 1954).

This example was selected since, as all who remember the experience of Singapore during the Second World War will appreciate, conditions there were still far from normal in 1949. There is, therefore, some excuse for the inadequate subsurface investigation but there are still all too many cases in which boulders have been mistaken for bedrock under no such extenuating circumstances. The possibility of boulders being present in soil strata overlying bedrock is one geological hazard that is common to both unglaciated and glaciated terrain. Residual soils may possess somewhat unusual properties due to unusual mineralogical composition, as the experience at the Sasamua Dam in Kenya so clearly demonstrated (Terzaghi, 1958). Of all the problems encountered with soil in civil engineering works, however, the varying characteristics of till ("glacial till" and "boulder clay" being common but somewhat inaccurate alternative names) have probably created more trouble than any other cause, possibly than all other soil problems combined, at least in the northern hemisphere. Accordingly, it may be helpful to review one of the most costly examples of neglect of available geological information on till, that which occurred during the construction of the St. Lawrence Seaway and Power project.

Tribute must first be paid to the magnitude and significance of this international civil engineering project, carried out and financed jointly by the United States of America and Canada. After half a century of political delays, the ceremonial first sod was turned on 10 August 1954. Less than 5 yr later, the Seaway was officially opened by the Queen of Canada and the President of the United States. The international power station had been delivering power since 1958, the first vessel having sailed through the completed Iroquois Lock on 22 November 1957, little more than 3 yr after the start of work. Total cost of the Seaway and the associated power development was over $1 billion, shared internationally. At one time more than 22,000 men were at work between Montreal and Lake Ontario. Over 500 professional engineers had worked in the offices merely of the four principal design agencies, and there were many smaller offices also involved. It must surely rank as one of the very greatest of all civil engineering projects.

The overall concept of the project will be familiar—the provision of a 27-ft waterway, capable of handling ocean going vessels, from the head of the navigable River St. Lawrence at Montreal to Lake Ontario, and thence into the Upper Lakes. This involved new locks and canal from Montreal to Lake St. Louis; locks at Beauharnois; two lift locks and a guard lock in the international section of the St. Lawrence, designed integrally with the international power house which utilizes almost the full fall from Lake Ontario to Lake St. Francis to generate over 2,000,000 hp; the Welland Ship Canal (opened in 1932) providing, with some deepening, the link between Lakes Ontario and Erie; all combined with much dredging of channels and the construction of a large number of ancillary works, some of which (such as the Long Sault Dam) are major structures in their own right by any normal standard (Legget, 1976).

By international agreement, the two lift locks in the International Section of the river were located on the United States side. All the earlier, smaller locks had been built by Canada and were located on the Canadian side, where geological conditions are favorable. The Eisenhower and Snell Locks are notable structures, with the same lock dimensions as for other Seaway locks (80 ft wide, 860 ft from pintle to pintle and a minimum of 30 ft of water over sills). They are located near the eastern end of the Wiley Dondero Canal, a straight channel almost 10 miles long, all but the western end being a major new cutting excavated through dry land, with a bottom width of 442 ft. This involved almost 20,000,000 cu yd of excavation, in advance of which extensive test boring, soil sampling and soil testing and a demonstration test pit were carried out. On the contract drawings for the canal, *till* is clearly indicated in the borehole records thereon plotted, although two tills are not differentiated. In the associated specification it is clearly stated that: "the materials to be encountered . . . will consist predominantly of compact to very compact glacial till." On the contract drawings (and specifications) for the power house works, the terms—sand, gravel and clay—are used in various combinations, many of the borings being indicated as "wash" borings, taken some years before construction started, the only mention of till being on a few borings taken along the line of the old Cornwall Canal in Canada, some distance removed from the location of the power house works.

It is doubtful whether the use or nonuse of the term till made much difference to the bids received. Bidding was keen from contractors with wide experience in earth moving who clearly wanted the jobs, to some extent, it has been suggested, for prestige purposes. Early unit prices for excavation of soil varied from 31¢/cu yd–58¢/cu yd; $1.26 was a later figure. Of the contractors involved with the canal, one went bankrupt, another defaulted, a third entered a claim for extra payment of $5,500,000 on a contract awarded for a total price of $6,500,000. Closely associated with the channel excavation was work on the United States half of the power development; this also involved large quantities of excavation. Here, on contracts with a total value of $89,100,000, claims for extra payment amounted to $27,600,000 (or 31% of the contract price); settlement was for $4,800,000 (or only 17% of the amount claimed). The chief complaints were that the till—for that is what the "sand and gravel" was—was "cemented," some requiring ripping or even blasting before being moved by scrapers; and that the marine clay was very difficult to handle with normal equipment because of its stickiness (St. Lawrence, 1955).

As is usual, contractors were supposed to have satisfied themselves as to the nature of the work on which they had tendered, an understanding clearly reflected in the small amounts given in settlement of claims. Quite apart from what might have been observed in the vicinity as to the character of the materials to be moved, the "glacial drift" (till) had been under renewed detailed geological study since 1952, early maps resulting from this work being on "open file" at the New York State Geological Survey in Albany (MacClintock, 1958). In the comprehensive paper summarizing this geological work, which includes observations made on the Seaway excavation, MacClintock and Stewart (1965) list about 150 papers dating back to 1843 dealing with the surficial geology of the St. Lawrence Lowland, at the heart of which lies the Seaway. As early as 1910, Fairchild had proposed two periods of glaciation over the area and thus suggested two tills (Fairchild, 1910). All papers after that date clearly indicate that till is the predominant soil beneath the upper stratum of marine clay, the well-known Leda Clay of the St. Lawrence Valley. The upper (Fort Covington) till was well recognized; the lower (Malone) buried, or paleo-till had been suggested, directly or indirectly, by several geologists. Since the lower till had been subjected to submersion and then to the pressure of the ice of the second glaciation, its compact nature was a certainty (see Fig. 13).

All this geological information was available before the bids for the Seaway and Power excavation were submitted but it does not appear to have been fully utilized. Some of those tendering might have preferred to rely upon the experience of other contractors, rather than upon the "academic" opinions of geologists. Let it be recorded, then, that immediately adjacent to the Wiley

Dondero Canal lies the Massena Power Canal, excavated at the turn of the century to lead water from the St. Lawrence to the new power station then constructed for the Aluminum Company of America. There were still alive (in 1957) men who had worked on this job and who could therefore have testified that the original contractor for this excavation also went bankrupt, apparently due to exactly the same difficulties with the same soils as were encountered on the Seaway and Power excavation and no more than 1 mile away (Nelson, 1957).

About 50 miles downstream is the Beauharnois Power Canal on the St. Lawrence, now a part of the main Seaway channel. It had been excavated between 1929 and 1932 in exactly the same type of marine clay as was encountered in 1956–57. And when the Welland Ship Canal was built in the 1920s, the same

**FIG. 13.—Members of "Friends of the Pleistocene" Examining Lower Till in Excavation for St. Lawrence Seaway U.S. Locks near Massena, N.Y., 1957**

type of problem in excavating tough till was encountered, leading to vast claims for extra payments and, indirectly, to the failure of a leading Canadian contractor. In a major account of this work, published in 1929, this statement appears:

> . . . the rock surface was overlaid by glacial till. Over the northern one-third the rock dropped much lower, and was overlaid by very hard boulder clay of the first glaciation, over this again being the ordinary till of the second glaciation (*Engineering*, 1929).

Admittedly, the Welland Canal is 250 miles upstream but the two areas had almost identical histories during and since the last glaciation.

Even though it was suggested at the time that a full account of the difficulties with the Seaway and Power excavation should be published, this does not appear yet to have been done. The claims for extra payments were so large, involving so much legal action, and had naturally aroused such strong feelings on the part of those involved, that this silence is understandable, but the profession is the poorer. This brief account has had to be based upon news reports of the time, personal recollections, discussions with some of those involved, and study of the few documents that are cited. Only one relevant United States paper has been traced and this concludes with a reference to "unsound and inadequate exploration data" in presenting, from the point of view of contractors involved, a review of physical properties of the "basal till" of which it is said that:

> the presence of a thick bed of tough, dense, highly cemented and consolidated basal till underlying the loose, sandy, superglacial tills came as a complete surprise to everyone concerned, and especially to the excavation contractor (Cleaves, 1963).

This experience on the St. Lawrence Seaway and Power project could provide invaluable information for the guidance of all who have to deal with till, especially its excavation. It is greatly to be hoped that, now the passage of time has brought the difficulties of the late fifties into perspective, somebody will be moved to embark upon the major task of reviewing all the relevant official documents and records of work so that a full and factual account of this matter may be prepared and published.

When one has seen millions of dollars being wasted, and even lost, due to neglect of geological information that was readily available for the asking, then one develops a "concern" (to use again that old Quaker word in its literal sense) such as is reflected in this paper. If the basis of this concern may be assumed to have been adequately exposed, then it is time to look ahead to explore, again and all too briefly, some of the ways in which geology can be of material assistance in geotechnical investigations. As will shortly be seen, the word geology is used here in its widest sense, including current geological processes that are now modifying the earth's surface, such as the action of rain. It will be assumed that there is an engineering geologist in the geotechnical team. When this is not the case, the expert services of an engineering geologist will be necessary for all but the most routine studies, recognition of this need calling for a sound general appreciation of geology on the part of the leader of the geotechnical team.

It will be noted that the term "engineering geologist" is used. Although there is no special branch of geological knowledge to meet the needs of engineering, special practical geological skills, tempered by an appreciation of the rudiments of civil engineering, are essential. The experience at the Broadway Tunnels showed what problems can result from reliance upon geological advice which, although scientifically sound as far as it went, has been given apparently without any real appreciation of the inter-relation between geology and civil engineering (Page, 1950). The remarkable standing of engineering geology in Czechoslovakia is well known. This owes much to outstanding teachers, such as Dr. Quido Zaruba with his thirty years of special training seminars, but official support for geotechnical investigations by engineering geologists stems in large measure from troubles experienced with the Nosice Dam on the Vah River in Slovakia, one of the first run-of-river water power plants to be built there after the Second World War (see Fig. 14). The geology of the site was studied by an eminent geologist who correctly delineated the bedrock strata that would be encountered in excavation. What was not appreciated was the importance of groundwater in civil engineering works, even though this was mentioned in his report. Artesian water, highly charged with sulphate salts, was encountered as excavation proceeded, necessitating a complete change in design and in construction methods, with a consequent delay of many months in the completion of the project. Thereafter competent engineering geologists were used for all official Czechoslovakian geotechnical investigations. [The unexpected ground water at Nosice was piped to a specially constructed spa where one may today "drink the waters" coming from beneath the dam to be seen below (Matula, 1961).]

**Regional Geology.**—A first requirement is a general appreciation of the regional

**FIG. 14.—Nosice Dam on Vah River, Czechoslovakia**

geology of the area in which is located the site under investigation. In many cases, this will already be known but if the region is unfamiliar a search must be made for all publications that may serve as an initial guide. Almost every country of the world today has an organization with some responsibility for its geology, usually a national geological survey, which will have publications of use. In this way, and even prior to any field work, a general impression of the local geology can be obtained, whether the region has been glaciated or not, typical depths to bedrock, the extent of bedrock weathering in warmer regions, general ground-water conditions. Bjerrum has told us from personal knowledge that:

> Terzaghi's work on a difficult dam project always starts with a study of the topography and geology of the entire area surrounding the site . . . the intimate knowledge of the geology of the whole area is as necessary for his work as the subsoil exploration which follows (Bjerrum, 1960).

For engineering projects covering very large areas, somewhat detailed engineering-geological surveys will usually be necessary. In the early days of the Tennessee Valley Authority (TVA), for example, a fine example of this type was developed (Eckel, 1934). In more recent years, for the extensive Snowy Mountains Hydro Electric project in Australia, involving an area of about 5,000 sq miles of rugged mountain terrain, active geological surveying had to be undertaken with only the most general background available from earlier studies, as the first major activity of the Authority (Moye, 1955). Investigations for the proposed Channel Tunnel between England and France necessitated regional geological studies on an international scale, even micropalaeontology being invoked to correlate geologically chalk strata on both sides of the 20-mile wide English Channel (Bruckshaw, et al., 1961).

Special regional geological features can be determined in this way, to serve as a guide and warning as more detailed site studies proceed. In the St. Lawrence Valley, for example, the extent of the Champlain Sea immersion is now well known and well recorded. Clays deposited in the water of this Sea are usually highly sensitive, susceptible to landsliding and often difficult to handle in excavation (Crawford, 1960). Once these features are appreciated, *any* clay encountered at a site anywhere within the 20,000 sq miles covered by the Sea must be initially suspect. In western Canada, bedrock at or near the surface of the ground will be found to be tough clay shale. Due to unloading of the shale following the recession of the last ice mass (about 11,000 yr ago), the shale exhibits the phenomenon of rebound, to such an extent that widespread geomorphological features have resulted (Peterson, 1957). Typical is the sloughing and sliding to be found along the banks of the South Saskatchewan River, and the unusual formation of valley ridges along the edges of incised valleys (Matheson and Thomson, 1972). The phenomenon appears to be related to some recent landsliding (Hardy, 1957; Thomson and Tacyshyn, 1977). A helpful account of unusual soils encountered in the New York City area has recently been published in the *Divisional Journal,* with full appreciation of the regional geology a key feature of the paper (Parsons, 1976).

To those who are well aware of these, and similar, features of regional geology, this attention to them in this review may savor of "the obvious of the obvious." But how many test drilling crews are still sent out into the field to "drill a few holes, and get some good undisturbed samples" without any prior consideration having been given to the overall geology of the site. The last few years have seen the publication of some masterly descriptions of how such engineering geological surveys should be made [see for example, (Fookes, 1969)]. Well recognized techniques have been developed for combining geological and general engineering indications on such maps. There has even been published by United Nations Educational, Scientific and Cultural Organization (UNESCO) an international tentative standard for the preparation of such engineering geological maps (Dearman, 1976). The preparation of specialized maps will be necessary for major projects, for all land-use planning and for all urban areas that can be so surveyed.

**Urban Geology.**—The steady growth of cities throughout the world, and especially of larger cities, will mean that a steadily increasing volume of geotechnical work will have to be carried out in urban areas (Legget, 1973). Superficially, it might be thought that city building sites can be studied in isolation, whereas exactly the opposite is the case. Geological structures, with all their vagaries, do not stop in some mysterious manner at municipal boundaries. They continue beneath the roadways, sidewalks, parking areas and buildings, frequently the more complex because of the activities of man, especially in older cities. Only rarely can urban geology be studied in situ. Usually it has to be investigated by means of historical records and by deduction from what is known of adjacent sites from borehole records or from records of adjacent excavations (such as the vital "as-constructed" drawings that should be made of every foundation). Only when all these sources of information have been fully explored and the best possible picture of the site geology has been developed, should any test drilling be planned (Jennings, 1976).

This is one area of geotechnical work in which familiarity breeds if not exactly contempt, at least carelessness. It can so easily be assumed that, because no unusual problems have been encountered at one building site, the sites immediately adjacent will be similarly satisfactory, thus needing only cursory investigation. This may be so; but just as often it may not be the case at all. The location of many cities near lakes or rivers inevitably means that in many cases small watercourses will have been covered up as urban development has proceeded. This is just one possibility that must always be kept in mind. In one Canadian city, the most complete record of the network of small watercourses that has gradually been covered up by the streets of the city, was built up in the depression years as a "make-work project" by a consulting engineer, whose office gladly shares this information with those who need it. Some cities have old mine workings beneath them, a relic of early development. Few cities have so good a guide to these potential hazards as the Atlas of old Coal Workings beneath the western Canadian city of Edmonton, privately produced by an interested consultant (Spence Taylor, 1971).

As but one indication of the amount of information on urban geology that is available, if one knows where to look for it, Appendix I lists a few of the current publications in this area of just one agency of the United States Government, the U.S. Geological Survey, by kind permission of Dr. V. S. McKelvey, the Director. State geological surveys, civic agencies and even local volunteer agencies such as naturalists' groups will often be found to have available useful publications on the local geology now hidden beneath city streets, the study of which may provide useful assistance in the planning of individual site studies.

So little thought is usually given to urban geology, so long removed from normal sight, that many citizens tend to forget there are geological formations beneath the streets of their city. Another potential danger may therefore usefully be mentioned. In considering site investigations in an unfamiliar city, there is a perfectly natural tendency to take a general look at the area, the local topography, the building types that are common, in general terms the "townscape," and to equate this with the corresponding appearance of a neighboring city that is well known and, if they are similar, to assume that the underlying geology will be the same. An interesting contrast of this sort has been drawn between the English cities of Oldham, Lancashire, and Stoke-on-Trent, Staffordshire, less than 40 miles apart and unusually similar in topography, in history, and in development, both susceptible to problems due to coal mining subsidence. Superficial judgment would conclude that the geological conditions beneath the two cities were similar, especially as both are underlain by carboniferous strata. In fact, subsurface conditions are quite different. Underlying strata include coal beds, but there the similarity stops. Due to fundamental differences in modes of geological deposition, the Oldham Coal Measure beds are much more variable in vertical sequence and in horizontal continuity than those at Stoke-on-Trent. Depth to bedrock in Oldham varies from nothing to 30 m, with sudden cliff-like drops in the bedrock surface beneath the overlying till. In Stoke-on-Trent a uniform glacial clay, always between 2 m–3 m thick, overlies the relatively horizontal surface of the bedrock. Accordingly, once these geological differences are appreciated, it can be seen that site investigations in Oldham will be more difficult, and the program of necessary exploration work necessarily more elaborated and costly, than in the apparently similar area of Stoke-on-Trent (Hassall, et al., 1973).

**Archival Material.**—One source of information about the subsurface of urban areas that can so readily be overlooked is the collection of historical records of development that most cities now possess, often in the collections of their public libraries. Increasing attention is now most fortunately being given to such archival material. Some cities now have a City Archivist; some have established municipal archives. The records thus preserved may be found to contain useful information about the urban geology now covered up by the streets of the city but earlier exposed to view. Not only can earlier natural features, such as buried watercourses, sometimes be traced in this way but of equal value on occasion may be records of early "man-made-geology."

In the early stages of the site exploration for the World Trade Center in New York City, e.g., early maps of the city dating back to 1783 were carefully studied. They showed that the shoreline of the Hudson River at that time was two city blocks inland from its present location. Simple timber structures, including rock-filled cribs, had then lined the shore. Long since covered up by miscellaneous fill, these old structures could have caused confusion in the test drilling program and difficulties during excavation had they not been located in advance (Kapp, 1969). In a similar way, old excavations and timber lock structures for early canals have been located in central Pittsburgh, with consequent benefit to site investigations (Gray, 1976: in an unpublished communication).

There is no pattern to the type of urban archival materials that may be found, especially for older cities, nor is there any certainty as to their location even within civic records. Each case must be investigated individually with no assurance of useful information resulting but with a probability that makes a search well worth while. The only group of such old records so far located is the collection of county histories that were prepared in 1876 in response to a Proclamation of President Ulysses Grant, following a request from the Congress, to mark the centenary of the United States. The discovery and use of one of these old histories was of assistance in the geological investigation of a damsite in the midwest (Hill, 1976: in an unpublished communication). It is hoped to present more information about these little known repositories of information in a paper now in preparation (Legget and Burn, 1977).

**Test Boring Programs.**—Only when all such possible sources of information have been thoroughly explored, and the best possible "picture" of the geology underlying the site has thus been obtained, should the work of test boring, test drilling and sampling be commenced. As has been so well said:

Because of the decisive influence of geological factors on the sequence, shape and continuity of soil strata, the first step in any subsoil exploration should always be an investigation of the general geological character of the site. The more clearly the geology of the site is understood, the more efficiently can the program for soil exploration be laid out (Terzaghi and Peck, 1948).

There are available helpful guides to the prosecution of this kind of work, notably Hvorslev's justly renowned report published by this Society (Hvorslev, 1949). Nothing need here be said about the conduct, supervision and recording of such exploratory operations save only to emphasize (again) their great importance, the necessity for the most meticulous records, and so the essentiality

FIG. 15.—Part of Transparent Plastic Model of Foundation Beds at Site of Proposed Auburn Dam, California, Showing some of the Extensive Subsurface Investigations: (Photo: U.S. Bureau of Reclamation, Auburn Dam Office, Louis R. Frei, Project Geologist)

of continuing *professional* supervision. The George's River Bridge, New South Wales, Australia, taking 5 yr to build instead of 2 yr, with corresponding increase in cost, stands as a constant reminder of what the neglect of this last requirement (and of geology) can mean (Allen, 1932).

In the lay-out of the pattern of test holes to be put down, geology has its part to play. A standard rectangular grid pattern is an obvious and convenient arrangement to adopt. It is statistically desirable—but unfortunately, from this point of view, geological structure and statistical regularity are incompatible. If there is any possibility, for example, of an old watercourse beneath an urban site, holes must be initially located in an attempt to penetrate this, further holes then being drilled at closer intervals once the general location has been detected. The same flexible approach must be used if there is any possibility of a contact between inclined strata of different types of bedrock underlying the site. Knowledge of the dip and strike of the bedrock to be encountered may well influence greatly the number and initial location of holes.

To go further would involve needless repetition. In summary, geology should be a principal determinant in the location of test holes at all sites, not only

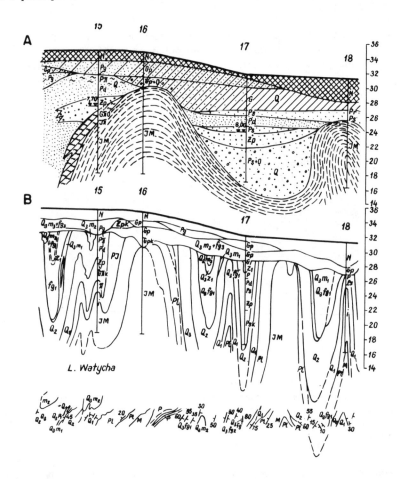

FIG. 16.—Subsurface Conditions along Part of New Expressway in Warsaw, Poland (a) As Deduced from Study of Borehole Records; and (b) as Actually Revealed by Excavation: [Illustrating care that must be exercised in interpreting borehole records: (Diagram from p. 79 of Bazynski, 1977, reproduced by permission)]

in regard to holes to be drilled into bedrock but also all holes penetrating overlying soils. Its utility will be evidenced not only in the preparation of the initial program but, and of equal importance, in the necessary adjustments and additions that close study of the first results may suggest. It is for this reason that geological advice throughout the full duration of a test drilling program must be available. This is obvious when the economics of possible mistakes in design due to inadequacies in exploration are considered. Many will know, however, of the difficulty so often experienced in obtaining authority for expenditure on such professional services, "merely to look at the cores brought up to the surface by a drilling crew of two or three men"—so runs an all too familiar argument. Yet if geotechnical investigations of building sites are to be effective, and are to give the best possible results, test boring is not something that can be left to the two or three men operating a drill-rig, excellent and devoted workers though they may be.

Plotting the results obtained from successive holes as the work proceeds is essential. With the facility possessed by all geotechnical engineers and geologists to visualize the three dimensional structure that is thus slowly revealed, such two-dimensional plots will be sufficient to guide the program to a successful conclusion. But for showing what is involved to those who are not trained in such thinking, the use of a model may be necessary or at least desirable. Various techniques are available that can be applied to models, from those that will fit on the top of a desk to such immense models as that of the complex foundation beds at the site of the proposed Auburn Dam in California which fills a fair-sized room (Frei, 1976), or that showing the geology beneath the Wolf Creek Dam (see Fig. 15).

Useful as they are, sometimes indeed essential, models of subsurface conditions involve some risks. Almost inevitably, the ratio of vertical to horizontal scale will have to be distorted so that the picture given by the model may not be quite a true one. And with a model constructed to any ordinary scale, the sticks used to denote individual borings will be many times larger than the cores actually made. This last distortion is so infrequently recognized in a wider context that it is worthy of comment. Consider the subsurface investigation for a simple bridge pier measuring 60 ft by 10 ft in plan. If two 3-in. diam boreholes were put down on its proposed site to a depth of 50 ft, this would be regarded as acceptable general practice. If, for simplicity, we assume that the prism of ground beneath the pier which will be stressed when the pier is loaded has its sides at 45° to the vertical, then a simple calculation will show that the volume of ground actually proven by the two borings is 0.0007% of the volume of ground that will be stressed, or one part in 150,000. (I am indebted to A. G. Stermac for this example). For man-made materials with completely controlled properties, this would be a modest degree of sampling and yet it is accepted without question in foundation studies for materials that are, by nature, heterogeneous. This unquestioning reliance upon the adequacy of test boring, probing so minute a fraction of the material being tested, is the highest of all possible tributes to geology since it can only be by unconsciously assuming a high degree of geological uniformity that such reliance can have validity. Figs. 16 and 17 are useful reminders of what geological conditions can actually be, confirming the impossibility of interpreting ground strata between adjacent boreholes, no matter how tempting this procedure may be.

**Groundwater.**—With the test boring program complete, and a reasonably assured picture of the geological structure beneath a building site thus obtained, some might think that the geotechnical investigation is complete. But ground water is also a constituent of the subsurface, at some depth. Its position must be determined, if reasonably near the surface, and, of equal importance, its range of level throughout the year. Observation of natural features such as springs,

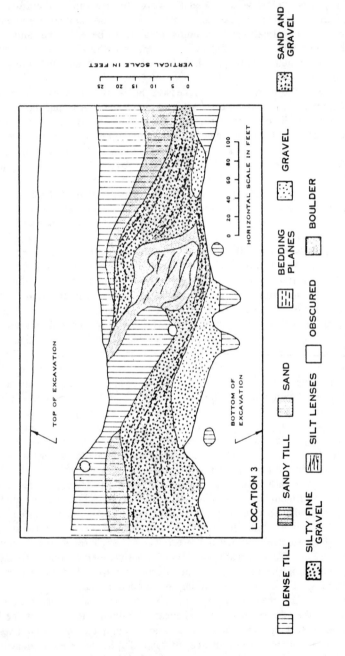

FIG. 17.—Stratification of Till as Exposed in One Section of Excavation for Closure Part of Main Cornwall Dyke of St. Lawrence Seaway and Power Project [Further illustrating the danger of interpolating between borehole records: (Diagram from Adams, 1960, reproduced by permission)]

if any are to be seen near the site, will assist with this but the only certain way is to have strategically located test holes lined with perforated casings and arrangements made for groundwater levels to be obtained regularly. It is desirable, although sometimes difficult, for this to be done for at least one full 12-month period. If construction starts before the year is out, special efforts should be made to continue observations as long as possible during construction. Groundwater is dynamic, usually moving horizontally and vertically in response to the rainfall cycle of the year. Other hydrogeological investigations may be necessary before a complete picture is obtained.

Possible complaints of flooded excavations can thus be anticipated and contract provisions for ground-water control included when necessary. Some project designs may be strongly influenced by ground-water conditions, as in the case of the Donzere-Mondragon water power and navigation scheme on the River Rhone in southern France (Henry, et al., 1961). Design of the headrace and tailrace of this run-of-river project had to be designed so as to preserve the local groundwater conditions, one feature being the provision of *Canaux de re-alimentation* (or canals that leak) in order to maintain the level of groundwater in the fertile land immediately downstream of the power house. In England, pumped groundwater is now being used to supplement low-water flow in the River Thames and other streams in order to obviate the need for costly balancing reservoirs, this imaginative project being the result of most extensive geotechnical and hydrogeological studies. In Canada, groundwater has to be intermittently pumped from a lake near one major water power installation in order to obviate the flooding of the sewerage system of the adjacent town (Legget, 1962).

At the present time, the New York Transit Authority is installing a system of deep-well pumps and underground pumping stations around the Newkirk Avenue station on their Nostrand Avenue line, as the only solution to a groundwater problem that may well be unique. When this line was built in 1915, the water-table was 16 ft below the base of rail. At that time, Brooklyn and adjacent areas were obtaining their public water supply through an extensive pumping program, drawing water from the previous water-bearing beds beneath Long Island. As pumping steadily increased, and more land was covered up, thus diverting normal infiltration, groundwater levels dropped steadily. Sea-water intrusion resulted. So serious did the situation become that pumping eventually had to be stopped; Brooklyn and much of the southwest end of Long Island now gets its public water supply from the surface sources of the New York Board of Water Supply. Pumping in the vicinity of the Newark Avenue Station, (by the New York Water Service Corporation) was stopped in 1947. By 1952 groundwater levels had risen to 2 ft 6 in. above base of rail level in the station, with results, such as leakage, that can be imagined. Various attempts have been made to correct the situation. The final solution is the permanent pumping installation now being constructed (O'Neill, 1977: in a private communication). This situation could not have been foreseen when this subway line was built. The costly solution now being constructed will, however, serve as a vivid example, to all who know of it, of the long-term significance of groundwater movements beneath civil engineering structures.

**Rainfall.**—Brief reference must be made to rainfall records as a further important tool in geotechnical investigations with strong geological overtones. A strong case can be made for the suggestion that no geotechnical investigation is complete without those responsible having seen during heavy rain the site being studied. Certainly no land-use planning of any sort should be finalized until not only thorough geological studies have been made of all the subsurface involved (and this is all too often not done), but also until the site has been examined during intensive rainfall, the heavier the better, by the planners responsible. Once such examinations have been made (uncomfortable though they may be) then the relevance to geotechnical studies of rainfall records, for as long as continuous period as possible, will be clear. The level of the water-table will be influenced but, often of even more importance, an estimate can be made of the extent of *groundwater depletion* (Thornthwaite, 1948). This is one branch of agricultural climatology that has yet to be fully appreciated in geotechnical work. Similarly, the pattern of the overall climate at the site may usefully be compared with that at better known sites, the simple concept of the *hythergraph* being a convenient means of so doing; the local hythergraph

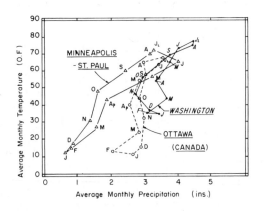

**FIG. 18.—Set of Typical Hythergraphs—Graphical Representations of Annual Climate at Specific Locations, Facilitating Comparison of Local Climates**

has a place in most geotechnical reports (see Fig. 18).

Rain is so potent a geomorphological agent that its relative neglect in geotechnical investigations is surprising. One of the worst landslides ever to start moving in Czechoslovakia was intimately related to heavy rainfall. The town of Handlova in central Slovakia is the site of an important thermal power station designed to burn local brown coal. The prevailing wind blew the exhaust from the station's tall chimney over a grassy slope to the south of the town, traditionally used over generations for the grazing of sheep. The exhaust included fly-ash, deposition of which on the meadowland affected the grass, some of which was dug up and the ground plowed for cultivation. During the year 1960, rainfall at Handlova totalled 1,045 mm as compared with a long term average of 689 mm. Apparently this excess precipitation was not noticed, nor was the fact appreciated that the tillage of part of the slope had removed Nature's own protective cover. But early in December 1960, the upper part of a large section of the sloping ground started to move downhill, 20,000,000 m$^3$ of soil and rock debris. The toe of the slope had advanced 150 m and 150 houses had been destroyed before valiantly built drainage tunnels and extensive pumping had brought the slide to a halt. Rainfall can be so very important (Zaruba and Menci, 1969) (see Fig. 19). All would do well to keep in mind Dr. Terzaghi's throw-away line: "On a planet without any water there would be no need for soil mechanics" (Terzaghi, 1939).

**Some Geological Hazards.**—In considering the assistance that geology can give in geotechnical investigations, it has been necessary to isolate geological matters from the other several disciplines which, together, make up Geotechnique. This is unfortunate in one respect since in geotechnical practice, geology must be in the background of all thinking about subsurface conditions, almost instinctively, as is indeed the case with so many geotechnical engineers. At the same time, there are some features of the subsurface which are essentially geological, the full significance may not be at first apparent. Brief reference to some of these seems desirable.

*Faults* are reminders of the dynamic character of geological processes. Geological study cannot always predict with certainty their presence beneath

FIG. 19.—Town of Handlova, Czechoslovakia, as Seen from Location about Half-Way up Slope that Started to Move in December 1960, as Described in Text

a site but areal studies can indicate the probability of their occurrence, while detailed geological examination can usually determine the relative activity of movement (if any) along the fault plane. It is not necessary to say more since site studies for several nuclear power stations have recently directed public attention to extensive studies of faults, but faults beneath dam foundations have been similarly studied for decades. The classic paper in *Geotechnique* "Potentially Active Faults in Dam Foundations" (Sherard, Cluff and Allen, 1974), provides a helpful guide (see Fig. 19).

Even small local faults if undetected can lead to difficulties. The city of Ottawa, underlain generally by Palaeozoic limestones, is underlain also by a network of small faults, many of which have been plotted in publications of the Geological Survey of Canada (Wilson, 1946). These records did not appear to have been studied prior to a recent small excavation contract, for which test holes had been put down on the usual grid pattern, spacing even at 50-ft centers missing the fault that ran across the site. Extra excavation of the gouge material and requisite backfilling with concrete cost an extra $50,000 (Legget,

1973). Another small fault in black shale, also plotted, runs beneath an important modern building in central Ottawa. Due to an unusual combination of local circumstances, pyrite in the shale along the upper part of the fault underwent complex chemical change in the course of time, causing serious heave of a basement floor (Penner, et al., 1970).

Particularly in sedimentary rocks, *seams* may be encountered, often parallel to bedding planes, frequently clay-filled. Failure along such a clay-filled seam was held responsible for the collapse of a lock wall at the Wheeler Dam and Lock on the Tennessee River in 1961 (Engineering News-Record, 1962). The Hawkesbury Sandstone, which underlies much of the city of Sydney, Australia, is notorious for its seamy character. Provision of a steel pipe lining to a 10-mile

FIG. 20.—Highway Bridge over Piscola Creek, Georgia, after Part of It Had Disappeared into Sinkhole that Developed Unexpectedly: [Sinkhole was satisfactorily sealed soon after the photograph was taken (Photo: Department of Transportation, State of Georgia)]

long, 10-ft diam water supply tunnel, between 1923 and 1928, was due in part to this character of the sandstone through which the tunnel was excavated (Haskins, 1953). The Sydney Opera House is founded on the Hawkesbury Sandstone. Changes from the initial design necessitated changes in foundation designs. In accordance with the local building by-law, core borings were sunk beneath each footing; a minimum depth of 12 in. to any seam was required, the nature of all seams encountered being studied by using scapers in the test holes (Mackenzie 1977: in an unpublished communication).

Since constituents of limestone are slightly soluble in water, cavities in limestones can constitute a real hazard in many areas, often described as *karstic* regions (after Karst in Yugoslavia) (see Fig. 20). Dramatic indeed was the collapse of a highway bridge in Georgia in 1976 when a 6-in. drill hole, being put down as a part of the investigation for a bridge replacement site, penetrated an underground cavity in limestone, over 90 ft below the surface. Water flowed down the hole, creating a cavity 45 ft deep into which the old bridge slipped (Moreland, 1977). This was an extreme example of man-made interaction with karst topography. Williams and Vineyard, in a careful study of karstic subsidences in Missouri, found that 46 out of a total of 97 catastrophic surface failures since the 1930's were due to the activities of man (Williams and Vineyard, 1976). Improved geophysical detection of hidden cavities is an urgent research need, even though useful work has already been done in this field (Kennedy, 1968).

*Weathering* is the more general process of rock metamorphosis under the long-term action of climate and allied factors. The depth of weathering above solid bedrock is often of vital importance in geotechnical investigations, especially so in tropical areas. Useful guides to the delineation and description of weathering are available (Dearman, 1974 and 1976). Especially significant in some areas, such as England and Czechoslovakia (and possibly even parts of North America), is the extensive rock disintegration that can result from paleo-permafrost conditions. One British example disclosed serious disintegration of the Chalk of southern England when excavation for a tunnel shaft approached bedrock (Haswell, 1969). Here also geophysical methods give promise of real assistance.

*Boulders* must be mentioned, again. They continue to be mistaken for solid bedrock despite at least a century of clear warning. In 1893 there was presented to this Society a short paper entitled "Borings in Broadway, New York" (Parsons, 1893), a percipient early example of urban geological studies. In the discussion it was quite bluntly pointed out—and this in 1893—that test borings sunk to a hard and resistant stratum "*do not indicate where rock is; they indicate where rock is not.*" The lesson has still to be learned. Geology can suggest the probability, or at least the possibility of boulders being present in soil. Well-developed cross-borehole geophysical methods can still further explore the possibility of boulders being present but excavation alone can provide the complete answer. If there is any possibility of boulders being present, suitable contract provisions can be made.

The study of Pleistocene Geology has provided geotechnical engineers in northern areas with invaluable information about the products of the multiple glaciations that have taken place in the last 1,000,000 yr and especially about *Till*. Difficulties with buried tills have already been mentioned. Usefully suggested as *Paleotills* (White, 1972 and 1974) they can cause remarkable problems in excavation if unrecognized in advance. A combination of geological studies and deep penetration testing can provide accurate predictions (Misiaszek, 1966). Tills are now of such widespread interest that useful interdisciplinary studies are available (Goldthwaite, 1971 and Legget, 1976).

Passing reference should be made to *Forensic Geology* recently recognized by the publication of the first book to bear this title (Murray and Tedrow, 1975), even though the study of the geological origin of soils for detection purposes goes back at least to 1874 (Marsh, 1874), with the shade of Sherlock Holmes ever in mind. The potential of geologically oriented geotechnical investigations in the study of crime holds interesting possibilities.

Finally, the problems that may be encountered with *residual stresses* in bedrock can be anticipated to a degree from geological studies and confirmed by appropriate observations in geotechnical investigations. Residual stresses causing rebound in western clay-shales can readily be foretold and explained. Stresses now being observed in limestones of northeastern North America, with the greatest principal stress in a horizontal plane, are not so readily explained (White, Karrow and MacDonald, 1974). Their magnitude is indicated by the rupture of a 6-ft thick reinforced concrete box tunnel wall at Welland shortly after construction [(Bowen, et al., 1976) see Fig. 21]. There is available a record of 70 yr of rock movement in the wheel-pit of the Canadian Niagara Power Company's plant at Niagara Falls (Hogg, 1959). Similar long-term rock movements have been reported from the Lockport, N.Y., area. Records are available also of similar movements in Australian sandstone at the Warragamba Dam (Nicol, 1964), and elsewhere in Australia (Moye, 1959). Similar experience in Penang has also been reported

FIG. 21.—Fracturing of Reinforced Concrete Tunnel Wall Due to Rock Pressure at Thorold, Ontario, Canada: (Photo: Acres Consulting Services Ltd., Niagara Falls)

(Newberry, 1970). There seems to be a possibility that these generally unrecognized stresses, now evidenced by resulting rock movements which can be observed and measured, may be in some way connected with plate tectonics, the concept of which has revolutionized the science of geology in the last two decades (Sbar and Sykes, 1973). If so, then Geotechnique might have another major contribution to make to the further advancement of the science.

**Geological Reciprocity.**—This would be no new thing, even though such a contribution might be more substantial than anything done up to now. From the earliest days, the work of the engineer has provided geologists with exposures previously unavailable (see, for example, Lyell, 1845). In more recent years, provision of geotechnical samples to geologists has facilitated notable scientific findings. Samples taken in the subsurface investigations for the Quinnipiac River bridge, across New Haven harbor, e.g., enabled geologists to determine, on the basis of Carbon 14 dating, the post-glacial changes in sea level at this location on the New England coast (Upson, Leopold and Rubin, 1964). Suites of samples taken at regular intervals along the first section of the Toronto Subway are now in the custody of the Royal Ontario Museum, available for study by geologists interested in the famous Toronto interglacial beds. When Steep Rock Lake, in northwestern Ontario, was drained in 1944–46, the significance of the remarkable sequence of varved clays then exposed in the lake-bed sediments was recognized and brought to the attention of a geological expert in this field, resulting in a valuable published account (Antevs, 1951; Legget, 1953 and 1958).

The methods and theories of Geotechnique have useful applications in geological studies, as was pointed out by Skempton as early as 1944 (Skempton, 1944 and 1953). Rominger and Rutledge (1954) in another pioneering paper, showed the significance of the Atterberg tests in Pleistocene geology; limited use only

FIG. 22.—Extra Excavation Adjacent to Interstate Highway 71 at Brookside Park, Cleveland, from which Fossil Devonian Fishes Were Obtained in Quantity: (Photo: Cleveland Museum of Natural History)

appears to have been made of this link since then, apart from the work of one eminent geologist (Boswell, 1961). Hubbert and Rubey paid due tribute to the work of Dr. Terzaghi in their well-known paper on Overthrust Faulting, in which they state that "it appeared that the phenomena of soil mechanics represented in many respects very good scale models of larger distrophic phenomena of geology" (Hubbert and Rubey, 1959).

It is, however, in the geology revealed by excavations that perhaps the greatest contributions to the science may be made. This was recognized in the Federal Highway Act of 1957 with a far-sighted provision (which is believed to be still in force) which makes it possible for Federal funds to be applied towards the salvage of historical, archaeological and paleaontological specimens when they occur within the limits of Federal highway projects. It was under this provision that the extra excavation was made adjacent to Interstate Highway 71 as it enters the city of Cleveland, Ohio, near Brookside Park (see Fig. 22).

Many must have blamed the obvious extra excavation at this location on "bad engineering" whereas it was a cooperative effort between the highway authorities and the Cleveland Natural Science Museum which resulted in the Museum obtaining what must be the world's most extensive collection of Devonian fossil fishes (Scheele, 1965).

A civil engineer, the late George B. Sowers, F. ASCE, Commissioner of Engineering, City of Cleveland had brought these unusual fossils to the attention of the museum nearly 40 years previously when he found them in an excavation. The construction project enabled the museum to obtain its first specimens.

Even more remarkable was the discovery of remains of five prehistoric civilizations in the course of the subsurface investigation for a new building in Zurich, Switzerland. Archaeologists were able to confirm that the site had been the location of a succession of early settlements of "lake-dwellers." The foundation design was drastically changed to span the historical site in which ancient piles were exposed to view during excavation, all of which was meticulously carried out by archaeologists while erection of the building superstructure proceeded above (Von Moos, 1977: in an unpublished communication).

All such cooperative efforts in the interest of science stem from an appreciation of the value of historical or geological exposures arising from engineering work. It is surely incumbent upon all geotechnical engineers to nurture such appreciation among their staffs so that unique opportunities for adding to the store of human knowledge, without interfering with necessary engineering work, shall not be lost.

**Geological Predictions May Be Wrong.**—It is only right that a note of warning should bring this review to a close. Predictions based on geological reasoning sometimes may be wrong. This is why the combination of the geological and geotechnical approaches to subsurface problems is so essential if sound and safe results are to be achieved. The geologist makes the best deductions he can from his observations, coupled with his knowledge of normal geological processes; the geotechnical engineer with desirable scepticism should always attempt to check the deductions by actual exploration techniques. In the search for a new site for the main settlement in northwestern Canada, near the Arctic coast, aerial photographs of the great Mackenzie River delta were studied to such good effect that 12 potential sites were selected in Ottawa before field work commenced. Examination of the sites in the field reduced the possibilities to four, of which one was outstanding—situated on what appeared to be sand and gravel alluvial fans emerging from Richardson Mountains, bedrock being clearly eroded no more than 1 mile from the proposed site. Test drilling was carried out to confirm the suitability of the site. Instead of sand and gravel, organic silt (naturally frozen) was encountered to depths of over 60 ft, eliminating completely a site that appeared to be, geomorphologically, excellent [(Legget, et al., 1966) see Fig. 23].

Far removed from this Arctic outpost is one of the best known geological features of United States cities, Beacon Hill in Boston. Boston Common, reserved as a public place since 1640, is a prominent feature. The historical significance of the Hill and other urban geological characteristics of the Massachusetts capital city have been interestingly discussed in the only geological bicentenary tribute yet seen (Kaye, 1976a). Beacon Hill, even without its three original "peaks," leveled by early excavation, has been well known as a drumlin; it has been repeatedly so described in the literature. C. A. Kaye, of the U.S. Geological Survey, has been observing excavations in its vicinity for the last 14 yr, as occasion permitted. During this period, 24 major buildings have been constructed on the slopes of Beacon Hill, with foundations extending as much as 55 ft below surface level. Construction of some of these foundations ran into difficulties, such as excessive groundwater, that did not tally with the assumed drumlin formation. Through a detailed study of all the geological conditions revealed by these engineering works, Kaye has been able to show that, instead of being a drumlin, Beacon Hill is almost certainly an end moraine of complex origin. His account of the use of excavation records to disprove "conventional wisdom" is an exemplar for all concerned with urban geology (Kaye, 1976b).

The foregoing paragraphs will show, it is hoped, that this presentation is

FIG. 23.—Alluvial Fans in Delta of Mackenzie River, Northwest Territories, Canada, Formed of Disintegrated Rock, Derived from Mountains in Immediate Background, in which No Sand and Gravel Were Found

not an over-enthusiastic starry-eyed promotion of geology as the "be-all and end-all" of subsurface investigations. Far from it. All that is hoped is that it will have provided a reminder that geology has a vital part—but only a part—to play in all geotechnical investigations, sometimes in a quite general way, but on many occasions as a rigorous concommitant of detailed soil and rock studies. It may been suggested that, although this has been recognized from the earliest days of civil engineering, there appears today to be at least a lack of *balance* between the practical initial approach to subsurface studies in civil engineering based on geology, and the attention given to theoretical and laboratory studies.

Consider how well this desirable balance has been achieved in the previous lectures in this series to which I can, in this way, add my tribute. Rewarding and valuable theoretical treatments were presented by John Lowe III, G. G. Meyerhof, F. E. Richert and Lymon Reese. S. D. Wilson described interesting examples of field instrumentation. The late Lauritz Bjerrum gave a wide-ranging review of his experiences with over-consolidated clays and H. B. Seed of his corresponding experiences with landslides due to soil liquefaction during earthquakes. Ralph Peck, P. C. Rutledge, T. W. Lambe and B. McClelland described outstanding examples of sound geotechnical engineering, and Arthur Casagrande used seven notable examples to illustrate his wise and philosophic comments on the calculated risk in earthquake and foundation engineering.

Do we find the same desirable balance in the current literature of geotechnical engineering, upon which practitioners of the art may be assumed to place reliance? We do not. Textbooks prior to the last decade had geology "built-in" to their contents, a good example being that by Sowers and Sowers (3rd ed., 1970). But a careful examination of 10 recent geotechnical texts, in English and French, published in three countries, revealed due and proper attention to geology as a starting point in one volume only of the 10. The excellent volume by J. K. Mitchell is, naturally, not one of the 10; its thorough treatment of geology is exemplary in its detailed treatment of *Soils in Soil Engineering* (Mitchell, 1976). Of the 10 books studied, five were found to make no reference to geology at all. Out of a total pagination of 3,329 pages, only about 50 were devoted to geology (and of these, 30 in one volume), or 1.5% of the total.

This is disturbing, but equally serious is the situation when periodical geotechnical literature is reviewed. *Geotechnique* was launched by a few enthusiasts in 1948. Dr. Terzaghi had this to say in his initial Foreword:

> Since soil mechanics has prepared the tools which are required to get the full benefit out of geological information, the practical value of the data secured in the field by the joint efforts of engineer and geologist will steadily increase. If *Geotechnique* becomes a clearing house for information with proper emphasis on such data, it may open up public sources of vital geotechnical information and earn the gratitude of every practicing civil engineer . . . (Terzaghi, 1948).

In the first few volumes the balance between theory and practice was maintained, the balance between what may be called "Case Histories" (accounts of projects or experimentation in the field with natural recognition of geology), and purely theoretical papers or reports upon laboratory work, equally naturally with no reference at all to geology. As the years have passed this balance appears to have been distorted in the direction of "theoretical" papers, valuable as these are in their place. This was, at first, merely a subjective impression but one which was found to be shared by others. Fig. 24 shows the result of a careful analysis of the contents of the first 25 volumes; it seems to confirm the impression. The analysis was also subjective but, making every allowance for some differences of judgment as to the classification of individual papers (the "border-line cases") there can be little doubt but that the *trend* in the character of the contents of this prestigious publication has been quite definitely away from geologically significant papers. The marked improvement in balance in the contents of Volumes 24 and 25 is to be welcomed. A few "practical" geotechnical papers do appear in the Proceedings of the (sponsoring) Institution of Civil Engineers but even with the inclusion of the papers now appearing in the *Quarterly Journal of Engineering Geology*, the balance is still not what it should be if these excellent British journals are to be at once a true reflection of the best of current practice and a guide to the way in which geotechnical engineering should be developing.

What of the *Journal of the Geotechnical Engineering Division* of ASCE, initiated in 1956? Despite the good work of successive Editorial Committees, the same sort of critical remarks about its increasingly theoretical bias are current. Fig. 24 shows the situation. There are surely grounds for even greater concern. If the present trend continues, there will probably be no "practical" papers at all in Volume 108 (1982). And as will be seen, inclusion of the modest number of geological papers now appearing in the *Bulletin of the Association of Engineering Geologists* does not redress the balance even as much as in the British case.

A number of explanations for this state of affairs can be advanced, always keeping in mind that geotechnical studies deal with two basic types of problem— the uses of soil and rock *in situ*, on the one hand, and their removal and possible modification for use as engineering materials on the other. It may

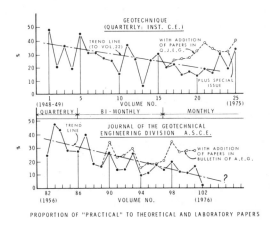

PROPORTION OF "PRACTICAL" TO THEORETICAL AND LABORATORY PAPERS

**FIG. 24.—Proportion of "Practical" to Theoretical and Laboratory Papers in Geotechnical Publications**

be that geotechnical engineers engaged on construction projects have no time to write up accounts of their work and that when site problems arise, so useful for professional review, legal constraints limit publication opportunities. As all who have written one know, the preparation of a concise paper on a "practical" subject is not easy, necessitating selection of material and condensation that are always difficult. On the other hand, the regrettable "publish or perish" syndrome requires the preparation of papers, usually "theoretical" in nature, for publication. Much of the research work arising from graduate study so reported is of great value, and is needed. It is the lack of a balancing amount of information from the field that is assuming serious proportions. Whatever the explanation, however, there appears to be little doubt that the situation is at least developing into one that should give cause for alarm to the profession, especially when thought is given to what the immediate future holds for civil engineering.

The very rapid increase in the population of the world, the greatest increases taking place in developing countries, means that many physical facilities around the world will have to be at least doubled before the year 2,000 A.D., now only 22 yr away. Much of this vast expansion will be in countries in which there is little geotechnical experience to serve as background and yet it is here that the need will be greatest. In all developed countries, the best sites for dams, bridges and other major structures have been progressively used up, so that site studies and site selection will inevitably become progressively more difficult. The energy situation (not a crisis that will pass but a situation that is here to stay) necessitates vastly increased attention being given to the use of underground space, as an energy saver, for all appropriate purposes. Attention now happily being given to environmental pollution is going to mean vast expenditures on sewerage systems and sewage treatment plants. These are but a few of the highlights clearly to be seen today. They all point in the direction of an increase in the volume and complexity of geotechnical work, in which geology must therefore be called upon to play its vital role.

What can, and what should be done? With due deference to the excellent work in using geology that is already being done by many geotechnical engineers in this and in other countries, may it be suggested, if only as a reminder, that:

1. Every geotechnical investigation must start with a review of the local geology, general in character if that is sufficient, but detailed when necessary.

2. The services of an engineering geologist should be utilized in such studies, when necessary, and always for detailed geological site studies.

3. No test boring and sampling program should be laid out except after due consideration of the geology of the site under study.

4. Test boring results should be correlated with previous deductions as to the site geology, during the progress of site investigation, and the location and number of borings steadily reviewed and if necessary revised in the light of such correlation.

5. When refusal is met with in a boring, diamond core drilling, preferably to a depth of at least 10 ft, should be put down and the cores so obtained geologically examined, as a protection against boulders being mistaken for bedrock.

6. In every foundation investigation, watch should be kept for features of possible geological or archaelogical significance and reports of any such occurrences so made that necessary precautions can be taken when construction starts.

7. All geotechnical investigations should include consideration of local rainfall records; sites under study, whenever possible, should be visited during the course of heavy rainfall.

8. Local hydrogeological features must be considered in all site investigations, by observation holes left in place, and in other ways not only at the time of the study but, to the extent that is possible, for at least one complete annual cycle so that groundwater level variations may be determined.

9. When a project goes into construction, geotechnical engineers should not regard their own work as complete until they have done everything possible to ensure that accurate *as constructed drawings* of all foundation features have

been prepared, with the most accurate delineation upon them of the geological conditions that are to be covered up by permanent works.

10. Every geotechnical report should start with reference to the local geology, however general, more detailed geological information being presented as may be necessary.

11. Geotechnical engineers should regard it as a part of their professional responsibility to share the results of their subsurface investigations for the common good, with due regard for necessary initial confidentiality, preferably in some cooperative manner.

12. When any geotechnical investigation reveals information of such an unusual character that it could be of benefit to the profession and so, indirectly, to the public, an account should be prepared for publication in the most appropriate medium.

13. The highest priority should be given to the preparation and publication of "case histories" of experiences with residual soils, since such information will be of special value in most developing countries.

14. When geotechnical engineers know of cases when valuable subsurface or allied information has been revealed by failures or problems on construction and which is temporarily restricted by legal constraints such as court action, they should record the fact privately and make appropriate efforts to ensure eventual publication once the legal restraints have been removed.

Does the need for attention to geology extend beyond the confines of geotechnical engineering? Surely it must, if only because, as must be observed yet once again, every civil engineering structure depends for its safety and stability upon the ground beneath it. Long ago, the Chief Geologist on the St. Gotthard Tunnel used words which Dr. Terzaghi often quoted:

> Every civil engineer is engaged in experimental geology without being conscious of the fact and without being spoiled by the recognition of the benefits which the science of geology has derived from his activities (Terzaghi, 1953).

Structural engineers in particular, need to have an instinctive realization that all their assumptions as to the stability of the foundations on which the structures they design are to be supported are ultimately dependent upon the geology underlying the site. And civil engineers who serve in construction will be immeasurably assisted in their practical problems by at least a good general grounding in geology.

Does it now follow, inevitably, that a basic part of the training of *every* civil engineer must be an introduction to the science of geology, preferably in a manner that will illustrate the relevance of geology to civil engineering? A good case can be made for the desirability of all undergraduate engineers being given at least the opportunity of gaining such an introduction, knowledge of the science of the earth in which they live being as culturally desirable, if not more so, than some of the elective subjects to be found in some current undergraduate curricula. For civil engineers an introduction to geology is imperative. And yet there are still a few civil engineering undergraduate curricula that do not include any geology as a mandatory subject. An introduction to

Pleistocene Geology, one would imagine, should be a prerequisite for any study of soil mechanics. It is often included, briefly, in the introductory parts of such courses, teaching in this case being in advance of textbooks.

Dr. Terzaghi had helpful words on this subject also, relative to the training of all civil engineers:

> The one ounce of geology is as essential as the yeast in the process of fermentation but it represents only a minute fraction of the vast domain covered by the sciences of the earth. Therefore, I believe that a two-semester course combined with field trips fully serves its purpose provided that the course represents the combined efforts of a geologist who appreciates the requirements of engineers and an engineer who has learned from personal experience that geology is indispensable in the practice of the profession (Terzaghi, 1957).

Like so much of this lecture, this is no new thought, not even with Dr. Terzaghi. As long ago as 1898 a paper was presented to this Society entitled: "Geology in its relation to Topography" by Professor John C. Branner of Stanford University (Branner, 1898). The entire paper and the stimulating discussion which it generated are well worth reading today. The paper includes such gems as this, with reference to the work of civil engineers: "To set a man at work on topography who knows nothing of geology is very like having someone perform a surgical operation who knows nothing of anatomy." And in the discussion, this from an eminent geologist of the time:

> The successful engineer of the future should know, not only how to locate his work, but how to locate it so that Nature will aid him in its building and take it under her protection. Too late he may know that Nature has resented his intrusion and in spite of his efforts is surely undoing his work (Kemp, 1898).

Make due allowance for the Victorian language, and these messages are still valid today.

What of all those members of the profession who have long since graduated without the advantage of any formal introduction to geology and yet who have come, through their own experience, to see the importance of geology in their work, be they designers or constructors? Most fortunately, geology is one science that can readily be studied, at least in its elements, by reading and personal observation. It was with good reason that it was described by Charles Kingsley, more than a century ago, as "the people's science" (Kingsley, 1877). So it should be today. It can be, and should be, the engineer's science! There are available a host of excellent introductory books, easy to read and to study, ranging from the convenient little *Golden Science Guides* (on Geology, Rocks and Minerals, Landforms, and Fossils) that can readily be carried in purse or pocket, through such admirable guides to field studies as the brilliantly illustrated pocket-sized *Field Geology in Colour* (Bates and Kirkaldy, 1976), and such delightful introductory texts as that by Gilluly, Waters and Woodford (1976) and Flint and Skinner (1974) to such a magisterial and fascinating volume as the 1,288 page *Physical Geology* of Arthur Holmes (1965), a small geological library in itself. One of the greatest of all contributions that geotechnical engineers could render to their profession would be for them to do everything possible to render geology familiar and meaningful to all civil engineers. Geotechnical work might then cease to be the Cinderella of the profession (sometimes described, as I have heard, as "messing about with mud"). It should be recognized as one of the most challenging of all branches of the profession if only because there are never any two cases identical, every problem requiring that exercise of judgment that is the hallmark of the true civil engineer. Much could be achieved if civil engineers would only regain the art of reading (Legget, 1976).

Let me conclude as I began, with the words of others to demonstrate that all I have essayed to do is to act as a sounding board for the firmly held views of some of our mentors and especially of him for whom this lecture is named. In 1973 Dr. Ralph Peck had this to say:

> We can never successfully divorce our thinking from the overwhelming influence of geology on our works . . . because nature is infinitely variable, the geological aspects of our profession assure us that there will never be two jobs exactly alike. Hence we need never fear that our profession will become routine and dull.

Almost 30 yr before, Dr. Terzaghi said this:

> In view of the variety and importance of the influence of geologic factors on the performance of soils in the field and on the methods for predicting their performance, the boundary between soil mechanics and engineering geology appears to be rather artificial. . . . Hence the time may come when it will be appropriate to combine soil mechanics and engineering geology into one unit, under such a name as Geotechnology" (Terzaghi, 1948).

We have, today, an even better name—Geotechnique—but have we yet forged the proper union between engineering geology and soil mechanics?

And, finally, these words:

> The satisfactory solution will one day be recompense for the work of those who, without separating mechanics from natural philosophy (geology), will best know how to adapt to the spirit of the first the material facts which it is the essential object of the second to discover and co-ordinate.

These are the words of Alexandre Collin, written in Paris in 1846 (Collin, 1846). Their challenge still stands.

APPENDIX I.—BIBLIOGRAPHY

A selection of titles of some publications of the U.S. Geological Survey that could be of assistance to civil engineers, typical of the many such publications of national and state (or provincial) geological agencies; published by permission of the Director of the United States Geological Survey.

Adams, V. W., Earth Science Data in Urban and Regional Information Systems—A

Review, *U.S. Geological Survey Circular* 712, 1975.

Anderson, J. R., Hardy, E. E., and Roach, J. T., A Land-Use Classification System for Use with Remote-Sensor Data, *U.S. Geological Survey Circular* 671, 1972.

Bailey, E. H., and Harden, D. R., Map Showing Mineral Resources of the San Francisco Bay Region, California: U.S. Geological Survey Misc. Inv. Map I-909, 1975, text, map scale 1:250,000.

Borcherdt, R. D., ed., Studies for Seismic Zonation of the San Francisco Bay Region, *U.S. Geological Survey Professional* Paper 941-A, 1975, 102 p.

Briggs, R. P., Pomeroy, J. S., and Davies, W. E., Landsliding in Allegheny County, Pennsylvania, *U.S. Geological Survey Circular* 728, 1975, 18 p.

Briggs, R. P., and Kohl, W. R., Map of Zones where Land Use Can Be Affected by Landsliding, Flooding, and Undermining, Allegheny County, Pennsylvania: U.S. Geological Survey Misc. Field Studies Map MF-685-D, 1975, scale 1:50,000.

Brown, R. D., Jr., Faults That Are Historically Active or That Show Evidence of Geologically Young Surface Displacements, San Francisco Bay Region; A Progress Report, Oct., 1970, U.S. Geological Survey Misc. Field Studies Map MF-331, 1972, 2 map sheets, scale 1:125,000.

Brown, R. D., Jr., and Lee, W. H. K., Active Faults and Preliminary Earthquake Epicenters (1969–1970) in the Southern Parts of the San Francisco Bay Region, U.S. Geological Survey Misc. Field Studies Map MF-307, 1971, one map sheet, 7 p. text, scale 1:250,000.

Brown, R. D., Jr., and Wolfe, E. W., Map Showing Recently Active Breaks along the San Andreas Fault between Point Delgada and Bolinas Bay, California, U.S. Geological Survey open-file map, 1973, 2 map sheets plus text and references, scale 1:24,000.

Bryant, B., Map showing Mines, Prospects, and Areas of Significant Silver, Lead, and Zinc Production in the Aspen Quadrangle, Pitkin County, Colorado, U.S. Geological Survey Misc. Inv. Map I-785-D, 1972.

Bushnell, K., Map Showing Areas that Correlate with Subsidence Events Due to Underground Mining of the Pittsburgh and Upper Freeport coalbeds, Allegheny, Washington, and Westmoreland Counties, Pennsylvania, U.S. Geological Survey Misc. Field Studies Map MF-693-C, 1975, scale 1:125,000.

Crandall, D. R., 1973, Map Showing Potential Hazards from Future Eruptions of Mount Rainier, Washington, U.S. Geological Survey Misc. Inv. Map I-836 (map with text), 1975.

Foxworthy, B. L., and Richardson, D., Climatic Factors Related to Land-Use Planning in the Puget Sound Basin, Washington, U.S. Geological Survey Misc. Inv. Map I-851-A (map with text), 1973.

Hack, J. T., Geologic Map for Land-Use Planning, Prince Georges County, Maryland, *U.S. Geological Survey Open-File Report* 75-208, 1975, scale 1:62,500.

Handman, E. H., and Hildreth, C. T., Depth to Bedrock, Hartford North Quadrangle, Connecticut, U.S. Geological Survey Misc. Inv. Map I-784-D, 1972.

Harris, L. D., Areas with Abundant Sinkholes in Knox County, Tennessee, U.S. Geological Survey Misc. Inv. Map I-767-F, 1973.

Hildreth, C. T., and Keune, C. H., Location of Wells and Test Holes, Hartford North Quadrangle, Connecticut, U.S. Geological Survey Misc. Inv. Map I-784-P, 1972.

Kohl, W. R., Map of Overdip Slopes that can Affect Landsliding in Armstrong County, Pennsylvania, U.S. Geological Survey Misc. Field Studies Map MF-730, 1976, Scale 1:125,000.

Langer, W. H., Thickness of Principal Clay Unit, Hartford North Quadrangle, Connecticut, U.S. Geological Survey Misc. Inv. Map I-784-E, 1972.

Langer, W. H., Thickness of Material Overlying Principal Clay Unit Hartford North Quadrangle, Connecticut: U.S. Geological Survey Misc. Inv. Map I-784-F, 1972.

Langer, W. H., Resources of Coarse Aggregate, Hartford North Quadrangle, Connecticut, U.S. Geological Survey Misc. Inv. Map M-784-G, 1972.

Leopold, L. B., Hydrology for Urban Land Planning—A Guidebook on the Hydrologic Effects of Urban Land Use, *U.S. Geological Survey Circular* 554, 1968.

Leopold, L. B., et al., A Procedure for Evaluating Environmental Impact, *U.S. Geological Survey Circular* 645, 1971.

Maberry, J. O., and Lindvall, R. M., Map Showing Relative Swelling-Pressure Potential of Geologic Materials in the Parker Quadrangle, Arapahoe and Douglas Counties, Colorado, U.S. Geological Survey Misc. Inv. Map I-770-D, 1972.

Maberry, J. O., Map Showing Relative Excavatability of Geologic Materials in the Parker Quadrangle, Arapahoe and Douglas Counties, Colorado, U.S. Geological Survey Misc. Inv. Map 770-H, 1972.

Maberry, J. O., Map Showing Inferred Relative Permeability of Geologic Materials in the Parker Quadrangle, Arapahoe and Douglas Counties, Colorado, U.S. Geological Survey Misc. Inv. Map I-770-I, 1972.

Nichols, D. R., and Wright, N. A., Preliminary Map of Historic Margins of Marshlands, San Francisco Bay, California, U.S. Geological Survey Open-File Map, One Map Sheet, Text and References, 1971, Scale 1:125,000.

Nichols, D. R., and Buchanan-Banks, J. M., Seismic Hazards and Land-Use Planning, *U.S. Geological Survey Circular* 690, 1974.

Nilsen, T. H., Preliminary Photointerpretation Map of Landslide and Other Surficial Deposits of the Mount Diablo area, Contra Costa and Alemada Counties, California, U.S. Geological Survey Misc. Field Studies Map MF-310, 1971, One Map Sheet, Scale 1:62,500.

Nilsen, T. H., and Turner, B., Influence of Rainfall and Ancient Landslide Deposits on Recent Landslides, Contra Costa County, California, *U.S. Geological Survey Bulletin* 1388, 1975, 18 p., 2 Plates, 5 Figs., 2 Tables.

Pessl, F., Jr., Langer, W. H., and Ryder, R. B., Geologic and Hydrologic Maps for Land-Use Planning in the Connecticut Valley with Examples from the Folio of the Hartford North Quadrangle, Connecticut, *U.S. Geological Survey Circular* 674, 1972.

Pomeroy, J. S., and Davies, W. E., Map of Susceptibility to Landsliding, Allegheny County, Pennslyvania, U.S. Geological Survey Misc. Field Studies Map MF-685-B.

Radbruch, D. H., and Wentworth, C., Estimated Relative Abundance of Landslides in the San Francisco Bay Region, California, U.S. Geological Survey Open-File Map, One Map Sheet, 1971, Scale 1:500,000.

Ritter, J. R., and Dupre, W. R., Maps Showing Areas of Potential Inundation by Tsunamis in the San Francisco Bay Region, California, U.S. Geological Survey Misc. Field Studies Map MF-480, 2 Map Sheets, 1972, Scale 1:125,000.

San Francisco Bay Region Land Use Maps, U.S. Geological Survey, Two Samples, U.S. Geological Survey Open-File Map, 1974, 2 Map Sheets, Scale 1:62,500.

Simpson, H. E., Map Showing Man-Modified Land and Man-Made Deposits in the Golden Quadrangle, Jefferson County, Colorado, U.S. Geological Survey Misc. Inv. Map I-761-E, 1973.

Slope Map of Part of West-Central King County, Washington, U.S. Geological Survey Misc. Inv. Map I-852-E, 1975.

Taylor, F. A., and Brabb, E. E., Maps Showing Distribution and Cost by Counties of Structurally Damaging Landslides in the San Francisco Bay Region, California, Winter of 1968–1969, U.S. Geological Survey Misc. Field Studies Map MF-327, One Map Sheet, 1972, Maps at 1:1,000,000 and 1:500,000 Scale.

Witkind, I. J., Map Showing Geologic Constraints on Placement of Sanitary Landfills in Henrys Lake Quadrangle, Idaho and Montana, U.S. Geological Survey Misc. Inv. Map I-781-H, 1972.

## Appendix II.—References

1. Adams, J. I., "Tests on Glacial Till," *Proceedings* 14th Canadian Soil Mechanics Conference, in *Technical Memo 69,* Associate Committee on Geotechnical Research, N.R.C., Ottawa, Canada, 1960, p. 37.
2. Allen, P., "The George's River Bridge," *Minutes of Proceedings,* Institution of Civil Engineers, London, England, Vol. 232, 1932, p. 183.
3. Antevs, A., "Glacial Clays in Steep Rock Lake, Ontario, Canada, *Bulletin,* Geological Society of America, Vol. 62, 1951, p. 1223.
4. Bates, D. E. B., and Kirkaldy, J. F., *Field Geology in Colour,* Blandford Press, Poole, United Kingdom, 1976, 215 pp.
5. Bazynski, J., "Studies of the Landslide Areas and Slope Stability Forecast," *Bulletin,* IAEG Krefeld, West Germany, No. 16, 1977, p. 77.
6. Boswell, P. G. H., *Muddy Sediments,* W. Heffer and Sons Ltd., Cambridge, United Kingdom, 1961, 140 pp.
7. Bowen, C. F. P., et al., (1976), "Rock Squeeze on Thorold Tunnel," *Canadian Geotechnical Journal,* Ottawa, Canada, Vol. 13, 1976, p. 111.
8. Branner, J. C., "Geology in Its Relations to Topography," *Transactions,* ASCE, Vol. XXXIX, 1898, p. 53 (see p. 54).
9. Brown, W. C., and Parsons, M. F., "Bosporus Bridge," *Proceedings,* Institution of Civil Engineers, London, England, Part I, Vol. 58, 1975, p. 506; discussion, Vol. 60, 1975, p. 503.
10. Bruckshaw, J. M., et al., "The Work of the Channel Tunnel Group," *Proceedings,* Institution of Civil Engineers, London, England, Vol. 18, 1961, p. 149; discussion, Vol. 21, 1962, p. 611.
11. "Clay Seam Wrecked Lock," *Engineering News-Record,* Vol. 168, January 4, 1962, p. 19.
12. Cleaves, A. B., "Engineering Geology Characteristics of Basal Till; St. Lawrence Seaway Project," *Case History Series,* No. 4, Geological Society of America, 1963, p. 51.
13. Collin, A., *Landslides in Clay,* translated by W. R. Schriever, University of Toronto Press, Toronto, Canada, 1846, 160 pp., see p. 144.
14. Crawford, C. B., "Engineering Studies of Leda Clay," *Soils in Canada,* R. F. Legget, ed., Royal Society of Canada Special Publications, No. 3, 1961, p. 200.
15. Deacon, G. F., "The Vyrnwy Works for the Water Supply of Liverpool," *Minutes of Proceedings* Institution of Civil Engineers, London, England, Vol. 126, 1896, p. 24.

16. Dearman, W. R., "Weathering Classification in the Characterisation of Rock for Engineering Purposes in British Practice," *Bulletin,* IAEG, Krefeld, West Germany, No. 9, 1974, p. 33.
17. Dearman, W. R., "Weathering Classification in the Characterization of Rocks; a Revision," *Bulletin* IAEG, Krefeld, West Germany, No. 13, 1976, p. 123.
18. Dearman, W. R., *Engineering Geological Maps,* M. Matula, chmn., Earth Sciences No. 15, UNESCO Press, Paris, France, 1976, 79 pp.
19. Eckel, E. C., "Engineering Geology and Mineral Resources of the T.V.A. Region," *Geological Bulletin No. 1,* Tennessee Valley Authority, Knoxville, Tenn., 1934.
20. Fairchild, H. L., and Cushing, H. P., "Geology of the Thousand Islands Region," *Museum Bulletin No. 145,* New York State Museum, Albany, N.Y.,1910, 194 pp.
21. Flint, R. F., and Skinner, B. J., *Physical Geology,* John Wiley, Inc., New York, N.Y., 1974.
22. Fookes, P. G., "Geotechnical Mapping of Soils and Sedimentary Rocks for Engineering Purposes," *Geotechnique,* London, England, Vol. 19, 1969, p. 52.
23. Frei, L. R., "Auburn Dam: Foundation Investigation, Design and Construction," *Field Trip Guide,* AEG Lake Tahoe Meeting 1975, (revised 1976).
24. Gilluly, J., Waters, A. C., and Woodford, A. O., *Principles of Geology,* 4th ed., W. H. Freeman and Co., San Francisco, Calif., 1975, 534 pp.
25. Goldthwaite, R. P., ed., *Till: A symposium,* Ohio State University Press, Columbus, Ohio, 1971, 402 pp.
26. Hardy, R. M., "Engineering Problems Involving Preconsolidated Clay Shales," *Transactions* Engineering Institute of Canada, Montreal, Canada, Vol. 1, 1957, p. 1.
27. Haskins, G., "The Construction, Testing and Strengthening of a Pressure Tunnel for the Water Supply of Sydney, N.S.W.," *Minutes of Proceedings* Institution of Civil Engineers, London, England, Vol. 234, 1932, p. 25.
28. Hassall, E. R., and Rankilor, P. T., "A Comparison of the Geological and Mining Aspects of the Urban Redevelopment of Oldham, Lancashire, with Stoke-on-Trent, Staffordshire," *Engineering Geology of Reclamation and Redevelopment* Geological Society of London, Engineering Group, University of Durham, United Kingdom, 1973, p. 69.
29. Haswell, C. K., "Thames Cable Tunnel," *Proceedings,* Institution of Civil Engineers, London, England, Vol. 44, 1969, p. 325.
30. Henry, B., and Mathian M. M., "Nappe de la Plaine Alluviale de Rive Gauche du Rhône Entre Donzére et Mondragon," *Compagnie Nationale du Rhône,* Lyons, France, 15 pp.
31. Hogg, A. D., "Some Engineering Studies of Rock Movement in the Niagara Area," *Case History Series,* No. 3, Geological Society of America, 1959, p. 1.
32. Holmes, A., (1965), *Physical Geology,* 2nd ed., T. Nelson and Sons Ltd., London, England, 1,288 pp.
33. Hubbert, M. K., and Rubey, W. W., "Mechanics of Fluid Filled Porous Solids and its Application to Overthrust Faulting," *Bulletin,* Geological Society of America, Vol. 70, No. 2, Feb., 1959, p. 115; and the following companion paper by the same authors.
34. Hvorslev, M. J., (1949), *Subsurface Exploration and Sampling of Soils for Civil Engineering Purposes,* ASCE.
35. Jennings, R. A. J., "The Problem Below," *Quarterly Journal of Engineering Geology,* London, England, Vol. 9, No. 2, 1976, p. 103.
36. Kapp, M. S., "Slurry Trench Construction for Basement Wall of World Trade Centre," *Civil Engineering,* Vol. 39, Apr., 1969, p. 36.
37. Kaye, C. A., "The Geology and Early History of the Boston Area of Massachusetts; a Bicentennial Approach," *Bulletin 1476,* United States Geological Survey, 1976a, 78 pp.
38. Kaye, C. A., "Beacon Hill End Moraine, Boston; New Explanation of an Important Urban Feature," *Urban Geomorphology,* D. R. Coates ed., Special Paper 174 of Geological Society of America, 1976b, p. 7.
39. Kemp, J. F., (1898), in discussion of Branner (1898), *Transactions,* ASCE, Vol. XXXIX, p. 86.
40. Kennedy, J. M., "A Microwave Radiometric Study of Buried Karst Topography,"

*Bulletin,* Geological Society of America, Vol. 79, No. 6, June, 1968, p. 735.
41. Kingsley, C., *Town Geology,* Daldy, Isbister and Co., London, England, 1877, 239 pp; see p. 4.
42. Legget, R. F., (1958), "Soil Engineering at Steep Rock Lake Iron Mines, Ontario, Canada," *Proceedings,* Institution of Civil Engineers, London, England, Vol. 11, 1958, p. 169.
43. Legget, R. F., (1962), "Experiences with Groundwater on Construction," *Journal of the Soil Mechanics and Foundations Division,* ASCE, Vol. 88, No. SM2, Proc. Paper 3092, Apr., 1962, pp. 1–17.
44. Legget, R. F., *Cities and Geology,* McGraw Hill Book Co., Inc., New York, N.Y., 1973, 624 pp. see p. 9.
45. Legget, R. F., (1976), *Canals of Canada,* David and Charles, Newton Abbot, United Kingdom, 1976, 261 pp.; see p. 129.
46. Legget, R. F., (1976b), "Do Engineers Read (Or Buy) Books?" *Scholarly Publishing,* Toronto, Canada, Vol. 9, 1976b, p. 337.
47. Legget, R. F., ed., (1976), *Glacial Till: An Inter-disciplinary Study,* Royal Society of Canada Special Publications No. 12, Ottawa, Canada, 1976, 412 pp.
48. Legget, R. F., and Bartley, M. W., "An Engineering Study of Glacial Deposits at Steep Rock Lake, Ontario, Canada," *Economic Geology,* Vol. 48, No. 7, Nov., 1953, p. 513.
49. Legget, R. F., and Schriever, W. R., "Site Investigations for Canada's First Underground Railway," *Civil Engineering and Public Works Review,* Vol. 55, 1960, p. 73.
50. Legget, R. F., Brown, R. J. E., and Johnston, G. H., "Alluvial Fan Formation near Aklavik, N.W.T., Canada," *Bulletin,* Geological Society of America, Vol. 77, No. 1, Jan., 1966, p. 15.
51. Legget, R. F., and Burn, K. N., "Archival Material and Site Investigation," paper in preparation.
52. Lyell, C., *Travels in North America,* Vol. 1, Wiley and Putnam, New York, N.Y., 1845, p. 253.
53. MacClintock, P., "Glacial Geology of the St. Lawrence Seaway and Power Project," pamphlet from New York State Museum and Science Service, Albany, N.Y., 1958, 26 pp.
54. MacClintock, P., and Stewart, D. P., "Pleistocene Geology of the St. Lawrence Lowland," *Bulletin 394,* New York State Museum and Science Service, Albany, N.Y., 1965, 152 pp.
55. McIldowie, G., "The Construction of the Silent Valley Reservoir, Belfast Water Supply," *Minutes of Proceedings,* Institution of Civil Engineers, London, England, Vol. 239, 1935, p. 465.
56. Marsh, G. P., (1874), *The Earth as Modified by Human Action,* 2nd. ed., Scribner Armstrong and Co., New York, N.Y., 1874, 656 pp., see p. 141.
57. Matheson, D. S., and Thomson, S., "Geological Implications of Valley Rebound," *Canadian Journal of Earth Sciences,* Vol. 10, No. 6, June, 1973, p. 961.
58. Matula, M., "Inziniersko-Geologiche Skusenosti z Vystavby Vodneho Diela Pri Nosiciach na Vahu," *Acta Geologia et Geographica Universitatis Comenianae,* Commenuis University, Bratislava, Czechoslovakia, No. 7, 1961, p. 39.
59. Misiaszak, E. T., "Engineering Properties of Champaign-Urbana Subsoils," thesis presented to the University of Illinois, at Urbana-Champaign, Ill., in 1960, in partial fulfillment of the requirement for the degree of Doctor of Philosophy.
60. Mitchell, J. K., *Soils In Soil Engineering,* John Wiley and Sons, Inc., New York, N.Y., 1976, 422 pp.
61. Moye, D. G., "Engineering Geology of the Snowy Mountain Scheme," *Journal of the Institution of Engineers,* Sydney, Australia, Vol. 27, 1955, p. 287.
62. Moye, D. G., "Rock Mechanics in the Investigation and Construction of T1 Underground Power Station, Snowy Mountains, Australia," *Case History Series No. 3,* Geological Society of America, 1959, p. 13.
63. Murray, R. C., and Tedrow, J. C., *Forensic Geology,* Rutgers University Press, New Brunswick, N.J., 1975, 217 pp.
64. Nelson, J. I., "St. Lawrence Blues," Letter in *Engineering News-Record,* Sept. 5, 1957, pp. 10 and 16.

65. Newberry, J., "Engineering Geology in the Investigation and Construction of the Batang Padang Hydroelectric Scheme, M1 Malaysia," *Quarterly Journal of Engineering Geology*, Vol. 3, 1970, p. 151.
66. Nicol, T. B. (1964), "Warragamba Dam," *Proceedings*, Institution of Civil Engineers, London, England, Vol. 27, No. 3, Mar., p. 491; see p. 542.
67. Nowson, W. J. R., "The History and Construction of the Foundations of the Asia Insurance Building, Singapore," *Proceedings*, Instution of Civil Engineers, London, England, Vol. 3, No. 4, July, 1954, Part I, p. 407.
68. Page, B., "Geology of the Broadway Tunnel, Berkeley Hills, California," *Economic Geology*, Vol. 45, No. 2, Mar., Apr., 1950, p. 142.
69. Parsons, J. D., (1976), "New York's Glacial Lake Formation of Varved Silt and Clay," *Journal of the Geotechnical Engineering Division*, Vol. 102, No. GT6, Proc. Paper 12118, June, 1976, pp. 605–638.
70. Parsons, W. B., "Borings on Broadway, New York," *Transactions*, ASCE, Vol. XXVIII, 1893, p. 13.
71. Peck, R. B., (1973), "Presidential Address," *Proceedings*, 8th International Conference on Soil Mechanics and Foundation Engineering, Moscow, U.S.S.R., Vol. 4, 1973, p. 156.
72. Penner, E., Gillott, J. E., and Eden, W. J., Investigation of Heave in Billings Shale by Mineralogical and Biogeochemical Methods," *Canadian Geotechnical Journal*, Ottawa, Canada, Vol. 7, 1970, p. 333.
73. Peterson, R., "Rebound in the Bearpaw Shale," *Bulletin*, Geological Society of America, Vol. 69, No. 9, Sept., 1958, p. 113.
74. Philbrick, S., "Kinzua Dam and the Glacial Foreground," *Geomorphology and Engineering*, D. R. Coates ed., Dowden, Hutchinson and Ross, Stroudsburg, Pa., 360 pp.; see p. 175.
75. Proctor, C. S., "The Foundations of the San Francisco Oakland Bay Bridge," *Proceedings*, First International Conference on Soil Mechanics and Foundation Engineering, Cambridge, Mass., Vol. 3, 1936, p. 183.
76. Rominger, J. F., and Rutledge, P. C., "Use of Soil Mechanics Data in Correlation and Interpretation of Lake Agassiz Sediments," *Journal of Geology*, Vol. 60, No. 2, Mar., 1954, p. 160.
77. Sbar, M. L., and Sykes, L. R., "Contemporary Compressive Stress and Seismicity in Eastern North America; an Example of Intra-Plate Tectonics," *Bulletin*, Geological Society of America, Vol. 84, No. 6, June, 1973, p. 1861.
78. Scheele, W. E., "Fossil Dig," *The Explorer*, Vol. 7, No. 3, 1965, p. 5; and articles in succeeding issues.
79. Skempton, A. W., "Notes on the Compressibility of Clays," *Quarterly Journal of the Geological Society*, London, England, Vol. 100, 1944, p. 119.
80. Skempton, A. W., "Soil Mechanics in Relation to Geology," *Proceedings*, Yorkshire Geological Society, Vol. 29, 1953, p. 33.
81. Sowers, G. B., and Sowers, G. F., *Introductory Soil Mechanics and Foundations*, 3rd ed., MacMillan Co., New York, N.Y., 1970, 556 pp.
82. Spence Taylor, R., (1971), "*Atlas of Coal Mine Workings of the Edmonton Area*," privately printed, Edmonton, Canada, 1971, 33 pp.
83. "St. Lawrence Seaway and Power Project," News Report in *Engineering News-Record*, Oct. 13, 1955; p. 31, June 13, 1957; p. 32, Apr. 18, 1957; p. 25, May 15, 1958.
84. Sherard, J. L., Cluff, L. S., and Allen, C. A., (1974), "Potentially Active Faults in Dam Foundations," *Geotechnique*, London, England, Vol. 24, 1974, p. 367.
85. Stevenson, P. C., and Moore, W. R., "A Logical Loop for the Geological Investigation of Dam Sites," *Quarterly Journal of Engineering Geology*, London, England, Vol. 9, 1976, p. 65.
86. Terzaghi, K., "Relation between Soil Mechanics and Foundation Engineering," *Proceedings*, Presidential address to First International Conference on Soil Mechanics and Foundation Engineering, Cambridge, Mass., Vol. 3, 1936, pp. 13–18.
87. Terzaghi, K., "Soil Mechanics—A New Chapter in Engineering Science," *Journal*, Institution of Civil Engineers, London, England, Vol. 12, 1939, p. 106.
88. Terzaghi, K., "Ends and Means in Soil Mechanics," *Engineering Journal*, Montreal, Canada, Vol. 27, No. 12, Dec., 1944, p. 608.
89. Terzaghi, K., Presidential Address to 2nd International Conference on Soil Mechanics

and Foundation Engineering, Rotterdam, The Netherlands, *Proceedings,* Vol. 5, 1948, p. 14.
90. Terzaghi, K., "The Influence of Modern Soil Studies on the Design and Construction of Foundations," *Proceedings,* Building Research Congress, London, England, 1951, Vol. 1, p. 139.
91. Terzaghi, K., Address at E.T.H., Zurich, *Proceedings,* 3rd International Conference on Soil Mechanics and Foundation Engineering, Zurich, Switzerland, Vol. 3, 1953, p. 78.
92. Terzaghi, K., "Influence of Geological Factors on the Engineering Properties of Sediments," *Economic Geology,* 50th Anniversary Vol., Part II, 1955, p. 557.
93. Terzaghi, K., Presidential Address at the 4th International Conference on Soil Mechanics and Foundation Engineering, London, England, Vol. 3, 1957, p. 55.
94. Terzaghi, K., "Design and Performance of the Sasamua Dam," *Proceedings,* Institution of Civil Engineers, London, England, Vol. 9, 1958, p. 369.
95. Terzaghi, K., "Storage Dam Founded on Landslide Debris," *Journal of the Boston Society of Civil Engineers,* Vol. 47, No. 1, Jan., 1960, p. 64.
96. Terzaghi, K., and Peck, R. B., *Soil Mechanics in Engineering Practice,* John Wiley, New York, N.Y., 1948, 566 pp; see p. 285.
97. Terzaghi, K., and Lacroix, Y., "The Mission Dam," *Geotechnique,* London, England, Vol. 14, 1964, p. 13.
98. Thomson, S., and Yacyshyn, R., "Slope Instability in the City of Edmonton," *Canadian Geotechnical Journal,* Ottawa, Canada, Vol. 14, 1977, p. 1.
99. Thompson, L. J., and Tannebaum, R. J., "Responsibility for Trenching and Excavation Design," *Journal of the Geotechnical Engineering Division,* ASCE, Vol. 103, No. GT4, Proc. Paper 12882, Apr., 1977, pp. 327–338.
100. Thorney, J. H., Spencer, G. B., and Albin, P., "Mexico's Palace of Fine Arts Settles 10 ft.," *Civil Engineering,* Vol. 25, No. 6, June, 1955, p. 357.
101. Thornthwaite, C. W., "An Approach Towards a Rational Classification of Climate," *Geographical Review,* Vol. 38, 1948, p. 55.
102. Upson, J. E., Leopold, E. B., and Rubin, M., "Postglacial Change of Sealevel in New Haven Harbour, Connecticut," *American Journal of Science,* Vol. 262, No. 1, Jan., 1964, p. 121.
103. Waddell, J. A. L., *Bridge Engineering,* Vol. 2, 1916, p. 1544, John Wiley and Sons, Inc., New York, N.Y.
104. "Welland Ship Canal IV," *Engineering,* London, England, Vol. 128, 16, Aug., 1929, p. 192.
105. White, G. W., "Engineering Implications of Stratigraphy of Glacial Deposits," *Proceedings,* 24th International Geological Congress, Montreal, Canada, Section 13, 1972, p. 76.
106. White, G. W., (1974), "Buried Glacial Geomorphology," *Glacial Geomorphology,* D. R. Coates, ed., State University of N.Y., Binghampton, 1974, 398 pp.; see p. 331.
107. White, O. L., Karrow, P. F., and MacDonald, J. R., "Residual Stress Relief Phenomena in Southern Ontario," *Proceedings,* 9th Canadian Rock Mechanics Symposium, Montreal, Canada, 1973, p. 323.
108. Williams, J. H., and Vineyard, J. D., "Geological Indicators of Catastrophic Collapse in Karst Terrain in Missouri," Transportation Research Record, No. 612, 1976, p. 31.
109. Wilson, A. E., "Geology of the Ottawa-St. Lawrence Lowland, Ontario and Quebec," *Memoir 241,* Geological Survey of Canada, Ottawa, Canada, 65 pp.
110. Zaruba, Q., and Mencl, V., *Landslides and their Control,* Elsevier, London, England, 1969, and Prague, 205 pp.
111. Zeevaert, L., "Foundation Design and Behaviour of Tower Latino Americana in Mexico City," *Geotechnique,* London, England, Vol. 7, No. 3, Sept., 1957, p. 115.

# GEOLOGY AND GEOTECHNICAL ENGINEERING
# (THE THIRTEENTH TERZAGHI LECTURE)[a]

**Discussion by Richard J. Proctor[2]**

This is both a tribute and a discussion—one the writer felt he had to write after reading Legget's superb paper. Those fortunate enough to know the author personally will immediately recognize his easy writing style and his marvelous command of English. His works read like novels. If you have not read the article, try to find the time. And if you want more enjoyment, browse through the author's book *Cities and Geology* (44).

A more worthy honoree would be difficult to find to present a Terzaghi Lecture. Karl Terzaghi and Robert Legget were friends; both were educated primarily as engineers, yet both had strong backgrounds in geology producing that all too infrequent and unique blend of engineer and scientist.

Legget's article contains brief, but wellchosen, case histories and discussions on the following topics: dams, bridges, tunnels, buildings, urban geology, test-boring Programs, ground water, and, under the heading Some Geological Hazards, are included: faults, karstic regions, weathering, boulders, till, and residual stresses.

In the context that the writer is a geologist who has worked with civil engineers for more than 20 yr, he wants to amplify only one aspect of the author's paper: Cooperation between engineers and geologists.

Legget points out the continual need for the professions of engineering and geology to merge their specialties to assure a safer and more economical end product. We know that this has not always been the case, as witness the St. Francis Dam failure in 1928, owing to lack of any geologic investigation, and, more recently, the Teton Dam failure owing largely to lack of communication between the field geologists and engineers and the design personnel at the headquarters office.

The importance of *geologist/engineer teamwork* has been the theme of many articles by both engineers and geologists. One of the earliest was by the "Father

---
[a] March, 1979, Robert F. Leggett (Proc. Paper 14444).
[2] Pres., Association of Engrg. Geologists, Arcadia, Calif.

of Engineering Geology" Charles P. Berkey (112), entitled "Responsibilities of the Geologist in Engineering Projects." Karl Terzaghi continually spoke and wrote of the importance of one profession to the other. Ralph Peck (71) succinctly put it: "We can never sucessfully divorce our thinking from the overwhelming influence of geology in our works." Other thought-provoking messages on cooperation and the often subtle differences between the two professions are worth quoting here:

> When geologists are called in to advise upon civil engineering work, they will have to act in conjunction with the engineers responsible for the work. Thus arises the need for cooperation between the civil engineer and the geologist, the practical builder and the man of science. Their cooperation may lead to a valuable partnership, and it often proves to be of considerable personal pleasure. This partnership is, in some ways, a union of opposites, for even the approach of the two to the same problem is psychologically different. The geologist analyzes conditions as he finds them; the engineer considers how he can change existing conditions so that they will suit his plans. From his analysis, the geologist cites problems that exist and suggests troubles that may arise; the engineer has to solve the problems and overcome the troubles. (115)

> . . . there has been a steadily increasing improvement in safety and economy in engineering planning and construction with the increasing utilization of professional geologic services. This improvement will continue as the geologist becomes more and more cognizant of his functions and responsibilities in the engineering organization and learns more and more about how his knowledge and discoveries dovetail with the practical problems of the engineer. (114)

> Obviously, a geologist to fit into such a group must have the same hard-headed practical drive toward the essential facts that characterizes a group of engineers . . . Only those men who approach their responsibilities in due humbleness of spirit and with transparent sincerity in the search of truth can gain the confidence of engineers whose best laid plans they may wreck by the discovery of some unexpected geologic difficulty. The passive and uncritical geologist may, however, be led by the enthusiasm of the group to view the natural difficulties with too much optimism and, therefore, fail for the very function for which he is hired.

A final quote from the author's Terzaghi Lecture sums the point (p. 343):

> Every structure built by man depends for its safety and stability upon the ground on which or in which it is built; upon geology.

### APPENDIX.—REFERENCES

112. Berkey, C. P., "Responsibilities of the Geologist in Engineering Projects," *Technical Publication No. 215,* American Institute of Mining and Metallurgical Engineers, 1929.
113. Bryan, K., "Problems Involved in the Geologic Examination of Sites for Dams," *Technical Publication No. 215,* American Institute of Mining and Metallurgical Engineers, 1929.

114. Burwell, E. B., Jr., and Roberts, G. D., "The Geologist in the Engineering Organization," *Application of Geology to Engineering Practice* (Berkey Volume), Geological Society of America, 1950.
115. Legget, R. F., *Geology and Engineering*, 2nd ed., McGraw-Hill Book Co., Inc., New York, N.Y., 1962.

## Discussion by Ronald J. Tanenbaum,[3] A. M. ASCE

Legget is to be congratulated as the conscience of the geotechnical engineering profession. The writer is in total agreement with the author's thesis concerning the critical role of fundamental geology in geotechnical engineering and the absolute necessity of a thorough background documentary study prior to planning and initiating a subsurface exploration program. However, the comments made by the author on p. 343 with reference to a paper (99) coauthored by the writer and Louis J. Thompson depicts a total misunderstanding of the content of that paper, thus requiring further discussion, since the misconception appears to pervade the minds of many other engineers.

It is obvious to all that, given the data from two borings, someone is going to attempt to interpret the subsurface conditions between them. Without this interpretation, it is virtually impossible to design foundations, determine cut and fill quantities, or perform any of the other numerous functions required in the design and construction of a project. All practicing engineers realize this.

A sane person does not seek the diagnosis of a plumber or biologist for his chest pains. He seeks a specialist. When it comes to interpreting subsurface conditions, the geotechnical engineer or engineering geologist is the specialist. The contractor might be classified as the biologist and the owner as the plumber. If someone must make an interpretation, or diagnosis if you wish (and certainly an interpretation must, or at least will be made), then logic dictates that it be made by a specialist, not a plumber.

The problem obviously lies with the meaning of interpretation. Making an interpretation does not imply precise prediction of subsurface conditions but rather an *educated* guess tempered by experience and sound engineering judgment combined with a thorough knowledge of fundamental geologic and geomorphic principles, specifically documentary data related to the site in question. Neither does interpretation imply a neglect of geology as inferred by the author on p. 344. Rather, it is the neglect of geology that prohibits interpretation.

When the author suggested that there is no certainty about geological conditions between adjacent borings even 5 ft apart, he is absolutely correct. It is for this very reason that the writer supports the inclusion of changed condition clauses in every contract, combined with adequate inspection of the work to identify, in a timely manner, those changed conditions.

The author suggests that the only way to ascertain subsurface conditions for a trench is to excavate the trench first. This would make sense if engineers designed each and every structure using 20/20 hindsight. In other words, before

---

[3] Asst. Prof., Dept. Civ. Engrg., Univ. of Pittsburgh, Pittsburgh, and Staff Engr., GAI Consultants Inc., Monroeville, Pa.

designing a building or dam foundation, all of the soil or rock beneath the structure should be excavated so that an interpretation can be made.

This concept is an obvious fallacy. The objective of a subsurface investigation is to provide the *best* interpretation of subsurface conditions and engineering properties while minimizing disturbance to in-situ conditions. If engineers were to adhere to the 20/20 hindsight hypothesis, then all subsurface and in-situ investigative techniques would be discarded and design based on pure guesswork. Such a hypothesis is unacceptable to the writer.

The author, on p. 343, also appears to object to the writer's belief that all subsurface information should be made available to contractors prior to bidding. It is interesting to note that on pp. 351 and 352 he reverses himself by citing, as an example of an excellent site investigation, the Toronto Subway where full disclosure of this information prior to bidding proved to be the most efficient and economical way of doing business.

The writer's objections to only a small portion of the author's paper should not be construed as a condemnation of the total effort. As a practicing registered professional engineer and educator in engineering geology and geotechnical engineering, the writer appreciates the author's refreshing reminder of the professional attitude that all engineers should maintain in their performance of site investigations. The Thirteenth Terzaghi Lecture should be required reading of all geotechnical engineers in the hope that it will help swing the pendulum away from the practice of geotechnical engineering as a business and back to the practice of geotechnical engineering as a profession, where it belongs.

## Closure by Robert F. Legget,[4] Hon. M. ASCE

The contributions of the two discussers are appreciated. Their commendation, and comments received privately but not submitted as formal discussions, encourages the writer to think that possibly it was timely to remind the geotechnical community of Terzaghi's guide lines.

All readers of the paper by Thompson and Tanenbaum (99), including the writer, will appreciate Tanenbaum's explanation of what his coauthor and he meant to say in their paper and in the closure to its discussion.

The writer cannot find on page 343 any words that either say or imply that all subsurface information available in advance of contracts should not be given to bidders. On the contrary, the writer has consistently advocated the fullest possible disclosure of all available subsurface information to bidders ever since his first publication in this field in 1939 (115).

Omission of the *Canadian Geotechnical Journal* in the comparative review of periodical geotechnical literature has been questioned. Those who know this Journal will appreciate the diffidence of the writer, himself a Canadian, in making what might have been thought to be an invidious comparison. If those who do not know the Canadian journal care to look at any recent issue, they will see at least an approach to that balance of practical and theoretical papers that the writer advocates.

---

[4]Consultant; formerly, Dir., Div. of Building Research, National Research Council of Canada, Ottawa, Canada.

Support for the view that more practice-oriented papers are desirable was endorsed in the pages of *Civil Engineering* by Appleton (116) and, more recently and more generally, by President Blessey (117). The writer trusts that positive response to these appeals will be reflected in the development of this Journal of the ASCE in particular, in keeping with the firmly held views of Terzaghi.

## Appendix.—References

116. Appleton, J. H., "More Practice-Oriented Papers in ASCE Journals?" *Civil Engineering*, Vol. 48, No. 6, June, 1978, p. 127.
117. Blessey, W. E., "An Open Letter from the President," *Civil Engineering*, Vol. 49, No. 10, Oct., 1979, p. 82.

# JOURNAL OF THE GEOTECHNICAL ENGINEERING DIVISION

## THE FIFTEENTH TERZAGHI LECTURE

Presented at the American Society of Civil Engineers Annual Convention and Exposition, Atlanta, Georgia

October 25, 1979

GEORGE F. SOWERS

## Introduction of Fifteenth Terzaghi Lecture
### By William F. Swiger

George Sowers has always been a civil engineer. As a child he accompanied his father, George B. Sowers, F. ASCE, on inspections of tunnel, foundation, and port construction. His undergraduate education was at Case Institute of Technology; his graduate work was at Harvard where he was a student of Karl Terzaghi and Arthur Casagrande.

George worked as a hydraulic and hydrologic engineer with the United States TVA and served in the United States Navy during WW II.

Since 1947 he has pursued a dual career: teacher and consultant, presently Regents Professor of Civil Engineering at the Georgia Institute of Technology, he pioneered geotechnical engineering education in the southeast. His textbook Introductory Soil Mechanics and Foundations is in its fourth United States edition, plus Spanish, international, and even pirated Taiwan editions. In addition, he has published more than 100 technical publications.

George is a cofounder of Law Engineering, one of the largest geotechnical environmental and materials engineering firms of the United States. He has been a consultant on geotechnical design throughout the United States and in many foreign countries.

He has been President of the Georgia Section ASCE and Chairman of the Geotechnical Engineering Division, ASCE. Presently he is the Vice President for North America of the ISSMFE.

He is married; his wife, Frances, has practiced hydraulic engineering. They have four children, two of whom are geologists.

# THERE WERE GIANTS ON THE EARTH IN THOSE DAYS

By George F. Sowers,[1] F. ASCE

## SYNOPSIS

Many examples of pre-Columbian earth construction have been uncovered in the Americas. Some reflect well-organized technologies comparable to those of the Old World of the same time period. Earth fills, stone and adobe structures, highways, and water management systems exhibit both well-conceived designs and organized effort for construction.

A few investigators speculate that such sophisticated works represent unrecorded technology transfer from the Old World; some pseudohistorians suggest visits from outer space. However, the evidence demonstrates that indigenous, ingenious responses to local social and economic problems and organized community efforts were responsible. Lack of both a formal logic-mathematics system and written communication inhibited synergistic development of the technologies. They were vulnerable to decay and were virtually obliterated by conquest.

## INTRODUCTION

Flying high above the earth, about 370 km south-southeast of Lima, Peru, and crossing a broad, arid valley 50 km from the sea, one can see patterns in the valley floor below. Even the earth resources satellites, 950 km (570 miles) high, can discern straight lines, some several kilometers or miles long. At lower altitudes, the lines become more distinct, up to 50 m wide, in strange criss-cross patterns on the 60 km (35 mile) long, 2 km (nearly a mile) wide braided-plain terrace 500 m (1,600 ft) above sea level.

---

[1] Regents Prof. of Civ. Engrg., Georgia Inst. of Tech., Atlanta, Ga. 30332; also Sr. Vice Pres., Law Engrg. Testing Co., 2749 Delk Road, S.E., Marietta, Ga. 30067.

Note.—Discussion open until September 1, 1981. To extend the closing date one month, a written request must be filed with the Manager of Technical and Professional Publications, ASCE. Manuscript was submitted for review for possible publication on February 5, 1980. This paper is part of the Journal of the Geotechnical Engineering Division, Proceedings of the American Society of Civil Engineers, ©ASCE, Vol. 107, No. GT4, April, 1981. ISSN 0093-6405/81/0004-0385/$01.00.

There is a cursory resemblance to a runway that one can identify approaching at low altitude, 600 m (2,000 ft) above the surface (Fig. 1). The Swiss science-fiction author, Eric Von Daniken (3), in *Chariots of the Gods,* asks, "What is wrong with the idea that the lines were laid out to say to the gods, 'Land here! Everything has been prepared as you ordered!' . . . What other purpose could they have served?" (Although Von Daniken's ideas are not taken seriously by archeologists, he has an amazingly large audience.)

With a closer look, the illusion of a landing field is dispelled by the criss-cross patterns. Some lines are oriented at random while others radiate from a common center like the points of a compass. Scattered among the straight lines are curves. Some are geometric, but most are cartoon-like outlines of familiar creatures, such as a monkey, a spider, or a hummingbird (Fig. 2). All are about 75 m–100 m across.

**FIG. 1.—Nazca Lines of Southwest Peru from 600 m (2,000 ft) above Ground**

Still closer, their construction is seen to be simple. The gravel-sized stones of desert pavement are covered with a dark brown coating, i.e., desert varnish (a dark brown, shiny coating of iron oxide and silica brought to the exposed surface of bed rock and gravel by capillarity and deposited by evaporation in a severe desert environment). When the stones are moved aside or overturned, the unexposed surface or light-colored protected sand is exposed. The patterns are merely strips of desert from which the pavement has been moved aside forming little brown ridges that outline the light colored lines and figures. A brief experiment suggests that a person can clear 1 $m^2$ in 10 min. Even one of the larger lines, 30 m wide and 300 m long, could be constructed in 100 man-days–200 man-days. Archaeologists are still debating the meaning of the lines (12,17). Most are oriented with prominent stars, the moon, or the sun at special times of the year. All probably have a religious significance. Considering the time that is required for desert varnish to develop, the figures are probably 500 yr–1,500 yr old. This is compatible with archaeological evidence of the pre-Inca civilization that inhabited this valley. It agrees with carbon-14 dating of wood stakes at the ends of the lines that were used to establish alignment by sighting (12).

These lines and figures are not unique. There are figures outlined in the southern California desert that are remarkably similar. High on a mountain ridge in northern Wyoming, there is outlined a lopsided wheel about 22 m (75 ft) diam made of light colored limestone cobbles and small boulders that contrast with the green grass. Because some of the spokes are aligned with sunrise at the vernal equinox, it also appears to have an astronomical significance.

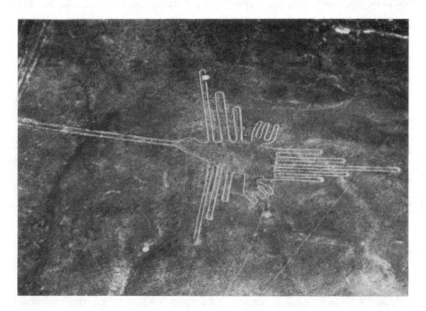

FIG. 2.—Nazca Humming Bird, 100 m (330 ft) Wide

The Indians of a century ago knew that it was sacred but were ignorant of its meaning or of who built it.

## Tradition of Western (European) Technology

The development of western (European) technology is reasonably well documented. First, there was the empirical or trial-and-error approach that marked the craftsmen of the Middle East and Egypt. Later, the Greeks added geometry and a formal logic for the behavior of physical objects. The Romans combined these with their genius for organizing and communicating throughout their empire. The Arabs added innovations in mathematics. The scientists and engineers of the Renaissance and Industrial Revolution resurrected these concepts from their suppression during the Dark Ages to establish the framework for engineering as we know it.

The members of the American Society of Civil Engineers are products of that tradition. With this background of empiricism, keen observation, logic, and mathematics, such a genius as Karl Terzaghi was able to conceive imaginative and innovative solutions to problems in soil and rock that had confounded engineers for milleniums. The engineering structures of this century are a testimony to this tradition.

It is sobering, however, to realize that until the twentieth century the largest structure known to western technology was the Pyramid of Cheops, built 4500 years ago. It was conceived and constructed without two of the tools of modern technology: Logic and formal mathematics. One contemporary religious writer suggests that it was built by Gods from other worlds—the Egyptians could not have undertaken such a feat. His idea is not new. The ancient Greeks attributed the large polygonal stones of the walls of Mycenae to the Cyclops—mythical one-eyed giants. Our term "cyclopean masonry," walls of very large boulders, is derived from their tradition. More than 3,000 years ago, Hebrew herdsmen explained monuments of giant stones that they found in the desert by exclaiming, "There were giants on the earth in those days" (8).

The traditions of the pre-Columbian Americas are not well documented. The early Spanish conquistadors found flourishing civilizations in Mexico and Peru, but destroyed many of the native works in a passion for gold and religion (4,10). Remains of some older, dying, or dead civilizations were gutted for building materials or looted by souvenir seekers.

The construction relics are still being uncovered; lost cities are coming to light in the jungles of Yucatan, Central America and the Andes of South America. Their scale is staggering. What was probably the largest structure in the world, the Pyramid of Tepanapa, 1.5 times the volume of the Pyramid of Cheops, is yet to be fully excavated. Countless other structural works on both continents are only now being analyzed by engineers.

It is the purpose of this paper to examine pre-Columbian structures and public works from the civil engineering point of view. It considers their features, their geographic and chronologic development, and presents hypotheses for their conception and construction. It explores whether they were conceived by occult forces, such as visitors from outer space (as suggested by Von Daniken) (3), inspired by lost mariners and early explorers from Europe and Africa who never reported back to the old world, or were the products of independent innovation. Most important, it relates the development and decline of pre-Columbian technology to the evolution of engineering today.

## Pits and Heaps

The earliest construction was a hole in the ground, either for disposal of excrement or burial of the dead. Both practices are common (but not universal) thoughout the world. The need may have been sanitation, or possibly the religious idea that both excrement and corpses have magic, malevolent power that must be kept from view. On the other hand, the Mayan farmers of Meso-America and the Creek Indians of the Southeastern United States buried their dead below the floors of their houses so that their spirits could rise and guide the future.

Where hard soil made excavation difficult, heaps of stones substituted for burial. An eskimo grave in Greenland is a crude cairn of small boulders over hard permafrost.

Both shallow pits and heaps of stones were used for shelter. The early Indians of the Mesa Verde area (Southwest Colorado, about 500 A.D.) found the ground was warmer than the outside air in winter and cooler in summer. They excavated elongated pits about 1 m deep and 3 m–6 m diam; where the soil was loose, the holes were lined with stones. Above ground, low rubble stone or laced twigs formed walls; poles, twigs, and grass formed a roof. Similar pit houses are found along the north coast of Peru, dated about 3000 B.C.

The more advanced Indian civilization of the American Southwest retained the pit house in the kiva, Fig. 3. This is a stonelined pit, about 3 m–5 m diam and 3 m–4 m deep with a roof of logs and earth. The kiva served as a fraternal-clan retreat for the men, isolated from the bustle of daily work. Although the details vary, most included a stone-lined fresh air duct at the base, a draft deflector at the inlet, a raised platform for a fire, and a small hole in the floor that served as a symbolic connection to the depths of the

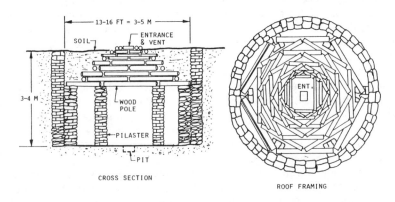

FIG. 3.—Details of Small Kiva, Mesa Verde, Southwest Colorado

earth. Kivas, some below ground, but others above (where the soil is shallow), play a vital part in the life of the Pueblo Indians today.

The Creek Indians also built earth covered ceremonial or clan lodges with provisions for a fire and a small floor hole. These date between 500 A.D. and 1200 A.D.

These excavations possibly reflect the instinctive feeling of security that a child exhibits when he retreats to a corner when under duress, or the dugouts made by older children to escape the adult world. The pit is not limited to primitive societies or children; the United States defense headquarters is far underground in the Colorado mountains.

## Mounds

Large heaps of earth and piles of stones were constructed by men in all parts of the world except Australia. Until the twentieth century, they were the largest structures ever built.

Trash heaps are one form. Despite the romantic fantasies of historians, some American Indian tribes merely dumped their trash, garbage, broken pottery, and fecal wastes in irregular piles immediately adjacent to their permanent camps. Refuse mounds up to 4 m (13 ft) high of shells, bones, and broken stone utensils from about 300 B.C. are found along the sea coast in Peru. Similar shell mounds occur along the South Atlantic and Florida Coasts in the United States dating from about 2000 B.C. Some became mounds as high as 6 m (20 ft). Although they are now rich sources for archeologists, they must have bred flies and rodents, and exuded rich odors when in use.

Burial mounds, constructed between 800 B.C. and about 150 A.D. by Woodland Indians, are widely distributed in the Eastern United States. Many are from 3 m–10 m (10 ft–33 ft) high with side slopes typically $30°–35°$ from the horizontal and have round tops. Most appear to have been built during a relatively short

FIG. 4(a).—Etowah Mound, Cartersville, Ga.

time period, representing only a generation or two, based on a comparison of possessions that accompanied the burials and on very limited carbon-14 dating. Long-term use of a site produced a succession of mounds instead of a single, growing burial. Burial mounds are also found in the coastal settlements of South America, dating from about 500 B.C. (11).

**Platform Mounds.**—Platform mounds for worship ceremonies or bases of temples appeared in the Eastern United States about 700 A.D. Similar structures in Meso-America, such as Tikal in Guatemala, date as early as 500 B.C. (1,2). Platform mounds, built by filling over old pit and wall houses with soil and debris are found on the north and central coast of Peru (11). The debris associated with them dates about 2000 B.C. by carbon-14. The mounds were constructed of local materials with uniformly sloping sides and flat tops. Most were built in stages over a long period of time, eventually becoming the largest structures of the pre-Columbian era.

The Etowah group, near Cartersville, Ga., 35 miles north-northwest of Atlanta, were built on a low terrace on the right bank of the Etowah River. It consists of three mounds, a village area, and a 1.3 m (4 ft) deep perimeter ditch enclosing a circular area about 450 m (1,500 ft) diam. The smallest mound has been completely excavated. It was constructed in stages between about 700 A.D. and 1200 A.D. and consists of a series of concentric truncated square pyramids with side slopes of 33°–37° with the horizontal. The top was flat and was similar in size for each stage. On each top was a simple wood temple with walls of vertical poles set in sockets in the ground. The bark-stained sockets remain and mark each platform level. At the death of a chief or priest, marking the end of a period of unified leadership, the wood structure was destroyed, and the leader was buried in the mound with weapons, jewels, and vestments. A new platform and temple were then constructed 1 m–3 m (3 ft–10 ft) above

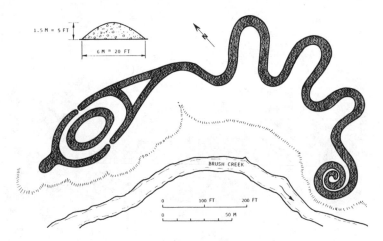

FIG. 4(b).—Serpent Mound, Southwest Ohio: Plan

the earlier one. The stages in construction are reflected in different soils and slightly different artifacts.

The largest mound of the Etowah group, Fig. 4(a), contains about 110,000 $m^3$ or 150,000 cu yd of soil. It has not been excavated. There are remains of a wood temple or council house on top. It appears to have been built in the same stage construction as the smallest mound.

Monks Mound of the Cahokia group, adjacent to a small creek about 25 miles east of St. Louis is the largest in the eastern United States. It is nearly 30 m or 90 ft high with 25° side slopes. Soil borings with undisturbed samples show it was built in about 14 stages, commencing about 900 A.D. and ending about 1150 A.D. (16). The stages range in thickness between 1 m (3 ft) and 5 m (17 ft). The total fill volume is about 750,000 $m^3$ or 1,000,000 cu yd.

**Protective Mounds.**—A few mounds are in the form of continuous embankments enclosing the remains of villages and ceremonial centers. At "Fort Ancient," near Cincinnati, Ohio, a perimeter mound 6 km (3.5 miles) long encloses a

40 ha (100 acre) segment of bluff overlooking a river. The mound varies from 3 m–5 m (10 ft–16 ft) high with side slopes of 30°–40°. Other fort-like or stockade mounds have been found in the upper Mississippi and Ohio River area.

**Effigy Mounds.**—A number of mounds are shaped like animals. Most are in southern Wisconsin or northeastern Iowa, although a few are scattered throughout the eastern United States. Most are low, 1 m–3 m (3 ft–10 ft) high, and cover areas 10 m to more than 100 m (33 ft to more than 330 ft) diam. Animals, such as the panther, bear, deer, buffalo, turtle, beaver, and dog predominate, but birds are also found. Most are on bluffs above rivers and are associated with small burial mounds. Those excavated appear to have been built in a single stage. Although there are some associated burials and nearby burial mounds, they appear to be clan symbols or ceremonial structures.

The Serpent Mound of southeastern Ohio, Fig. 4b and 4c is the largest and

FIG. 4(c).—Aerial View of Serpent Mound

best known of the effigy mounds in the United States. It lies on a bluff overlooking a small river. It is made of gravelly clay, following a pattern laid out with a row of stones and ashes after removing the topsoil. The approximate volume is 2,000 $m^3$ or 2,500 cu yd.

**Mound Construction.**—Those mounds that have been excavated or core drilled all exhibit evidence of similar construction. They were placed in level layers 2 cm–8 cm or 1 in.–3 in. thick. In some cases the edges of the layers were confined by logs secured by stakes: simple forms. These forms were removed and reset as the work progressed, but occasionally left to rot in place as shown by remains of bark and rotted wood. Side slopes were typically 33°–37° with the horizontal with no obvious correlation between height, soil texture, and slope. Although the sides were usually dressed to a common straight slope, the only protection was vegetation which developed naturally. The occasional eroded gully was sometimes filled with stones.

The tools for construction have been pieced together from fragments remaining in the mounds. The soil was loosened with a pointed digging stick, hardened by fire. The same sticks were also used for planting corn. I have seen them in use today in remote rural Mexico, Honduras, Peru, and Bolivia. Excavation was by hoes made from shoulder blades of deer lashed to poles. The earth was carried in baskets or leather bags on the backs of the workers, about 20 kg or 45 lb at a time.

Experiments by various investigators, cited by Erasmus (6) suggest that a team of one digger and three basket carriers could excavate 3 m$^3$ (4 cu yd) and move and place the soil 200 m (650 ft) from the source in one day. At this rate, the largest Etowah mound required about 100,000 man-days and the largest Cahokia mound 1,000,000 man-days. The Incas of South America were required to devote about 90 days/yr to community labor. Reed et al. (16) estimate that the North American Indian may have contributed 40 days. At this rate, Etowah required 2500 man-years and Cahokia 25,000 man-years. Both were built over periods of about 250 years. Thus the required average annual labor force would have been only 10 persons and 100 persons, respectively. Cahokia, however, was built in 14 stages averaging 18 years (probably 14 generations), thus Cahokia required 1,800 man-years/generation. On the same basis Etowah required only 200. The large volumes are not unreasonable for the modest labor forces available, when the religious and secular leaders organized the people for the common effort.

## Platforms and Pyramids

In the southeastern United States the earth platform temple base possibly evolved from the burial mound. In Meso-America and Peru the temple base developed independently. The earliest were flat platforms of earth and boulders faced with walls of rubble masonry and topped with flat stones (11). On top were stone monoliths and remains of small buildings that probably were worship related. The earliest appeared about 1800 B.C. in Peru and Mexico.

During later periods, some of the platforms were enlarged by building new walls around them, and then raised in increments of 3 m–5 m (10 ft–16 ft). The platform became a series of terraces or a step pyramid, with very steep sides, and a level top which was the focus of religious rites.

Still later, the platforms were augmented by faces of rubble, adobe brick, and later of trimmed stone, battered at slopes of 60°–70° with the horizontal. These truncated pyramids first appeared about 1800 B.C. in Peru. Their construction climaxed with the Pyramid of the Sun at Pachamac, about 30 km (18 miles) south of Lima, between about 200 A.D. and 1000 A.D. The pyramid is of stones and adobe brick, about 220 m (700 ft) square, flat topped, and about 23 m (75 ft) high.

The largest earth-stone pyramids were constructed in the Valley of Mexico. At Cholula, about 50 km (30 miles) east of Mexico City is the Pyramid of Tepenapa. It is possibly the largest structure built before the 20th century, constructed between 350 A.D. and 600 A.D. The base is 402 m$^2$ (132 sq ft) and it is 70 m (231 ft) high. Although it is not as tall as the Pyramid of Cheops, its volume of 3,770,000 m³ or 5,000,000 cu yd is approximately 1.5 times greater (Fig. 5). At the time of the Spanish Conquest, there was a small temple on the flat top. It was replaced by a Roman Catholic Church which is still in use.

The pyramid is constructed of light-colored tuff boulders about 0.6 m (2 ft) diam in a matrix of dark volcanic ash and soil. Exploratory tunnels into the heart of the pyramid have found five (or possibly seven) smaller concentric pyramids with plastered faces. A painting of a giant grasshopper on one face is still grotesquely colorful.

Deep in the jungles of northern Guatemala, along the humid Gulf Coast of southern Mexico and in the tangled underbrush of the Yucatan Peninsula are the most magnificent structures of pre-Columbian North America, i.e., the Mayan temple pyramids (14,20). Commencing about 500 B.C. and ending between 1200 A.D. and 1400 A.D., complex religious centers resembling cities were constructed in the tropical lowlands. A wide variety of massive structures remain, many still choked by the jungle and awaiting archeologic study. Enough have been examined to demonstrate the evolution of advanced construction technology over a wide area (Fig. 6). Although each has its imposing structures, Tikal,

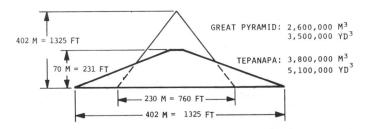

FIG. 5.—Comparison of Pyramid of Tepanapa, Cholula, Mexico with Pyramid of Cheops, Egypt

Guatemala represents one of the earliest and longest lived of these areas (1,2,20).

Temple I, Fig. 7 was completed about 700 A.D. near the climax of Tikal's 1000 yr history. It is 40 m² (130 sq ft) at the base, with the overall slope of its nine terraces 65°. It has a narrow truncated top, 30 m (100 ft) above the Plaza. On top is a small temple with a soaring crest or comb rising another 14 m (45 ft). Although other temple pyramids at Tikal are higher, and many in Yucatan, such as at Uxmal and Chichen Itza, are broader, Temple I remains an impressive example of the best Mayan construction. A stair leads up to the temple, so steep that one climbs it today with the aid of a chain. The pyramid surface is of well-fitted soft limestone blocks laid in courses.

Surface repair of the Tikal pyramids reveals that they were built of boulder-sized soft limestone rubble in a matrix of limestone chips and cemented with lime mortar. Tunnels into the pyramids reveal smaller, older pyramids within (1,2). Large carvings on the older pyramid surfaces, hidden by the larger, later enveloping construction, can be seen in exploration adits, excavated by archaeologists (Fig. 8).

The quarries from which the soft limestone blocks were cut and the rubble obtained are still in use, furnishing fresh blocks for restoration of the structures

(Fig. 9). Deep, narrow slots were first cut, using tools of chert or flint. The rock was then broken apart by prying or wedging in the slots.

## ADOBE

Adobe is a universal construction material used throughout the Americas as well as in the Old World. Technically it is unburned soil (clay) brick, but the term has been applied to other forms of dried soil construction.

Sun-dried brick is a form of artificial stone used in arid regions where local rock is difficult to shape or is absent. Based on data from sites in the western United States, a wide variety of soil materials have been used, the major

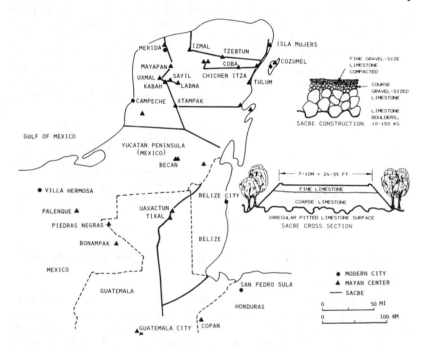

**FIG. 6.—Mayan Centers and Roads**

requirements are sufficient cohesion to provide strength when dry and enough inert filler such as sand, fine gravel, and even straw, grass, or twigs to minimize shrinkage. Based on unpublished tests by the writer, unconfined compressive strengths range from 100 kN/m$^2$–nearly 700 kN/m$^2$ (15 lb/sq in.–nearly 1000 lb/sq in.). The tensile strength is typically 1/6 the unconfined compressive strength. Adobe resembles weak, plain concrete in its strength and ability to absorb repeated strain and tension. It resists static loads in walls up to 15 m (50 ft) high but crumbles under the dynamic loads of earthquakes.

Adobe balls and cones were used as masonry units in coastal Peru as early as 1000 B.C. Rectangular adobe blocks appeared shortly afterwards sometimes interlayered with stones. The largest adobe brick structure in the Americas was the pyramid of Pachamac, mentioned previously (11).

The Pueblo Indians of the United States southwest used both adobe and stone for construction depending on material availability. The Casa Grande of Arizona was built between 1350 A.D. and 1400 A.D. as a part of a village compound of the Hohokam, whose descendents may be the Pima Indians (9,25). Lenticular pancakes, typically 0.35 $m^3$ (1.3 cu ft) diam of firm silty clay were laid in very irregular noncontinuous courses and level tiers, 1.2 m (4 ft) wide and 0.5 m–0.75 m (1.5 ft–2.5 ft) high. Each tier was shaped to make flat vertical surfaces and then allowed to dry before adding the next, producing cold joints in which hand prints are still visible. Although all the soils in the walls were obtained from the site, there are significant variations between tiers, reflecting the different calcium carbonate percentages in the soil horizons excavated. The insides of the walls (and possibly the outside) were plastered with adobe and then painted.

FIG. 7.—Temple I, Tikal, Guatemala

Archeologists (25) have estimated that the 41,000 cu ft of walls required about 50,000 man-hours of labor, based on the excavation, mixing, and placing of each soil pancake. A village of 20 families could have built it in a year or two after allowing for their farming and hunting. Similar adobe construction has been found in other southwestern United States pueblos.

## MASONRY WALLS

The temple platforms and pyramids are massive, requiring much labor, but limited ingenuity. Walled structures, which also evolved from simple heaps of stones, required advanced technology to provide stability against overturning by wind, earthquakes, and security against enemy attack.

The first walls in the Americas were of cobbles, stone slabs, or adobe with clay binder for mortar, about 2 m high. They appeared along the coast in Peru about 2500 B.C. Similar cobble and mud walls were built along the sandstone cliffs of the southwestern United States to enhance the shelter provided by the sandstone overhangs.

Wall construction evolved rapidly in the southwestern United States, using flat sandstone flags about 30 cm$^2$ (1 sq ft) by 3 cm–5 cm thick, laid in courses, with the stones in successive courses overlapping to form a bond. The earliest construction used rough flags selected for uniform size. In later construction,

FIG. 8.—Stone Face 2.5 m (8 ft) High Revealed in Exploratory Tunnel into Pyramid at Tikal, Guatemala

the broad surfaces and one edge of the stone were chipped to provide a rectangular outer face of uniform thickness. A good example is Pueblo Bonito, Chaco Canyon, N.M., built about 1000 A.D. The higher walls consisted of outer casings of bonded and faced flat stones with cores of rough flat stones crudely interlayered (bonded) with the facing. An understanding of the greater stress at the base of the wall is reflected in the greater width at the base, 1 m (3.3 ft) tapering to 0.3 m (1 ft) at a height of about 8 m (25 ft). The masonry was laid with stone chips and clay mortar to level the rough split stone surfaces. Lateral stability was provided by cross walls; the rooms were typically 2 m–4 m wide and long.

Coursed masonry walls were used more sparingly in Meso-American construction. Some of the temple complex of Teotihuacan, Mexico, included coursed masonry of tuff blocks, 25 cm × 50 cm, on the outer face. There are a few Mayan (700 A.D.–1200 A.D.) walls of soft limestone, generally laid without mortar. The blocks were of varying sizes, from 20 cm to about 50 cm high, but uniform in each course. The typical length to width ratio of the blocks ranged from 1.1–1.4.

Masonry walls were constructed extensively in the Peru-Bolivia highlands. Most of the earlier walls, 200 B.C.–600 A.D. were laid with more or less rectangular

FIG. 9.—Ancient Quarry in Soft Limestone, Tikal, Guatemala Furnishing New Blocks for Restoration Work

blocks of unequal height, as in Fig. 10. This required notching of the corners to fit successively higher courses together. The fitting was very precise and most of the walls were laid without mortar. At Tihuanaco, Bolivia, some of the stones were held together by I-shaped copper cramps, forced into dumbbell shaped notches cut between adjoining stones. The use of the wall was reflected in the quality of construction in the later Inca masonry, 1200 A.D.–1500 A.D. Terrace walls in rural areas were built of roughly trimmed rectangular blocks of varying sizes, laid in courses that were not necessarily level nor of equal height. Military structures such as the Fort of Sacsahuaman near Cuzco (Fig. 10) were made of huge, more or less rectangular blocks, laid in irregular courses, with blocks of unequal size notched to fit together. The fitting of these limestone blocks was very precise, with either very thin clay mortar or no mortar at all. The surface texture of the stones suggests that they were shaped by pounding with small hard stones, followed by surface grinding with stones and sand. According to Inca tradition, recounted by de LaVega (10), the blocks were dragged from the nearby quarry by hundreds of men utilizing twisted vine ropes and either wood skids or rollers. Earth ramps were constructed to place the upper blocks. The construction of Sacsahuaman, completed about 1480 A.D., required 50 yr utilizing tax labor, about 120 days/yr/person, and a legendary labor force of more than 20,000. One large block, the "Weary Stone," was never trimmed or lifted in place. According to Inca legend (10), it was so weary after being dragged to the fort site that it wept blood from two holes in one corner. More likely it was the laborers who wept—3,000 were killed when it fell from a slope during hauling.

**FIG. 10.—Inca Wall of Notched Rectangular Blocks: Fortress of Sacsahuaman, Cusco, Peru**

Because of the irregular coursed, notched corners, and close fitting, the walls have demonstrated unusual resistance to earthquakes compared with the uniform coursed, mortared walls built in the same areas by the Spanish conquerors. Some historians have speculated that the notched form represents a conscious effort for greater stability. However, if this were true, it would be used on the most important structures, i.e., religious and governmental buildings. However, uniform, continuous, coursed (unnotched) masonry was reserved for the most sacred buildings in Cuzco, such as the Convent of the Virgins of the Sun and the Temple of the Sun. The writer believes that the notched masonry reflects a compromise between the ordered appearance of uniform coursed masonry and the labor required to produce uniformity from rock that did not split readily.

**Mortar.**—Mortar is employed to bed and cement the masonry units of walls and to plaster over a rough wall to provide a smooth surface, often for decorating. A low to moderate plasticity clay was the most widely used material throughout the Americas. It is durable if protected from wetting.

When the masonry units were so irregular that thick mortar was required, shrinkage became a problem. Pebbles and stone chips were forced in the mortar joint to steady the stones, reduce the mortar volume, and minimize shrinkage. This practice was widespread in the United States southwest and in Meso-America.

In the United States southwest, caliche, a calcareous silt or clay, is abundant near the ground surface and has been employed for mortar probably because of availability. It has the property of hardening somewhat, in alternating damp-and-dry conditions, and becoming slightly water resistant.

FIG. 11.—Composite Masonry Wall with Rubble Core, Uxmal, Yucatan, Mexico

The Mayans discovered that burned limestone would reharden slowly into soft rock upon wetting and subsequent exposure to air. Lime was produced by burning limestone fragments on firewood racks (20). The mortar was a mixture of lime, limestone dust, soil (if abundant), and probably volcanic ash. Although the Egyptians and Persians used gypsum mortar, lime mortar apparently did not appear in the Old World until a few hundred years B.C. The Mayans used it possibly as early as 500 B.C. for masonry construction as well as plaster.

**Composite Construction.**—The Mayans made extensive use of massive composite wall construction. The interior of the wall was coarse, rubble masonry, cemented with clay or lime mortar and chips of stone into a homogeneous mass. A veneer of stone masonry, precisely fitted and coursed, covered the surface, providing the illusion of a masonry block wall (Fig. 11). The practice can be seen in early structures at Tikal, about 200 B.C. and some of the latest at Chichen Itza and Uxmal, 1200 A.D. to 1400 A.D.

**Wattle and Daub.**—Wattle and daub is a form of reinforced adobe or a wood-soil-plaster composite. The reinforcement is a grid of wood poles 5 cm–10 cm (2 in.–4 in.) diam spaced 1 m–2 m (3 ft–7 ft) apart. Between them is an open matting or wattle made of parallel (but sometimes loosely woven) reeds or thin poles 1 cm–2 cm or (3/8 in.–3/4 in.) diam, 1 diam or 2 diam apart. Low plasticity clay mortar is applied to both sides to form a wall 8 cm–15 cm or (3 in.–6 in.) thick. Such a contemporary house in Yucatan is seen in Fig. 12. The deteriorating clay daub exposes the underlying wattle. The age of this form can be judged from its representations, in a Mayan carving on a structure named the "Nunnery" (although its use is not known) of Uxmal. Similar houses can be seen today throughout the Mayan areas, in other parts

FIG. 12.—Wattle and Daub House, near Chichen Itza, Yucatan

of Meso-America (and in Kenya and adjacent areas of East Africa).

**Retaining Walls.**—Retaining walls served two purposes in the pre-Columbian Americas: (1) Agricultural terraces; and (2) support of steep slopes. The civilizations of the Andes Highlands of Bolivia and Peru built extensive terrace walls to make level land on steep slopes and to conserve runoff. Most were of untrimmed, small boulders laid in crude courses (Fig. 13).

Walls to support steep slopes were a part of the structures of Monte Alban in Mexico and Mayan Copan in Honduras. They appear to be incidental to the structures to which they are attached and not built specifically for retaining.

Pueblo Bonito, Chaco Canyon, N.M., is built at the toe of a sandstone cliff. Soft shaley layers near the base of the cliff were weathering and eroding, allowing the massive sandstone above to tilt outward from a vertical joint crack (Fig. 14). The pueblo builders constructed a wall at the toe of the cliff to retain the weathering rock and to prevent further deterioration, probably about 1000 A.D. For more than 800 yr it functioned. In 1941 the cliff collapsed demolishing the wall and a portion of the pueblo.

## FOUNDATIONS

Foundations in the Americas were usually simple. Although the bottom of a high masonry wall was sometimes wider than the top, there was no enlarging of the base in contact with the ground to make a footing. At the sites of the larger Mayan and Mexican structures, rock was usually shallow; therefore, foundation bearing was no problem. The same is probably true of the western South American structures; however, foundations are not mentioned in archeo-

**FIG. 13.—Inca Agricultural Terrace, Abandoned 400 yr Ago but Still Intact, Machu Pichu, Peru**

logical reports. Some of the precisely fitted Inca masonry walls rest on crudely laid rubble below the ancient ground level. This suggests that a trench was dug, filled with stones, and the wall placed on top. The trench width and depth, however, are not significantly greater than the wall thickness.

Wood columns in most North American construction were placed in holes about 1 m deep, and backfilled with soil to provide lateral support during construction and (probably without intent) some vertical support for the eventual load. Stone columns of the Mayans and Mixtecs usually sat on the stone platform of the temple complex; separate foundations were not required.

The Pueblo Indians in the vicinity of Chaco Canyon, N.M. used prefabricated stone disks at the bases of the columns for the Great Kivas. Three or four, about 1 m diam and 4 cm thick, were stacked like pancakes under each column.

Wood piles served as both foundations and unsupported columns in areas of wet soil. The Spanish explorers found Indians in Lake Maricaibo on the north coast of South America living in pile-supported houses over the water. They named the area Venezuela: "Little Venice." There are similar pile supported houses in the area today; the only change is the substitution of wood-plank siding and corrugated iron roofs for the indigeneous pole and thatch.

Similar wood pile supports were used in the Valley of Mexico where the Aztecs built Tenochtitlan, the first Mexico City, in a marshy arm of Lake Texcoco (14). Excavations for modern construction have uncovered wood pile foundations for houses, and to retain soil fill for roadways. Although there are no records of how the piles were driven, the writer observed men driving wood piles in the loose sands at Acapulco, Mexico in 1940, Fig. 15. They raised a wood hammer about 25 cm (10 in.) diam and 60 cm (2 ft) long, by

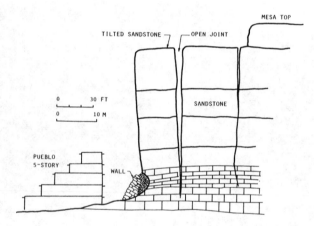

FIG. 14.—Retaining and Erosion Prevention Wall Pueblo Bonito, Chaco Canyon, New Mexico

handles, and dropped it about 60 cm on top of the 10 cm diam wood pole while standing on a wood platform fastened to previously driven piles.

## Transportation Routes

Transport of people and commodities developed late in the pre-Columbian Americas. There is evidence of loosely organized trade in the finding of native copper in mounds in Georgia, which probably came from the only known source in Michigan, far away.

Similar trade in mica and flint between different areas of the eastern United States is inferred from grave artifacts. The trade routes, which were also used by the early European explorers, were water courses and loose networks of foot paths, 0.6 m wide, which probably evolved from hunting trails. There was little need for more. The only beasts of burden were men, women, and occasionally dogs. The largest boats were only canoes or rafts. In the Andes of South America, the llama, a mini-camel, still carries loads of about 50 kg (100 lb) but has never pulled a load. Coastal ocean commerce developed late, primarily in the Yucatan area and to a small extent along the North Pacific Coast of the United States and Canada.

Surprisingly, systems of wider "highways" are found in many areas. Because these routes focus on major settlements and religious centers, archeologists speculate that they were required for processions or perhaps for armies moving in compact units.

The best known are the Inca roads in Peru and Ecuador (22,23). One 5,400 km or 3,200 miles long followed the high valleys in the Andes, from the administrative capital at Cuzco, Peru northward into Ecuador and southward into Chile. A second road followed the coast linking the irrigated valley civilizations across the sandy deserts. They were connected by east-west roads. In the plains they were merely trampled paths, marked by rows of stones 3

FIG. 15.—Pile Driving by Hand, Acapulco, Mexico, 1949

m–6 m apart, and swept clean of sand regularly. The roads were essentially straight, up and down the mountainsides, and across streams 5 m–7 m (16 ft–23 ft) wide. In the mountains they included causeways across low wet areas, retaining walls, paving, stone stairways up steep slopes (Fig. 16) and parallel marker stones. Suspension bridges spanned narrow canyons and tunnels pierced steep obstructions. At intervals of 7.5 km (4.5 miles) there were distance markers. Relay stations were built 2.5 km (1.5 miles) apart for messengers who ran the interval, passed messages and perishable food to the next runner, and then rested. At larger intervals, about a one day walk or 20 km–30 km (12 miles–18 miles), there were rest stations for official travelers and supply warehouses for small military parties. Possibly the litter parties that carried the nobility and their military escorts determined the road width. There were no vehicles and only occasional llama pack trains. Parts of the Inca road system are still in use, while others lie in disuse because of the steepness of the pathways or because they are lost under the desert sands.

The Mayans likewise had an extensive road system called "Sacboebs" connecting many of their cities and religious centers (Fig. 6). Many can be traced today as lines of different growth in the jungles of Guatemala and the scrubby forests of Yucatan. The ground surface in much of the Mayan area is extremely rough, deeply pitted from the solution of the soft limestone, making cross country travel difficult. Causeways 8 m–10 m (25 ft–33 ft) wide were constructed with level surfaces and in straight lines.

The causeways were built of small limestone boulders, topped with limestone chips and limestone sand to form a smooth surface, anticipating the ideas of McAdam, the Scottish engineer, centuries later and an ocean apart. A stone

FIG. 16.—Inca Road near Machu Pichu, Peru

cylinder, 4 m long and 0.65 m (13 ft × 2 ft) diam weighing 5 ton, found adjacent to a road suggested a roller to Mexican geotechnical engineers (18). They found that 15 men could easily push it across freshly placed stones and compact the surface. (Some archeologists are skeptical of this explanation of the stone's use.) Most of the causeways have become overgrown with vegetation after centuries of disuse. A few sections have been restored by the Mexican Highway engineers; it is likely that many others remain to be discovered.

Archeologists of the United States National Park Service, studying air photographs of Chaco Canyon, N.M., for prehistoric irrigation channels, were surprised by faint linear features radiating from some of the ancient pueblos (5). The lines go straight across topographic features but are best defined on the uplands and mesas (Fig. 17). They were located on the ground and found to be cleared pathways 6 m–9 m (22 ft–30 ft) wide. Across bed rock, they were defined by lines of cobble stones, and in some cases by leveling of high points in the rock. Some terminated in staircases down the cliffs. Across sandy ground the pathway was a shallow excavation. Despite seven centuries of neglect, these pathways or "roads" are still evident (Fig. 18). Their use is not known; archeologists speculate that they were paths from the cultivated fields and springs to the pueblos.

Although much of the transport of people and commodities into the Aztec Capital of Tenochtitlan was by canoe, there were major causeways across the

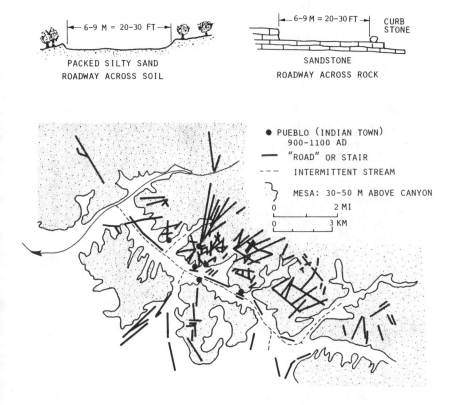

FIG. 17.—Road or Pathway System, Chaco Canyon, New Mexico (5)

marshes connecting it with the mainland (Fig. 19). Closely spaced pole piles formed parallel retaining walls (Fig. 20). Possibly reeds or canes were interwoven with the piles to produce a tighter wall with fewer piles. Earth fill from the adjacent marsh was placed between the pile walls forming an earth causeway with a canal for canoes on each side. There were occasional openings in the causeways to allow water circulation and passageway for boats. According to the Spanish conquerors, these were bridged with timbers which could be removed for military defense (4). Cortes, his soldiers, and Indian allies were nearly trapped by the Aztecs who opened the bridges during his first sortie into the city.

## WATER AND SOIL MANAGEMENT

Water and soil for agriculture are twin essentials to community living. In many regions of the Americas, highly seasonal rainfall was unfavorable to continuous settlement and agriculture. In coastal Peru, where the sea was a source of food, lack of any rain made life nearly impossible. In mountainous areas, steep slopes eroded and slumped when the permanent vegetation was destroyed by farming. Despite shortcomings of climate, some of the most advanced civilizations in the Americas developed in such areas. Water and soil management were the keys to their success.

Terracing, in steep terrain, with the aid of low rubble walls, retained both soil and runoff. There are remains of such rudimentary terraces in the south-

FIG. 18.—Chaco Canyon Road or Path across Sandstone Mesa

western United States, some possibly 2000 years old. In the Tehuacan Valley, southeast of Mexico City, terraces have been found in cultures dating from 900 B.C.

In South America, terracing in the Andean highlands and in coastal settlements probably began about 500 B.C. (11,23). Because of the continuity of community life in the Andean highlands, the terrace systems were improved and enlarged until the end of the Inca empire in the sixteenth century. There are many that remain in use today. Many others are in good physical condition despite four centuries of neglect, such as at Machu Pichu (Fig. 13). Others have crumbled; their remains are a testimony to better water and soil management than is practiced today in the same region. The terraces in some areas are as high as 2 m or 6 ft constructed on slopes typically 20°–30° or 2.7(H)–1(V)–1.7(H)–1(V) but occasionally steeper than 1–1 (Fig. 13). The rubble walls at Machu Pichu were backfilled with soil, plus a filtered underdrain of sand over gravel to prevent soil saturation (and also to prevent hydrostatic pressure from overturning the wall) (15). Unfortunately, the technical knowledge and the political system that made it possible to build and maintain these terraces were destroyed by the European conquerors.

**Water Diversion Canals.**—Diversion of water from streams and springs to farm lands, and to terraces by canals or ditches, represents a logical step beyond terracing. Unlined ditches have been found in the southwest United States built by the Hohokum nearly 2,000 years ago (9). The builders diverted streams

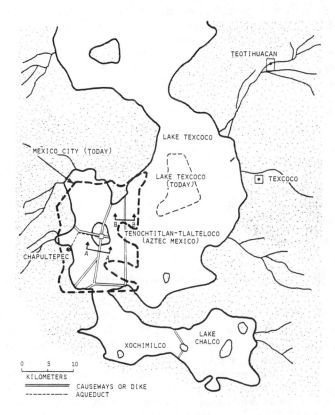

FIG. 19.—Causeways and Dike, Aztec Tenochtitlan (Mexico City), Mexico

using temporary dams of woven fiber matting supported by short poles driven into the stream bed. In Chaco Canyon, N.M., there are stone rubble structures built in dry gullies to divert thunderstorm runoff into cultivated fields (24).

In Coastal Peru, where there is no rainfall, agriculture 2,500 years ago depended on diverting water from the rivers which flowed from the mountains into the flat terraces alongside the arid flood plain. In the Andean highlands with seasonal rainfall, multiple use was made of the occasional dry weather springs. The water was collected in masonry channels, diverted first to fountains, and baths, and then into trenches leading to terraced fields (Fig. 21).

The Hierve el Agua Spring, about 20 km east of Milta in Mexico's Oaxaca Valley, is a combination terrace and canal system (Fig. 22) (7). Warm water, rich in magnesium and calcium bicarbonates, flows from several small springs on a steep hillside. The carbon dioxide is released in the air, precipitating the minerals as travertine, a soft limestone. Ordinarily, alkaline waters cannot be used for irrigation because the minerals inhibit plant growth. At this site the builders constructed an elaborate system of minicanals using the precipitating travertine (Fig. 23). The lengthy network prolonged the air exposure, enhanced precipitation, and reduced the mineral content. When the travertine built up, the soft deposits were recut forming new channels on top of the old. Today, more than 2,000 years after the system was begun, the minicanals are perched on travertine ridges 1/2 m–1 m high and 1/4 m–1/2 m wide. These ridges also help form terraces on the steep hillside (Fig. 24). At intervals along the terrace-canals there are pot-like enlargements. The archeologists believe that the water was handdipped from these to water the plants on the terraces (7). Other small canals with small cisterns for dipping water are found throughout ancient Mexico; some were derived from springs or small streams but a few utilized shallow well water.

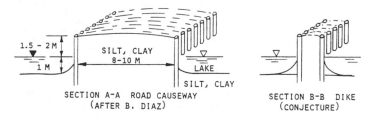

FIG. 20.—Cross Section of Road Causeways and Dikes of Tenochtitlan (4)

**Water Storage.**—In the Southeastern Gulf Coast of Mexico, the lowlands of Belize and Guatemala and the Yucatan Peninsula, there is little stream flow despite abundant rainfall. The runoff disappears into cavities in the soft, porous limestone. The occasional large open sinkholes, or cenotes, were the only natural source of water. These became the focal points for many Mayan Temple cities. The large cenote of Chichen-Itza, Yucatan was a sacred well. When the ground-water level dropped during dry weather, jewelry and sometimes humans were sacrificed into it to bring the water back.

At other Mayan Temple cities artifical reservoirs were constructed in the limestone to store surface runoff. Some were probably old solution pits in the rock, others were carved in the soft limestone. The pores of the rock were sealed with lime-limestone dust mortar and occasionally clay. Some at Uxmal and Sayil were covered to vinimize evaporation and included paved rainfall catchments (Fig. 25). At Tikal, there were a number of open reservoirs with areas of 1,000 $m^2$–10,000 $m^2$ and 2 m–5 m deep.

The "Mummy Lake" in Mesa Verde, Fig. 26($a$) and 26($b$), is a combination masonry and earth-storage reservoir in a runoff-collecting system on the sandstone mesa top. Storm water was diverted into it by blocking the collector ditch, which also fed water to small agricultural terraces (19).

**Drain Systems.**—Surface water drainage systems were incorporated in many of the temple city complexes of Meso-America and in the Andean Highlands where locally intense rainfall could accumulate on impervious masonry surfaces.

At Copan, Honduras, and Monte Alban, Oaxaca, Mexico, there are extensive stone-slab culvert systems, draining to nearby gullies and streams (13). Grooved stone bars were used as runoff channels at Copan. Similar grooved stone water conduits can be seen on the Andean highlands. It is possible that matching bars were joined to form closed water pipes.

There is some speculation that the stone drains of Monte Alban were also used for drainage of human waste. If so, they probably accummulated decomposing solids during periods of low rainfall because there are no springs on the mountain top and rainfall is highly seasonal.

FIG. 21.—Spring, Baths, and Water Diversion System at Tambumachy, Cuzco, Peru

At Dainzu, Southeast of Oaxaca, Mexico, there are extensive conduit systems of grooved stones and stone slab culverts. There are also burned clay pipes about 10 cm OD (Fig. 27). They brought spring water into the temple complex from a hill above.

A ribald sense of humor can be seen in roof drains for a temple in "Old Chichen," a part of Chichen Itza. The roof downspouts are in the form of a penis; when it rains, they discharge in a realistic manner.

**Valley of Mexico.**—The valley of Mexico (Fig. 19) was once occupied by a shallow lake between the surrounding volcanos. The early inhabitants excavated canals in the marshy lake margins and piled the excavated peaty, silty clays between to form low rectangular artifical island called *chinimpas* (14). Several crops per year were possible in the warm sunshine with the organic soils and their adjacent source of water. Human wastes were collected from nearby settlements to enhance the natural fertility. The present "Floating Gardens of Xochimilco" (the remnant of an ancient south arm of the lake) still produce flowers and vegetables, although they are now mostly a tourist attraction.

The Aztec capital of Tenochtitlan (now Mexico City) in the western marshes of Lake Texcoco was an adaptation of the chinimpas (plus some older islands) to urban development. The island city was connected to the mainland by causeways (Figs. 19 and 20). Lake Texcoco was becoming saline because it had no outlet. The causeway isolated the western part of the lake, with its large fresh water inflow from springs and streams compared with its surface area from which water evaporated. By limiting the area of the cross-canal links to the more saline larger eastern area of the lake, the Aztecs were able to maintain freshwater around the city.

FIG. 22.—Hierva el Agua Canal and Terrace System Near Oaxaca, Mexico

Windstorms generated sufficient waves and water pileup on the main lake that the causeways were overtopped and parts of the Aztec City flooded on a few occasions. Shortly before Cortez's conquest, the Aztecs constructed an outer dike for the sole purpose of protecting the city and causeways from flood damage and salt-water contamination. Today, modern Mexico covers much of the old causeway and water control system, and Lake Texcoco has shrunk to a shallow salt pond.

## Significance of Pre-Columbian Construction

The pre-Columbian civil works, with their unusual variety and size, are a tribute to those ancient engineers who conceived their designs and supervised their construction. The civilizations are dormant and the names of those responsible for these achievements are lost. However, when these works are evaluated by modern standards, many are examples of what is good in engineering.

The first test of the value of engineering is its contribution to society's primary physical needs, i.e., shelter, water, food, communication, and defense. Enough examples of engineering works that meet each need remain unswallowed by jungle or unmelted by weather to demonstrate that they rival comparable structures of Europe and the Middle East of 2,500 yr–1,000 yr ago. As will be considered later, the works of the Americas exhibit neither the evolutionary growth of technology from one location to another nor the development with passing time that can be seen in the Old World (although this conclusion may be partially blamed on less archeologic study).

Structures that expose man's religious and spiritual feelings also fulfill a fundamental need. In size and grandeur the temples and pyramids of the Americans compare with those of the Egyptians, Babylonians, and Assyrians. In construction, they are more advanced, with their composite masonry core and fitted mosaic casings.

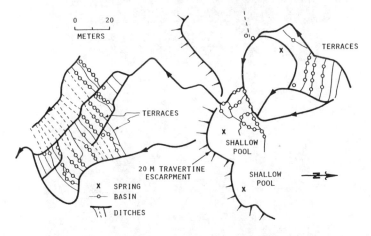

FIG. 23.—Plan of Hierva el Agua Canal and Terrace System (7)

A third test of technology is the adaptation to available materials. In this the pre-Columbian Americans excelled. Because long, heavy timber was scarce in the Andean highlands, the Valley of Mexico, Yucatan, and the arid southwestern United States, local stones and soil were exploited. As today, the soil was dirt cheap—a versatile, strong, durable material if handled correctly, a pile of mud if not. Even a child knows this instinctively and empirically; some modern engineers have forgotten it.

Soil was made into walls, and was used for the mortar for walls of stone, and the plaster that smoothed the surfaces. Reinforced with sticks and reeds, it became floors, roofs, and the walls of huts in areas where soil was too scarce for adobe block construction, or where moisture softened the blocks.

The geotechnical engineer can find a deeper meaning in this use of simple earthy materials. Many religions teach that man becomes a part of the earth when he dies, therefore, he must be made of earth. The ceremonial underground pit or kiva of the Southwest Indian, is thought to represent the womb of the earth from which man springs. The civilized child sometimes finds comfort in a close space among rocks, a shallow pit in the ground, or a dark corner of a room. The earth, or rock mound that supports a temple, reflects man's (and all life's) dependence on the earth. The pit grave and the burial mound signify his return on death.

Thus, the geotechnical engineer can feel that he is working with something more fundamental than a mere construction material—he builds with life itself!

FIG. 24.—Detail of Raised Minicanal Constructed of Travertine

Unfortunately to some modern engineers, soil has a dirty connotation. However, to the Geotechnical Specialist, soil is a primary material for construction and dirt is something swept off the floor, or which he washes from his child's face.

The final test of technology is its ingenuity in transmitting an idea into a reality. The civilization of the Western world was born in Asia and Africa and matured in Europe and the post-Columbian Americas. The initial development of technology was slow, with similar ideas emerging in widely spaced localities. Writing fostered the long distance communication of ideas and transmission of knowledge from one generation to the next. After the development of writing, technology developed at a rapidly increasing rate through the synergism of independent (nuclear) technologies.

Development in the Americas was different. Man appeared only recently, compared to Africa and Asia, as recently as 12,000 years (or perhaps as long as 50,000 years ago). The settlements were dispersed and their life was dependent on hunting and food gathering. Primitive technologies arose spontaneously in widely separate areas, about 3,000 years ago, much later than in the Old World. However, they evolved at least as rapidly as those in the Old World during comparable time spans.

**FIG. 25.—Paved Rainfall Catchment 10 m diam above Cistern, Sayil, Yucatan, Mexico**

There were also other major differences in the development in the Americas. The lack of a logic system and mathematical computation inhibited an orderly understanding of the mechanics of technology. The lack of written language inhibited the synergism between geographically isolated areas as well as the transmission of technology to succeeding generations.

With these differences between the Old and New Worlds, many historians have speculated on the origin of the technologies that produced the ingenious works that are being uncovered.

**Old World Contacts.**—There were legends of both the Incas in Peru and the Aztecs in Mexico that describe bearded, light-skinned gods landing from large ships. However, the stories were embellished with such fanciful details that they could be entirely imaginary. There are a few scattered finds of pre-Columbian European artifacts. Recent small-craft crossings of the Atlantic demonstrate that skilled seamen, such as the Phoenicians, could have crossed to the Americas, probably because of bad navigation. The chances of a safe return were small, which would explain no records of such voyages. There are fragmentary stories of Norse expeditions to "Vineland" but conflicting evidence of its location. Certainly there are no records of major expeditions that might have brought Old World skills to the Americas.

The evidence of comparative technology is also weak. There are some similarities, e.g., weaving, burned clay pottery, adobe, and pyramids. However, a detailed comparison shows that the similarities are superficial. The American stone pyramids are far steeper than those of the Old World. Most were built box-in-box and of rubble with only a veneer of faced stone. Their primary use was in religious ceremonies; burial was secondary.

Some of the most important Old World advances are absent such as the wheel and metallurgy (except for Western South America, remote from Atlantic

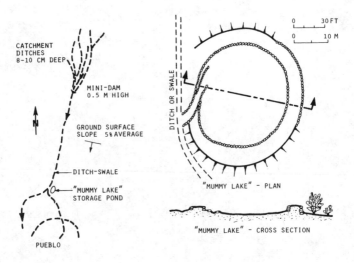

FIG. 26(a).—Plan and Cross Section, Mummy Lake Reservoir, Mesa Verde, Southwest Colorado (19)

crossings). Most important, there was neither a system of logic nor written communication. By way of contrast, there are some New World ideas that may have predated those in the Old, e.g., lime mortar and composite masonry using a rubble core with only a thin facing of finished stone work.

The absence of Old World diseases, such as smallpox, in the Americas before Columbus is also evidence of little or no communication. Large segments of the indigenous people were wiped out by Old World infections after contact with the early explorers.

The available evidence suggests that although contact was possible, there was no significant transfer of technology from the Old World to the New.

**Creatures from Space.**—The idea of extraterrestrial communication is intriguing, but entirely lacking in evidence. *The Chariots of the Gods* strains at comparing the weird stylized costumes of the Mayan gods with the modern space suit and suggesting that the Nazca lines were landing fields for space craft. We can understand the ancient Hebrews explaining still older megalith construction (lavstones) by stating, "There were giants in the earth in those days." There is little excuse for such a viewpoint today (21).

The extraterrestrial hypothesis as voiced by Von Daniken is developed by a logic of elimination: (1) the huge structures built today are the product of modern technology; (2) the ancient people possessed no such technology; (3) they did build some extraordinary structures; and (4) because they had no modern technology, they must have received supernatural help.

This conclusion is based on the unstated assumption that the ancients could not have contrived an innovative, admittedly primitive technology to produce their structures. It is a form of intellectual snobbery that cannot admit the

FIG. 26(b).—Mummy Lake Dike, Mesa Verde, Southwest Colorado

potential for achievement without the tools and gadgets which we now possess.

**Indigenous Ingenuity.**—The alternative to the hypotheses of Old World influence and extraterrestrial intervention for the New World construction achievements is indigenous ingenuity. The word "engineer" is derived from ingenious, the capacity to contrive something new. It is related to genie, the mythical being with magic powers to create. Confronted with a problem, the engineer (the ingenious one) contrives a plan to transform the available materials into a solution. The plan becomes a reality through the organization of labor.

The available evidence supports this hypothesis. Each of the widely separated pre-Columbian civilizations exhibited a logical technical evolution within its geographic nucleus. The problems were common among the different civilizations. However, each evolution was somewhat different, reflecting different materials and different ideas, e.g., the Maya built their temple bases of limestone rubble bonded with lime mortar and cased with cut stone; the Mexicans at Cholula used volcanic stones in a soil-volcanic ash matrix; and the peoples of the Andean highlands fitted large stones together.

Although there were rapid technological developments between 2,500 and 500 years ago, many were forgotten before the Old World conquest. Moreover, the technology of one nuclear civilization did not appear to influence that of its near neighbors. Each invented its own, more or less independently. When a society deteriorated, its technology was largely forgotten. The lack of an effective means of recording and transmitting ideas was a major deficiency of the pre-Columbian Americas. In my opinion, this loss confirms the indigenous nature of their origin.

**For Today.**—A study of the past is of greatest value in its message to the

FIG. 27.—Drain Pipe, 10 cm OD, Dainzu, Oaxaca, Mexico

present and future. To the geotechnical engineer, our pre-Columbian heritage can reinforce our dedication to the earth. The Western Indians revered it as their mother or grandmother, the Judeo-Christian tradition maintains that we were created from it, our childish instinct is to play in it, and as engineers, we build with it. Thus, our work with the earth as a raw material and a foundation for our creations is more fundamental than child's play or an intellectual exercise for the laboratory.

The arrested development of pre-Columbian technology is a warning that communication and interchange of ideas is vital to development. Concepts that remain within a nuclear society eventually die with it. We continue to waste effort in "reinventing the wheel."

Finally, these achievements remind us that the most important factor in achievement is man's innate ingenuity—the essence of engineering. Modern technology has enabled us to build bigger, faster, and even better. However, the complexity and beauty of our scientific tools causes some of us to fall in love with them instead of their use. The past remains us that man has produced magnificently with little except his ingenuity and labor.

The ancient Hebrews were closer to the truth than they realized when they conclude, "There were Giants on the Earth in Those Days." There were giants of ingenuity, fulfilling their potential.

Karl Terzaghi was such a giant. He generated new ideas 50 years ago that have transformed engineering. We can express this ingenuity if we aren't seduced by our own tools. We too can be giants, and society will be the better.

## Acknowledgments

Many have contributed observations, data, and ideas from which this paper has been compiled. Particularly, R. E. Gray, F. ASCE, G.A.I., A. Rico Rodriguez, Departmento Geotechnica, S.O.P. Mexico, T. Lyons, Chaco Research Center, New Mexico, and R. G. Vivian, Arizona State Museum made obscure data available and provided personal insights. Warren Bennet conducted a brief demonstration on desert varnish on the Mojave Desert, Calif. Teresa Groves, Librarian, Marlene Barnes, and Chris Shattuck, secretaries and Robert Alexander, Draftsman, of Law Engineering provided the technical assistance for preparing the manuscript. My wife, Frances, accompanied and encouraged me through jungle, highland, and desert in pursuit of the elusive ancients.

## Appendix I.—Bibliography

deLanda, D., *Yucatan before and after the Conquest,* Translation with Notes, W. Gates, ed., Dover Publications, Inc., New York, N.Y., 1978, (Manuscript 1566).

Innes, H., *The Conquistadors,* Alfred A. Knopf, Inc., New York, N.Y., 1969, 336 p.

Lanning, E. P., *Peru before the Incas,* Prentice-Hall, Inc., Englewood Cliffs, N.J., 1967, 216 p.

Spiden, H. J., *A Study of Mayan Art,* Dover Publications, New York, N.Y., 1975, 285 p. (originally *Memoirs,* Vol. 6, Peabody Museum of American Archeology and Ethnology, Harvard University, Cambridge, Mass., 1913).

## Appendix II.—References

1. Adams, et al., *Tikal Reports 5–10,* University Museum, University of Pennsylvania, Philadelphia, Pa., 1961.
2. Coe, W. R., *Tikal: A Handbook of the Ancient Mayan Ruins,* University Museum, University of Pennsylvania, Philadelphia, Pa., 1967, p. 124.
3. Von Daniken, E., *Chariots of the Gods,* Putnam's G. P. and Sons, New York, N.Y., 1969.
4. Diaz, B., *Conquest of New Spain,* Translated from Spanish by J. M. Cohen, 1963, Penguin Books, Harmondsworth, Middlesex, England, 1963, (Manuscript 1576).
5. Ebert, J. I., and Hitchcock, R. K., "Chaco Canyon's Mysterious Highways," *Horizon,* Autumn, 1975, p. 49.
6. Erasmus, C. J., "Monument Buildings, Some Field Experiments," *Southwestern Journal of Anthropology,* Vol. 21, No. 4, 1965, pp. 277–299.
7. Flannery, K. V., and Kirkby, A. T., "The Use of Land and Water Resources in

the Past and Present Valley of Oaxaca, Mexico," *Memoirs of the Museum of Anthropology*, Vol. 1, No. 5, University of Michigan, Ann Arbor, Mich., 1973.
8. Genesis 6:4.
9. Haury, E. W., "The Hohokum," *National Geographic*, May, 1967, pp. 670–696.
10. de LaVega, G., *The Incas, Royal Commentaries*, Translated by Jolas, Discus-Avon-Prion Press, New York, N.Y., 1961, (Manuscript 1610–1616).
11. Mason, J. A., *The Ancient Civilizations of Peru*, Penguin Books, Baltimore, Md., 1961.
12. McIntyre, L., "Mystery of the Ancient Nazca Lines," *National Geographic*, May, 1975, pp. 716–728.
13. Paddock, J. ed., *Ancient Oaxaca*, Stanford University Press, Stanford, Calif., 1966, p. 36.
14. Peterson, F. A., *Ancient Mexico*, Putnam's G. P. and Sons, New York, N.Y.
15. Polo, N., *Macu Picu*, Editorial de Cultura Andina, Cuzco, pp. 279–298.
16. Reed, N. A., Benett, J. W., and Porter, J. W., "Solid Core Drilling of Monks Mound: Technique and Findings," *American Antiquity*, Vol. 33, Apr., 1968, pp. 137–148.
17. Reiche, M., *Mystery on the Desert*, Nazca, Peru.
18. Rico, R. A., and Castillo, H. Del, *La Ingenieria de Suelos en las Vias Terrestres*, Vol. 1, Editorial Limuasa, S. A. Mexico, 1974, p. 134.
19. Roh, A. H., "Prehistoric Soil and Water Conservation on Chapin Mesa, Southwestern Colorado," *American Antiquity* Vol. 28, No. 4, Apr., 1963, pp. 441–455.
20. Stierlin, H., *Mayan Architecture*, Office Du Livre, Fribourg, Switzerland, 1964.
21. Thiering, B., and Castle, E., et al., *Some Trust in Chariots*, Popular Library, New York, N.Y., 1972.
22. Von Hagen, V. W., *Highway of the Sun*, Revised Edition, Plata Publishing Co., Ltd., Chur, Switzerland, 1975.
23. Von Hagen, V. W., *Realm of the Incas*, 2nd ed., Mentor Books, New York, N.Y., 1961.
24. Vivian, R. G., "Prehistoric Water Conservation in Chaco Canyon," *Final Report, National Science Foundation Grant GS 3100*, Arizona State Museum, Tucson, Ariz., 1972.
25. Wilcox, T. R., and Schenk, L. O., *The Architecture of Casa Grande and its Interpretation*, Archeological Series 115, Arizona State Museum, Tucson, Ariz., 1977.

# THERE WERE GIANTS ON THE EARTH IN THOSE DAYS[a]
## Discussion by N. J. Schnitter,[2] F. ASCE

The writer would like to take the opportunity to expand somewhat the author's treatment of pre-Columbian water and soil management and draw the attention to the importance of dam building on the American continent long before the arrival of the Spaniards, let alone the Anglo-Saxons.

As a matter of fact, the oldest pre-Columbian dam found so far was begun already between 750 and 600 B.C. on the Lencho Diego creek near the southern end of the Tehuacan valley, some 260 km (160 miles) southeast of Mexico City (36). Thus, the Purron Dam is considerably younger than the oldest dams in the world known at present, i.e. the Jawa earth backed masonry dam in Jordan from the end of the fourth millennium B.C. (30), and the Kafara rockfill dam in Egypt from the middle of the third millennium B.C. (29). The Purron Dam is about the same age as the famous Arim earth dam near Marib in Yemen (27), the Urartian dams east of the Van lake in eastern Turkey (28), and the Assyrian masonry dams northeast of Mosul in Iraq (34).

While the first stage of the Purron Dam was a simple embankment of just 3 m (10 ft) in height, its first heightening to 7 m (23 ft) occurred around 600 B.C. This shows already a systematic structure of dry masonry cells, that are filled with compacted sandy soil (upper section in Fig. 28). Its upstream face consisted of a masonry wall, while the downstream face was lined with stones. The crest measured 400 m (1,310 ft) in length and the reservoir volume amounted to some 1,400,000 m$^3$ (1,140 acre-ft). The base width of 100 m (330 ft) was grossly overdesigned, but facilitated the subsequent heightenings according to the progressive silting up of the reservoir. The final dimensions attained after more than 1 1/2 millenia were a dam height of 19 m (62 ft) and a fill volume of 370,000 cubic meters (480,000 cu yd), with a gross storage capacity of 5,100,000 cubic meters (4,140 acre-ft).

Slightly younger than the first stages of Purron is the dam found near Xox-

[a]April, 1981, by George F. Sowers (Proc. Paper 16190).
[2]Executive Vice Pres., Motor-Columbus Consulting Engineers Inc., CH-5401 Baden, Switzerland.

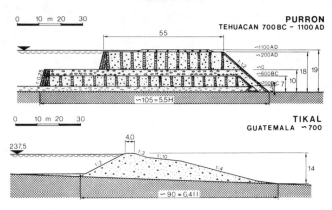

FIG. 28.—Cross Sections of Purron Dam in Tehuacan Valley (Top), and Palace Reservoir Dam in Tikal, Guatemala (Bottom)

ocotlan, a few km southwest of Oaxaca, Mexico. It dates back to the fifth century B.C. (35). It consisted of two 40 m (130 ft) long walls of boulders which towered up to 10 m (33 ft) high. Each was laid in lime mortar, which in plan formed a V pointing upstream, and between which was located the outlet to an irrigation canal. The many water and soil retention dams on the slopes of the nearby Monte Alban religious center, are of more recent origin, which formed part of the rainfall collection and drainage systems mentioned already by the author (p. 410).

In connection with his mention (p. 409) of the open Mayan reservoirs at Tikal, Guatemala, one might note that the dams impounding them were much more important than indicated by the author. From the maps given in Ref. 2, the section of the "Palace Reservoir" dam reproduced in the lower part of Fig. 28 could be drawn which shows a still existing structure which is 14 m (46 ft) high. The crest length amounts to 83 m (270 ft), its volume is 16,000 cubic meters (21,000 cu yd), and its storage capacity was about 0.05 million cubic meters (40 acre-ft). Considerably larger storage capacities were attained in the elaborate system of excavated ponds and canals around Edzna, near the westcoast of Yucatan, Mexico (31).

While these Mayan achievements date from the early centuries of the Christian Era, more recent samples of pre-Columbian dam building were found near Maravilla, west of the well known religious center of Teotihuacan, Mexico (26,32). One earth dam was 11 m (36 ft) high and 530 m (1,740 ft) long, and the other was a 25 m (80 ft) long diversion dike of a construction, similar to that shown in Fig. 20. Finally, a most interesting Incan structure existed across the outlet of lake Huinso, 150 km (90 miles) east of Lima, Peru (35). The 7 m (23 ft) high and 40 m (130 ft) long dam consisted of a rockfill retained on the downstream side by a 5 m (16 ft) thick masonry wall. This dam also contained seven superimposed openings to closely control the withdrawal of water.

Appendix.—References

26. Armillas, P., Palerm, A., and Wolf, E. R., "A Small Irrigation System in the Valley

of Teotihuacan," *American Antiquity*, Vol. 21, 1956, pp. 396–399.
27. Bowen, R. L., and Albright, F. P., *Archaeological Discoveries in South Arabia*, J. Hopkins Press, Baltimore, Md., 1958.
28. Garbrecht, G., "The Water Supply System at Tuspa (Urartu)," *World Archaeology*, Vol. 2, 1980, pp. 306–312.
29. Hellström, B., "The Oldest Dam in the World," *La houille blanche*, France, 1952, pp. 423–430.
30. Helms, S. W., "Jawa," Cornell University Press, Ithaca, N.Y., 1981.
31. Matheny, R. T., "Maya Lowland Hydraulic Systems," *Science*, Vol. 193, 1976, pp. 639–646.
32. Millon, R., "Irrigation Systems in the Valley of Teotihuacan," *American Antiquity*, Vol. 23, 1957, pp. 160–166.
33. O'Brien, M. J., Lewarch, D. E., Mason, R. D., and Neely, J. A., "Functional Analysis of Water Control Features at Monte Alban, Oaxaca, Mexico," *World Archaeology*, Vol. 2, 1980, pp. 342–355.
34. Thompson, R. C., and Hutchinson, R. W., "The Agammu of Sennacherib on the Khosr," *Archaeologia*, Vol. 79, 1929, pp. 114–116.
35. Tang-Sie, A., "Represa de Huinso, un ejemplo de ingenieria incaica," *Aguas de regadio*, Lima, Peru, 1964, pp. 58–60.
36. Woodbury, R. B., and Neely, J. A., "Water Control Systems of the Tehuacan Valley," *The Prehistory of the Tehuacan Valley*, Vol. 4, Univ. Texas Press, Austin, Tex., 1972, pp. 81–153.

# THE SIXTEENTH TERZAGHI LECTURE

Presented at the American Society of Civil Engineers Annual Convention and Exposition, Hollywood Beach, Florida

October 30, 1980

**GERALD A. LEONARDS**

# INTRODUCTION OF SIXTEENTH TERZAGHI LECTURE

### By John A. Focht, Jr.

Past President Ward, Terzaghi recipient Frank E. "Bill" Richart, previous Terzaghi lecturers, ladies and gentlemen, we are gathered here to hear the sixteenth Terzaghi lecture. This lectureship was established in 1960 to honor both Karl Terzaghi, father of our profession, and its most capable and accomplished practitioners. The lecturer tonight is another distinguished engineer, Dr. Gerald A. Leonards. Jerry is a Canadian by birth and graduated from McGill University in 1943. He worked in several positions in Canada before entering Purdue University where he earned his M.S. in 1948 and PhD in 1952. He must have liked Lafayette because he has been on the staff at Purdue ever since. He advanced rapidly becoming Associate Professor in 1955 and Professor in 1958. He was Head of the Civil Engineering Department from 1965–1968 and is proud that the students voted him the Best Civil Engineering Teacher in 1976.

Jerry has served as member and Chairman of several Geotechnical Engineering Division Committees, including the Executive Committee. He was also active on the U.S. National Committee for Rock Mechanics and the Highway Research Board. In 1965 he received the highest HRB award, its Best Paper Award, entitled "Use of an Insulating Layer to Attenuate Frost Action in Highway Pavements."

Professor Leonards is an active consultant on a wide variety of geotechnical problems around the world. His publications are extensive including the editorship of the text "Foundation Engineering." Many of his papers contain pioneering ideas and concepts on consolidation, deformation of sands, compacted clays, pavements, and piles. He won the 1965 Norman Medal for his paper on "Flexibility of Clay and Cracking of Earth Dams." Since 1975 he has been the invited lecturer at conferences and seminars in Bangalore, India, Haifa, Israel, New York, Berkeley, Lima, Peru, Minneapolis, and Boston. It gives me great pleasure to invite Prof. Jerry Leonards to deliver the sixteenth Terzaghi Lecture on "The Investigation of Failures."

# INVESTIGATION OF FAILURES

## By Gerald A. Leonards,[1] F. ASCE

**ABSTRACT:** The problems associated with determining what actually happened after an unexpected instability has occurred are illustrated by case records. Examples are taken from failures of cut slopes in, and embankments on, soft clays and from the failure of a large pile foundation during proof loading. It is then shown how more information can be gleaned from post-failure investigations than has generally been the case up to now, and the important lessons learned are highlighted. It is proposed that a National Center for Investigating Civil Engineering Failures be created to build up concentrated expertise and to develop rewarding methodologies for investigation of failures, to make the lessons learned cumulative and accessible to the profession, and to provide focal points for rewarding research to reduce significantly the frequency of unexpected failures.

## INTRODUCTION

In the sense used herein, the term "failure" connotes an unacceptable difference between expected and observed performance. Included are rupture or erosion of the ground, unexpected cracking that is unsightly, or reduces load-carrying capacity, or both, movements that impair the function of a structure, and monitoring systems that do not convey intended warnings in time. When a sudden rupture occurs, herein called an "instability," the result is usually spectacular and often incurs heavy property damage and possibly loss of life. It is commonly believed that investigation of instabilities offers the ultimate opportunity to gain new insights, to evaluate analytical tools, and to improve the design process. This, however, is not always the case, for the following reasons:

1. The geometry of the structure, the loading conditions, the soil stratigraphy, and the ground-water regime prior to failure must be inferred from the original site investigation and extrapolation from adjacent ground conditions, from the plans and specifications, and from examination of the post-failure conditions. In many cases, this information is incomplete, and the very fact that adjacent material did not fail implies that conditions in the failure zone may have been different.

2. It is always difficult and frequently impossible to reconstruct the *sequence* of events that led to failure; thus the factors which initiated the instability seldom can be positively identified.

---

[1] Prof., School of Civ. Engrg., Purdue Univ., W. Lafayette, Ind.

Note.—Discussion open until July 1, 1982. To extend the closing date one month, a written request must be filed with the Manager of Technical and Professional Publications, ASCE. Manuscript was submitted for review for possible publication on March 11, 1981. This paper is part of the Journal of the Geotechnical Engineering Division, Proceedings of the American Society of Civil Engineers, ©ASCE, Vol. 108, No. GT2, February, 1982. ISSN 0093-6405/82/0002-0187/$01.00.

3. As the original conditions are irreparably altered (and cannot be fully reconstructed), it is not possible to test different hypotheses of failure on the prototype.

4. The time and expense required to investigate all aspects of a failure are often incompatible with the objectives of the parties involved. The aim may be restricted to collecting and interpreting evidence favorable to the parties seeking redress for damages, or the need to reconstruct the facility quickly may take precedence, or politics may interfere with the investigative process.

5. There exists no rational methodology for investigating failure. As a result, lessons from past experiences are not cumulative. In each new instance, the investigator is compelled to develop his own strategy to minimize bias in collecting data and in analyzing and interpreting the results.

The first three of the previously mentioned factors are inherent in unexpected instabilities and can be mitigated only by monitoring prototype structures under controlled loadings, including contrived failures. While such studies deserve the strongest possible encouragement, they are seldom feasible and, as will be shown, are rarely successful. On the other hand, there seems to be no abatement in the frequency of unexpected instabilities. Detailed investigations of these failures often are made and it is imperative that lessons of permanent value be learned from them.

It is my aim to illustrate by case records the problems associated with determining what actually happened by investigating the debris after an unexpected instability and to show, by example, how more information can be gleaned from post-failure investigations than generally has been the case up to now. First, attention is focused on slides in the cut slopes for the Kimola Canal in Finland; in this respect, three planned failures of embankments on soft clay also are reviewed. Then, the failure of a pile foundation beneath a large oil storage tank at Fawley, England is presented, analyzed, and explained. Gaps in the existing mechanisms for investigating failures are identified, and it is proposed that the void be filled by creating a National Center for Investigating Failures. The center would be empowered to collect data and make independent measurements, but its reports would not be published until after all claims are settled; thus, it would supplement but not replace any of the existing mechanisms for investigating failures. A library documenting past experiences would be generated, and would be open for scrutiny by qualified parties, and then studies could be made to discriminate between more advantageous versus less rewarding approaches. Gradually, in this way sound methodologies for investigating failures would be developed: the lessons learned would be cumulative, and would provide focal points for the initiation of rewarding research. Immeasurable benefits would accrue both to the profession and to the public at large.

## KIMOLA CANAL, FINLAND

The Kimola Canal is located approximately 120 km northeast of Helsinki, Finland, and was constructed to facilitate the floating of logs and timber to market. It consists of two segments with zero gradient but at different elevations (13 m apart), connected by a tunnel. The soil deposits in the canal area are plastic ($w_L \simeq w \simeq 55$; $w_P \simeq 25$; $S_t \simeq 8$ to 16), slightly overconsolidated post-glacial clays overlying interbedded sandy and silty tills. The canal is in a valley that runs approximately NE-SW. Bedrock outcrops in ridges more or less parallel to the valley alignment.

When the first phase of the work on the lower canal was completed in the Spring of 1963 (Fig. 1), slopes 4–6 m high, cut to an inclination of 1:1.5, began to fail. In April–May, 1963, eleven separate failures were observed between stations 50 and 57 (700 m), and by the middle of October some of these had widened and some new ones had appeared. The location and date of these failures are shown in Fig. 2. The numerous small slides were of no economic importance as the failed material would be removed in the next phase of excavation; however, they called into question the $\phi = 0$ method of stability analysis used in the design of the canal. A comprehensive investigation was initiated which included studies of effective stress-strength parameters, installation of piezometers, and examination of different techniques for stability analysis.

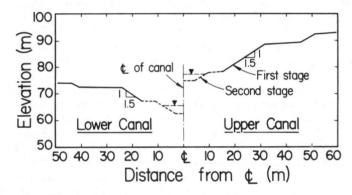

FIG. 1.—Representative X-sections of Upper and Lower Segments of Kimola Canal, Finland

In 1964, no slope failures occurred on the upper canal, although 1:1.5 slopes had been excavated to depths as much as 12 m. Immediately after the upper canal was excavated to grade, with cuts up to 15 m high, a large failure (30,000 m³ of clay) took place during the period March 30 to April 4, 1965 between stations 43 + 60 and 45 + 70 (Fig. 3). At this one location, a steep smooth rock wall without till cover dips towards the canal. At one end of the slide, some 50% of the back wall of the failure was bounded by smooth rock sloping towards the canal. Small (1,000–2,000 m³) slides have subsequently taken place in the upper canal every spring and fall (Fig. 3), although no other large failures have been observed. In fact, so many small slides have occurred since Fig. 3 was prepared that at present the cut slopes on the upper canal are hardly recognizable.

On November 3, 1965 a major slide occurred on the lower canal between stations 51 + 80 and 54 + 20. Suddenly, at a location where excavation to grade was completed nine months earlier, 90,000 m³ of clay slid into the canal. The slide blocked more than 200 m of the canal and cut the Kouvola-Heinola road (Fig. 2). By coincidence, the slide occurred at a location where piezometers

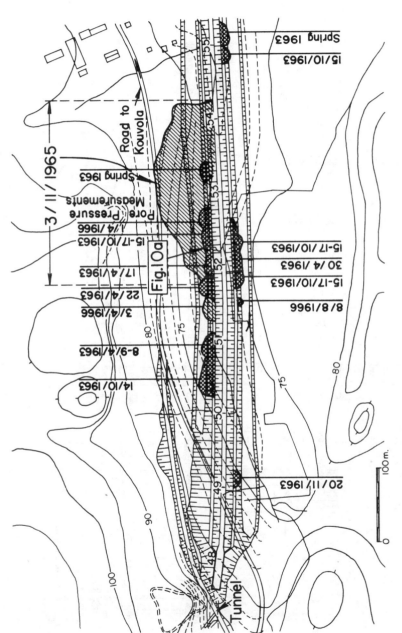

FIG. 2.—Plan View of Slides on Lower Kimola Canal (26)

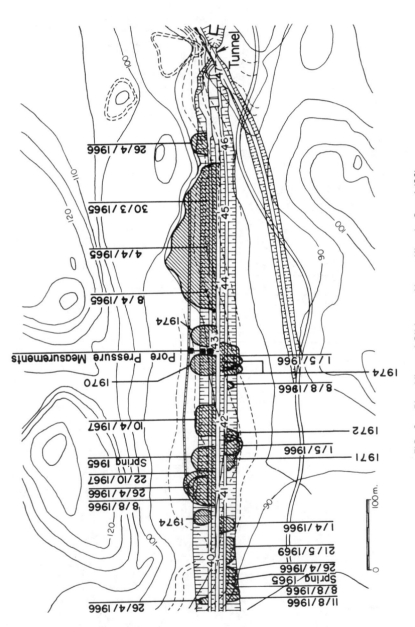

FIG. 3.—Plan View of Slides on Upper Kimola Canal (26)

had previously been installed. Two eyewitnesses, 350–400 m from the slide, heard a thunder-like clap and saw a surge rising in the canal. Within half a minute they arrived at the edge of the failure but movement had already stopped. They believed that the entire clay mass slid at the same time and gave "five seconds" as the duration of the failure. A cross section of the slide showing the material that slid into the canal is sketched in Fig. 4. Since November, 1965 no significant slope failures have occurred on the lower canal.

An outstanding doctoral thesis by Esko Kankare (26) contains complete documentation of all these events, and a summary of his detailed studies is readily available (27). Kankare made stability analyses of the large slide on the lower canal (at Sta. 52 + 70) considering critical circles for both the upper slope and the full slope (Fig. 4), using undrained as well as effective stress stability analyses. The undrained shear strengths were calculated from the results of conventional vane borings, and the $c'$-$\phi'$ strength parameters were obtained from isotropically consolidated, drained triaxial compression tests (with the consolidation pressure never exceeding the preconsolidation pressure). In-situ

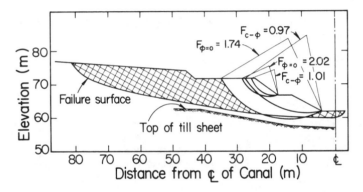

FIG. 4.—X-section of Nov. 3, 1965 Slide at Sta. 52 + 70 on Lower Canal (26)

pore-water pressures measured a few days before the slide occurred were used to calculate effective stresses. Similar stability analyses were made at Sta. 43 + 20 on the upper canal, where no failure had occurred. The results are shown in Table 1 and are noted on Fig. 4. Based on these studies, Kankare concluded the following:

1. The $\phi = 0$ analysis is unreliable for predicting the stability of cut slopes in such highly plastic, slightly overconsolidated clay deposits.

2. The $c'$-$\phi'$ analysis, using *measured* pore water pressures, accounted for observed behavior at the two locations analyzed.

3. The failure surface finally observed (Fig. 4), which involved so much more material than that predicted by the $c'$-$\phi'$ analysis, was caused by the development of a rapid retrogressive slide propagating along a slightly inclined sensitive clay bed at the level of the sliding surface.

The large slide at Sta. 52 + 70 on the lower canal (hereafter termed "the slide" at Kimola) has become a *cause célèbre* among geotechnical engineers debating the pros and cons of undrained versus effective stress analyses for slope stability investigations. For example, Andersen (1) conducted anisotropically consolidated, undrained shear tests on the Kimola clay (with $\sigma'_v = p'_o$ and $K_o = 0.5$) in triaxial compression, direct simple shear, and triaxial extension, and found $\tau_{max}/p'_o = 0.42, 0.33$, and $0.25$, respectively. He also found a 20% reduction in undrained shear strength when the strain rate was reduced from 1.0 to 0.001 percent/hr. Andersen concluded that by taking the effects of anisotropy and strain rate into account, undrained analysis gave a safety factor of nearly 1.0 on a failure surface close to the actual one, although it was acknowledged that "the failure mechanism is not yet fully understood, and will be subject to research at NGI in the future."

Bjerrum (8) identified four factors that complicate the evaluation of long-term stability of cuts in soft clays: (1) Undrained shear strength anisotropy; (2) undrained strain rate effects; (3) reduction in effective stress due to redistribution of pore-water pressures with time and associated swelling of the clay; and (4) cumulative creep strains exceeding the critical strain at which the clay structure

**TABLE 1.—Summary of Kankare's (27) Stability Analyses at Stations 52 + 70 (Nov. 3, 1965) and 43 + 20 (May 26, 1967)**

| Location (1) | | Factor of Safety | |
|---|---|---|---|
| | | $\phi = 0$ (2) | $c', \phi'$ (3) |
| Station 52 + 70 | Upper slope | 2.02 | 1.01 |
| (slide) | Full slope | 1.74 | 0.97 |
| Station 43 + 20 | Upper slope | 1.64 | 1.16 |
| (no slide) | Full slope | 1.71 | 1.17 |

breaks down, thereby reducing the strength sufficiently to induce failure. Bjerrum (7) had already proposed an empirical correlation with plasticity index (PI) to account for undrained anisotropy and strain rate effects in the case of embankments of soft clay, and in 1973 he indicated this same correlation could be applied to "end-of-construction" stability calculations of cuts. To deal with creep effects, Bjerrum proposed that Hvorslev's (24) shear strength parameters, which he called effective friction and effective cohesion, were mobilized at particle contact points; effective friction resulted from mineral to mineral contacts and effective cohesion was due to interparticle attractions. He reasoned that as the strain required to mobilize effective cohesion is much smaller than that required for effective friction (53), an increment of applied shear stress is first taken up largely by effective cohesion. As the frictional resistance is stable while the cohesive bonds are viscous, with time creep in the cohesive bonds transfers the stress to effective friction. If the effective friction can support the stress transfer, creep strains will decay with time until all the stress is carried by effective friction; if it cannot, the shear strains eventually will accelerate and failure will ensue.

The slide at Kimola was cited by Bjerrum (8) to illustrate the uncertainties involved in calculating long-term stability of cuts:

The Kimola canal was designed with a safety factor of 1.5 (Kankare, 1969a and b) based on the uncorrected vane strength. If the shear strength values are corrected in accordance with the procedure outlined above [empirical correlation with PI], the safety factor drops to 1.3, which thus should represent the conditions immediately after the excavation was completed. The failure occurred 9 months later. Most likely, the failure occurred within or at the end of the period of pore-pressure redistribution. The failure thus occurred as the result of the combined effect of a reduction in strength due to 'softening,' and a creep due to the shear stresses exceeding the available effective friction.

As one of the earliest and strongest advocates of effective stress stability analyses, (5,4), why was Bjerrum so reluctant to use an effective stress analysis to account for pore-pressure redistribution and softening of the clay slopes along the Kimola canal? It was because a major portion of the slip surface was a plane (Fig. 4) following approximately the same inclination as the natural stratification of the clay, i.e. parallel to the till layer. Experiences with such cases in Sweden had been disconcerting: conventional stability analyses in terms of effective stresses, introducing $c'$, $\phi'$ parameters obtained from drained shear tests, gave factors of safety on the order to 1.5 to 2.5 for slides at Säveån (12), Sköttorp (49), Guntorp (19), and Göta (50).

The clincher was a landslide that occurred in April, 1959 in a natural slope along the River Namsen at Furre, in central Norway. It was reported (23) that an essentially cohesionless soil mass slid as a unit on a thin seam of sensitive normally consolidated clay of low plasticity. Using pore-water pressures interpreted from field measurements in the adjacent unfailed sections, the back-calculated value of $\phi'$ was 7°. Isotropically consolidated undrained triaxial tests (with pore-pressure measurements) gave values of $\phi'$ ranging between 25° and 28°. This large discrepancy led Bjerrum to pursue the development of empirical adjustments to the undrained shear strength. It has been suggested (35) that at least part of the anomaly might be explained in terms of effective stresses if the following were recognized.

1. The effective sttress strength parameters are stress-path dependent (38,59).
2. $c'$ and $\phi'$ are strain rate dependent (38).
3. Constraints in laboratory strength tests (2,52) involve strain energies different from those released in-situ.
4. The accumulation of strain due to long-term creep can reach a critical strain level at which the clay structure breaks down; this is accompanied by a decrease in $c'$, $\phi'$ and an increase in compressibility, which can induce failure in undrained shear even though drained conditions had existed previously for a long period of time (9,35,36).

Kenney and Uddin (28) utilized the pore pressure data published by Kankare (1969a) to calculate the safety factor at selected times during the nine month period preceding failure. Their results are reproduced in Fig. 5, from which it was concluded that, "The case provides us with valuable assurances that in the case of slopes in weak clay soils the effective stress approach in estimating slope stability yields correct results when laboratory-determined shear-strength properties are used together with field-measured values of pore-water pressure." It is implied that there are no "strain-rate" nor "critical strain" effects in the sense proposed by Bjerrum, that isotropically consolidated (below $p'_c$) drained triaxial compression tests correctly reflect anisotropy in the $c'$, $\phi'$ parameters, and that an "initial" slide occurred whose boundaries approximately coincided with the critical circle in the $c'$-$\phi'$ analysis. This conclusion was strongly endorsed by Morgenstern (45) and by Janbu (25) in the General Report on Slopes and Excavations to the International Conference of Soil Mechanics and Foundation Engineering in Tokyo.

There is no doubt that a safety factor close to unity can be calculated for the slide at Kimola, using a critical slip circle along which pore-pressures are interpolated from readings on piezometers (Fig. 5) a few days before the slide,

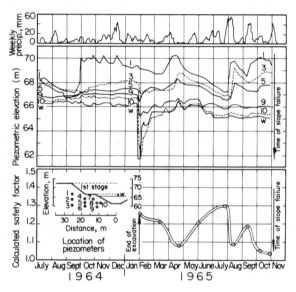

FIG. 5.—Piezometric Levels and Safety Factors at Sta. 52 + 70 on Lower Canal (28)

and using $c'$ and $\phi'$ obtained from triaxial compression tests. However, if this is to be interpreted to mean that the procedure "yields correct results," then it also should explain all other events that were experienced on the canal:

1. There is a difference of more than an order of magnitude between the volume of soil that slid and that predicted from the effective stress analysis (Fig. 4). Both Kankare (26) and Kenney and Uddin (28) attribute this to a rapid retrogressive slide that took place after an initial slip at the toe. The eye-witnesses' account tends to refute this explanation but, as it is not possible to reconstruct the *sequence* of events, it never can be proved whether or not a rapid retrogressive slide occurred. Considering the many other locations where small slides were observed, why has only one large slide occurred on the lower canal, and how can it be foretold when and where a large slide will occur?

2. On the upper canal no failures occurred at the end of construction. By Fall, 1966 more than a dozen slides had occurred (Fig. 3). Effective stress analyses near Sta. 43 + 20, where a line of piezometers was also available (Fig. 3), gave a safety factor of 1.17 on May 26, 1967 (Table 1); this seems to support the utility of effective stress analyses as no failures were observed in the vicinity of Sta. 43 + 00 to this date. However, slides occurred at this location in 1970 and 1974 and, as previously noted, so many toe slides have since occurred that the cut slopes on the upper canal are hardly recognizable. If this is due to subsequent development of less favorable pore-water pressures, why has no further sliding been observed on the lower canal since 1965? Moreover, all of these slides extend only to the toe of the first-stage excavation although the calculated safety factor for this case is the same as for the entire slope.

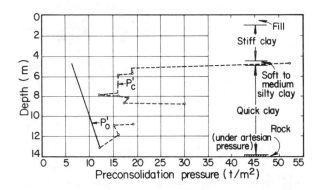

FIG. 6.—Preconsolidation Pressure versus Depth at Eugene 86 in Drammen, Norway

It is hard to escape the conclusion that important factors involved in the Kimola canal slides have not been identified in either the previously described undrained nor effective stress stability analyses. Once it is accepted that not all observed events have been explained adequately, the way is open to identify other possible contributing factors. These, in turn, may illuminate facets that heretofore were incorrectly interpreted, or even *unrecognized*. This turned out to be the case with the Kimola canal.

### Characteristics of Sedimentary Clay Deposits

Conditions during deposition of sedimentary clays vary. Sometimes these variations can be spectacularly rhythmic, forming "banded" or "varved" clays. The mechanisms responsible for the formation of varves are now well understood (51). As these deposits are readily recognized, attempts are made to deal with the special problems they raise (32,44). I am convinced that less marked nonrhythmic banding is characteristic of sedimentary clay deposits and that, in many cases, it is responsible for large discrepancies between predictions from conventional stability analyses and observed behavior. The first indication of the importance of this concept became apparent in the summer of 1964, when I made a study of building settlements in the town of Drammen, Norway. As shown in Fig. 6 (33), erratic scatter was observed in the variation of preconsolidation pressure ($p'_c$) with depth in a clay deposit which, until then, was thought to be relatively homogeneous (6). Because I obtained the samples and conducted the tests, random variations in test procedure and sample disturbance were not accepted as a plausible explanation. Fortunately, at the time, following the lead of Calvert and Veevers (13) and of Hamblin (21), x-radiography was being used at the Norwegian Geotechnical Institute to study structures in Norwegian clays (55). Advantage was taken of this opportunity

FIG. 7.—X-radiogram of Quick Clay at Eugene 86 in Drammen, Norway (55) (Actual Size, Approximately 10 cm)

and the radiograph shown in Fig. 7 was obtained. The variation in tone is due largely to differences in bulk density (or water content). It is evident that there are substantial differences in the soil profile within distances as small as a centimeter, or less, which can account for the erratic distribution in $p'_c$ and thus also in shear strength. The inclination of the "bedding" planes in Fig. 7 roughly parallels the slope of the underlying rock surface.

As the nonrhythmic banding shown in Fig. 7 is believed to be characteristic of sedimentary clays, they all contain continuous thin layers, or "seams," parallel to the depositional bedding planes whose strengths can vary erratically. The weakest seams are preferred planar surfaces that can control the location and extent of sliding. They may be sufficiently thin that conventional techniques for measuring shear strength would not assess their actual resistance. Thus, whenever a slip surface is observed to be partly planar, it is unlikely that the shear strength along this surface was correctly assessed; accordingly, the calculated safety factor (SF) should differ from unity. If it happens to equal one it could mean there are other compensating errors, and not necessarily that the method of analysis is correct! The realization of this fact prompted a careful review of potential errors in previous stability analyses of the slide on the Kimola canal.

TABLE 2.—Effects of Stress Path on Effective Angle of Shearing Resistance ($\phi'$) for Undisturbed, Normally Consolidated Haney Clay (59)

| Stress path after $K_o$ consolidation (1) | Effective Friction Angle, $\phi'$ | |
|---|---|---|
| | Failure condition, $(\sigma_1 - \sigma_3)_{max}$ (2) | Failure condition, $(\sigma_1'/\sigma_3')_{max}$ (3) |
| Drained tests | | |
| Triaxial compression | 26.2 | 26.2 |
| Plane strain compression | 29.4 | 29.4 |
| Triaxial extension | 34.3 | 34.3 |
| Plane strain extension | 34.7 | 34.7 |
| Undrained tests | | |
| Triaxial compression | 21.4 | 29.8 |
| Plane strain compression | 25.2 | 31.6 |
| Triaxial extension | 33.8 | 33.8 |
| Plane strain extension | 34.3 | 34.3 |

## Kimola Canal Failures Revisited

At the time it was published, Kankare's thesis was the best documented case record of a slope failure that had come to my attention. Nevertheless, important uncertainties in the main input parameters were not adequately explored.

**Effective Stress Strength Parameters.**—It has long been recognized that the undrained shear strength of soft clays is markedly anisotropic (11,30) and strongly dependent on strain rate (14,15). It seems evident that such behavior in undrained shear merely reflects the response in terms of effective stresses; i.e., $c'$ and $\phi'$ also should be affected by stress path and strain rate. Table 2 shows how important these effects may be. Even in drained shear, the difference in $\phi'$ values between triaxial compression and plane strain extension is 8.5°, while the corresponding difference in undrained shear is nearly 13°! In undrained shear, the value of $\phi'$ is also dependent on the failure criterion adopted, mainly because the post-peak pore pressures are increasing at a sufficiently rapid rate that $\sigma_1'/\sigma_3'$ continues to increase even though $(\sigma_1 - \sigma_3)$ is decreasing (35). This means that if failure is induced by any mechanism that causes structural breakdown of a soft clay, including creep, values of $\phi'$ determined at peak

TABLE 3.—Results of Triaxial Compression Tests on Kimola Clay (26)

| Station (1) | Borehole, in meters from the center line (2) | CIU-TESTS ||||||||| CID-TESTS ||||
|---|---|---|---|---|---|---|---|---|---|---|---|---|---|
| | | Level, in meters (3) | $n^a$ (4) | $\bar{w}^b$ (5) | $(\sigma_1 - \sigma_3)_{max}$ |||| $(\sigma_1'/\sigma_3')_{max}$ ||| Level, in meters (11) | $n$ (12) | $\bar{c}'$, kilograms per square centimeter (13) | $\bar{\phi}'$ (14) |
| | | | | | $\bar{c}'{}^c$, in kilograms per square centimeter (6) | $\phi'{}^c$ (7) | $n$ (8) | $\bar{c}'$, in kilograms per square centimeter (9) | $\bar{\phi}'$ (10) | | | | |
| 43 + 00 | 33 left  | 81.8 ... 69.9 | 28 | 43.6 | 0.174 | 22.1 | 24 | 0.117 | 27.3 | 87.5 ... 79.5 | 15 | 0.122 | 27.6 |
| 50 + 20 | 38 right | 67.5 ... 61.7 | 11 | 69.2 | 0.031 | 22.1 | 8  | 0.022 | 26.7 | | | | |
| 52 + 70 | 14 left  | 66.5 ... 58.0 | 36 | 64.7 | 0.147 | 20.5 | 30 | 0.098 | 28.2 | | | | |
| 52 + 70 | 24 left  | 70.9 ... 65.6 | 15 | 50.9 | 0.070 | 25.7 | 13 | 0.107 | 27.8 | | | | |
| 52 + 70 | 24 right | | | | | | | | | 68.8 ... 61.7 | 16 | 0.049 | 27.7 |
| 55 + 20 | 24 right | 66.5 ... 57.9 | 8  | 52.0 | 0.068 | 22.2 | 7  | 0.062 | 26.3 | | | | |
| 55 + 70 | 16 left  | 66.1 ... 57.5 | 26 | 48.7 | 0.098 | 25.7 | 25 | 0.083 | 28.7 | | | | |
| 55 + 70 | 22 left  | 66.4 ... 60.2 | 8  | 58.1 | 0.061 | 21.7 | 20 | 0.084 | 27.0 | | | | |
| 62 + 80 | 16 left  | 62.4 ... 54.9 | 23 | 72.6 | 0.105 | 19.0 | 25 | 0.084 | 27.8 | | | | |
| Average | | | | | 0.100 | 22.2 | | | | | | | |

[a] Number of samples.
[b] Average water content.
[c] Effective stress, strength parameters for each borehole, determined by drawing the best fit straight line defined by $n$ sets of $1/2\,(\sigma_1 - \sigma_3)$ and $1/2\,(\sigma_1' + \sigma_3')$ test results on undisturbed samples.

$\sigma_1'/\sigma_3'$ will not be mobilized in-situ. For the Haney clay from British Columbia (Table 2), this difference in $\phi'$, determined from undrained triaxial compression tests, was 8.4°.

Results of triaxial compression tests on the Kimola clay are given in Table 3. Values of $c'$, $\phi'$, determined at $(\sigma_1 - \sigma_3)_{max}$ vary from borehole to borehole, although tests from a substantial number of samples in each borehole were "averaged" to obtain the values cited. See Leonards (35) for a simple explanation of the effect of variation in soil properties on the interpretation of failures. It is important to recognize that a calculated safety factor equal to, say, 1.5 does not necessarily mean that the probability of failure is very low. Conversely, a calculated safety factor of 1.5, based on *mean strengths*, may have been correctly determined although failure had occurred.

Moreover, $\phi'$ at $(\sigma_1 - \sigma_3)_{max}$ in undrained shear is (at Sta. 43 + 00) 5.5° less than in drained shear, compared with 4.8° for corresponding tests on the Haney clay (Table 2). Accordingly, it can be concluded that the $c'$-$\phi'$ parameters

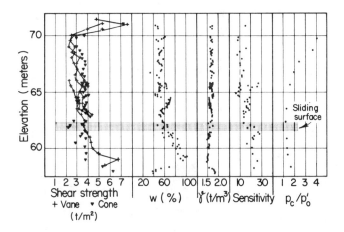

**FIG. 8.—Geotechnical Profile at Sta. 52 + 70 on Lower Canal (26)**

for the Kimola clay are also strongly dependent on stress path. To obtain a safety factor of one on a critical slip circle, the strength parameters used were $c' = 0.05$ kg/cm$^2$ and $\phi' = 27.7°$ (drained triaxial compression test results). As the stress path on much of the failure surface is not simulated by a state of triaxial compression, and as the variance of the strength parameters is large, the value of $\phi'$ used in the analysis may not be at all appropriate. Thus, although the calculated safety factor was equal to one it does not follow, a priori, that the analysis was correct.

**Pore-water Pressures.**—Fig. 5 shows the location of piezometers and the piezometer readings at Sta. 52 + 70, which is essentially at the midpoint (in plan) of the large slide on the lower canal. Equipotential lines were drawn and the pore pressures along a potential slip surface were interpolated. While this approach is reasonable insofar as the phreatic surface and the water level in the canal are concerned, if excess pore pressures were developing due to breakdown of clay structure along a weak seam, pore pressures along the actual slip surface could be much higher than the interpolated values. Moreover, these excess pore pressures could propagate from zones of high shear strain along the planar slip surface and induce a rapid progressive (not retrogressive) failure although there would be no surface evidence that this was occurring. Such behavior is especially troublesome when the slip surface passes through clay layers of markedly different stiffnesses.

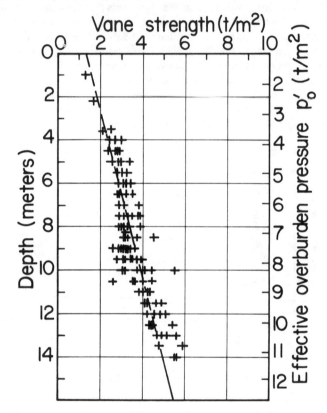

FIG. 9.—Composite of Results of Vane Borings at Sta. 52 + 70 on Lower Canal (26)

The foregoing analysis has two important implications:

1. Pore-water pressures measured in a slope adjacent to an actual slide may not be representative of the pore pressures in the sliding zone.

2. If the effects of creep are to be investigated, of if the danger of an impending slide is to be assessed from pore pressure measurements, then: (1) The piezometers must be capable of responding to rapid changes in pore-water pressure; (2) automatic read-out in response to a rapid change in pressure is required; and (3) a sufficient number of piezometers must be installed to assure a high probability that some of them will reflect effects in the actual slip zone. These demanding requirements severely restrict the use of effective stress stability analyses, even when field measurements of pore-water pressures are available.

Fig. 5 shows how rapidly negative excess pore pressures due to excavation equilibrate with the natural ground-water regime. The response was immediate, and within a week, virtual equilibrium was achieved. This response is very much faster than heretofore believed (4,8), and is characteristic of excavations in soft clays. It is important to recognize not only that undrained conditions are not critical, but that the reduction in shear strength due to dissipation of negative excess pore pressures is essentially complete in a matter of a few weeks, or less. Consequently, delayed failures of longer duration must be due to other causes.

**Weak Seam at Kimola.**—The preceding analysis was intended to establish that a calculated safety factor of unity, using conventional effective stress analysis, could result from compensating errors and does not preclude an alternate explanation for the cause of sliding. In retrospect, there is strong evidence pointing to the presence of a weak seam at Kimola that controlled the depth, extent, and timing of the slide:

1. The geotechnical profile (Fig. 8) shows that, at the depth of the sliding surface, the clay has a lower shear strength, is more sensitive, and becomes more nearly normally consolidated than at any other depth in the profile.

2. A composite of vane shear strengths (Fig. 9) taken before the slide at Sta. 52 + 70 shows anomalously low strengths at depths of 9 and 11 m, i.e.,

FIG. 10.—(a) Slide Between Sta. 51 + 70 and 52 + 57 on Lower Canal (photo by S. Korhonen, Nov. 21, 1963); (b) Slide Between Sta. 50 + 75 and 51 + 07 on Lower Canal (photo by E. Martti, Apr. 4, 1963)

at depths which are in the vicinity of the failure zone (it is unlikely the van sensed the strength of the weakest thin seam).

3. The long, planar sliding surface is more or less parallel to the underlying till sheet (Fig. 4).

4. A photograph of an impending slide between Sta. 51 + 70–52 + 57 is shown in the foreground of Fig. 10a (its location in plan is shown in Fig. 2). Noting the large tension crack, it is evident that the slide is in an advanced stage. The absence of a scarp, and the mud wave at the toe are characteristic of planar displacements and not rotation about a slip circle. A photograph of an adjacent slide is shown in Fig. 10b. Again, the near-vertical scarp and the broken up soil mass are not characteristic of rotation along a slip circle.

5. Numerous small slides occurred all along the upper and lower segments of the canal (Figs. 2 and 3). Why have no other large slides developed? A possible answer is suggested by the ground surface contour lines shown in Figs. 2 and 3. It may be that the geometry of the dipping weak seam is more

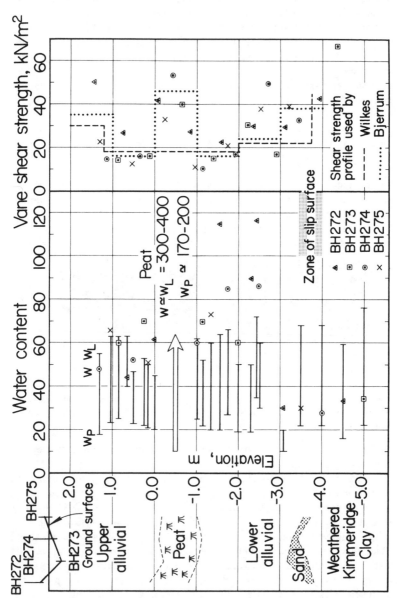

FIG. 11.—Geotechnical Profile at King's Lynn Trial Embankment, England (62)

favorable to sliding in the vicinity of Sta. 52 + 70 and 44 + 60 than at any other locations.

Fig. 10a is very revealing. Once a small tension crack develops, stability is impaired and failure along a planar slip surface is impending. The actual demise of the slope may take place months later and be associated with heavy rainfall that soaks the surficial soils and fills the cracks with water, as was the case at Kimola. Analysis of such slides using slip circles, pore-water pressures measured outside the slip zone, and $c'$, $\phi'$ parameters determined from drained triaxial compressions tests, surely is not representative of the real conditions and hardly could be used to judge the validity of the approach, although the calculated safety factor does happen to equal one.

If weak seams are characteristic of sedimentary clay deposits, there should exist other published case records (in addition to the previously cited landslides in Sweden) in which slip surfaces were observed to be at least partly planar [or the evidence indicates planar slip surfaces but they were not thus identified]. Three such cases, all of them planned failures, are especially relevant in this connection, and will be reviewed briefly.

## King's Lynn Embankment

The trial embankment at King's Lynn, England was described by Wilkes (62). The main features of the geotechnical profile are summarized in Fig. 11.

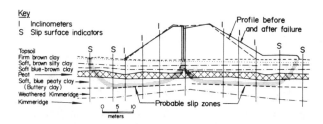

FIG. 12.—Slip Zones Beneath King's Lynn Trial Embankment (62)

The section at which failure was induced had been instrumented with 37 piezometers, 6 inclinometers, 4 slip surface indicators, 9 settlement gages, and 36 heave and thrust gages.

In the design stage, stability calculations were made using the vane strengths represented by the plotted points in Fig. 11, and it was found that with an embankment height of 6.25 m, failure would occur on a slip circle just reaching the peat layer.

The embankment was constructed on a slope of 1:1.4 and cracking was observed when the height reached 6.36 m. Failure occurred at 8:00 a.m. on the morning following its completion to a height of 7.05 m (including 0.33 m of settlement in the alluvium); suddenly, a heave of about 1.5 m occurred at the toe and a scarp of 1.25 m developed on the crest. The slip surfaces interpreted by Wilkes are shown in Fig. 12. The actual slip zone was much deeper than predicted and its major portion was planar rather than circular in shape. Using the shear strength profiles noted in Fig. 11, undrained stability analyses were carried out (by Wilkes and by Bjerrum at the Norwegian Geotechnical Institute) on slip circles and on the observed failure surface. The results are summarized in Table 4. Undrained triaxial compression tests on samples from the peat gave $S_u = 16$ kN/m$^2$; thus, Bjerrum also made calculations using this strength value for the peat. (The units used in the original papers are followed to preserve accuracy and authenticity of the case records.) The results are shown in parentheses in Table 4. The difference in calculated safety factors along the observed slip surface is due to the different shear strengths used by Wilkes and Bjerrum along the planar portion of the slip surface. Wilkes used 22 kN/m$^2$ while Bjerrum used 38 kN/m$^2$. Examining Fig. 11, Bjerrum's strength value appears to reflect the measured data reasonably well, yet the safety factor on the observed slip surface was found to be much higher than one. The use of $S_u = 22$ kN/m$^2$ along the planar portion of the observed slip surface, as Wilkes did in his analysis, is hardly justified by the available data but it is necessary to back-calculate a safety factor of one using the observed slip surface.

**TABLE 4.—Summary of Undrained Stability Analyses, King's Lynn Trial Embankment, Based on Field Vane Tests (62)**

| Analyses (1) | Factor of Safety | |
|---|---|---|
| | Critical slip circle (2) | Observed slip surface (3) |
| Analysis by Bjerrum | 1.38 (1.05) | U.C. Berkeley Program[a] 1.72 (1.49) |
| Analysis by Wilkes | Little and Price 1.02 | Janbu 1.0[b] |

[a] Slope 8R, based on Spencer's method.
[b] Subsequently, in his thesis, Wilkes corrected this value to 1.18.

Thus, it is pertinent to ask what might have been done during the initial site investigation in order to predict the observed behavior of the trial embankment?

The first prerequisite is to take the position that thin weak layers or seams are characteristic of alluvial deposits, and to plan the site investigation so that if any are extant there is a high probability they will be identified. This requires the use of inexpensive probing tools that can sense thin layers, such as cone penetrometers or piezometer probes (57) with tip diameters on the order of 10 mm. Suspected weak zones then can be sampled, and examined in the laboratory prior to extrusion, using x-rays (29) or gamma-ray scanners (20). Once identified, the weak seams can be extruded and their shear strengths tested separately. This approach to site investigations in soft clays should be adopted in every day practice without further delay.

It is of more than historical interest to note that the Eau Brink Cut, which was constructed in 1821 to eliminate a broad meander in the Great Ouse River, lies immediately west of the trial embankment. In 1943, a slip occurred in this cut due to erosion near the toe (54). Borings were made and the position of the shear surface was found "with some certainty." Using undrained analysis, the calculated safety factor (SF) was unity along a critical slip circle. The "actual slip surface" was interpreted to be a circle (although the evidence for this is not conclusive) that was considerably different from the critical slip circle.

Moreover, on the actual surface, the SF was 1.3. It was concluded that undrained analysis gave a correct safety factor but did not lead to the correct slip surface, an anomaly considered tenable in many quarters to this day. It is time to lay this notion to rest and to conclude that the shear strengths along the actual slip surface were not evaluated correctly.

### James Bay Test Embankment

Dascal and Tournier (18) reported on a test embankment that was constructed

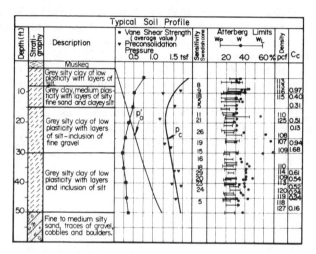

**FIG. 13.—Geotechnical Profile at James Bay Test Embankment, Site R-7, Quebec (18)**

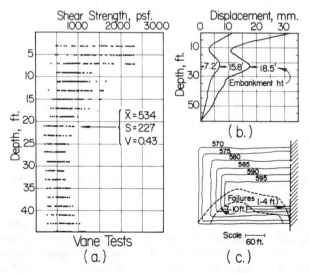

**FIG. 14.—(a) Composite of Field Vane Strengths; (b) Lateral Deformation versus Depth at Different Embankment Heights; and (c) Plan View of Scarps after Failure, James Bay Test Embankment (18)**

on the soft, sensitive marine clay along the Rupert River in the James Bay (Canada) area. A representative geotechnical profile at site R-7 is shown in Fig. 13. A composite of the vane shear strength profiles is shown in Fig. 14a. While the scatter appears large, the variance, $V$, of the undrained shear strengths is not unusual; nevertheless, the data pose a formidable challenge to the designer selecting an appropriate shear strength profile for use in deterministic stability analyses. Only a probabilistic approach can deal with the situation in a rational manner.

The embankment was constructed in less than 2 months and was instrumented with 13 piezometers, 12 settlement plates, 8 lines of surface monuments, and one inclinometer at the toe of a 1:2 slope. Although noticeable localized lateral deformations were observed at a depth of about 22 ft (6.7 m) beneath the bottom of the embankment as the fill height approached 18 ft (5.5 m) (Fig. 14b), failure (as evidenced by surface discontinuities) occurred abruptly at a fill height of 25 ft (7.6 m). In a few minutes, a two-step scarp (Fig. 14c) with a total height of about 10 ft developed on the crest, and a heave of nearly 6 ft (1.8 m) was recorded at the toe. The critical and "observed" slip *circles* given by Dascal and Tournier are shown in Fig. 15. The difference between the positions of the two circles is considerable, with the critical circle falling outside the bounds of the observed displacements.

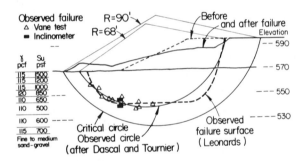

**FIG. 15.—Slip Surfaces Beneath James Bay Test Embankment (18)**

At each depth, mean values of the vane shear strength (Fig. 14a), both uncorrected and corrected according to Bjerrum (7), were used to make undrained stability analyses. The results are given in Table 5. To examine the influence of the granular fill, calculations were made assuming $\phi_{fill} = 0$ and 35°. In this case, the effect of the strength of the fill was of no importance.

From investigations of an earlier nearby test embankment at Matagami, Quebec, Dascal et. al (17) had concluded that similar discrepancies between calculated and observed behavior were due to progressive failure effects. The results shown in Table 5 were considered to reinforce this previous conclusion, and a modification to Bjerrum's correction factor was proposed in which a progressive failure effect was added to those of time and anisotropy, although Bjerrum's chart is empirically based and already included a progressive failure effect. It is agreed that progressive failure plays a role in the stability of slopes and embankments (35), but it is of importance here to consider also the possibility that a thin, weak layer is located at a depth of 22 ft (6.7 m) below the ground surface, a possibility previously proposed (43) for the companion embankment at Matagami. This is suggested by the lateral spreading at this depth indicated by the slope indicator and by the composite plot of vane test results shown in Fig. 14, which shows the minimum value to be at this depth. Considering the depositional history, it is expected that the weak zone pervades an area larger than the zone influenced by the embankment loading. Based on the aforementioned considerations, the "observed" failure surface was redrawn (Fig. 15). An undrained stability analysis on the revised slip surface, using a shear strength of 200 psf (95.8 kPa) on the planar portion, gave a safety factor of 1.3 if the planar portion is horizontal and 1.2 if it dips down 5° (an inclination approximately equal to that of the ground surface). These safety factors are comparable to the ones listed in Table 5 for the observed slip circle.

All of the safety factors previously cited for the R-7 embankment are unreliable, for the following reasons:

**TABLE 5.—Results of Stability Calculations, James Bay Test Embankment**

| Vane shear strength (1) | FACTORS OF SAFETY | | | |
|---|---|---|---|---|
| | $\phi_{fill} = 35°$ | | $\phi_{fill} = 0$ | |
| | Uncorrected (2) | Corrected[a] (3) | Uncorrected (4) | Corrected[a] (5) |
| Critical slip circle | 1.21 | 1.18 | 1.17 | 1.12 |
| Observed slip circle | 1.36 | 1.29 | 1.27 | 1.21 |

[a] As a representative value of the plasticity index is approximately 15, the correction factors actually should be slightly greater than one.

1. The critical slip circle lies outside the bounds of the observed displacements, and thus is inappropriate.
2. The "observed" failure surfaces, whether circular or composite, were assumed to be defined by the outer bounds of the observed movements. As a dual scarp developed, the slide was probably retrogressive.
3. In the case of a composite slip surface, when the shear strength along a planar portion is much smaller than on the remainder of the surface, the usual assumptions made to effect the stability analysis are considered to be unreliable.
4. Incompatibility between the strain distributions induced by the loading and the relative stiffnesses of the underlying clay strata aggravates the effects of progressive failure, which makes selection of the "operating" shear strength in the clay crust both difficult and arbitrary. At R-7, the strong tendency for lateral deformations induced tensile strains, which further exacerbated the problem. The sudden failure, without prior evidence of surface cracking, is a consequence of the brittle response to tensile straining.

In earlier times, embankments on soft clay designed with safety factors on the order of 1.5 were less likely to fail than is the case today. The reason is that shear strengths were determined by testing samples of low quality and the ordinary method of slices was used for stability analysis. Now that better quality samples are routine, and more "exact" methods of stability analysis are in common use, there is much less in reserve to accommodate other errors, such as an incomplete assessment of the soil stratigraphy or of the mechanism of failure. If these latter errors are not eliminated, then higher factors of safety should be used for design, but not necessarily in relation to the PI of the clay. Otherwise, it must be recognized that embankments on soft clay, designed according to present practice, involve a high risk of failure.

The R-7 test embankment underscores an earlier conclusion (34) that the same "fundamentals of soil mechanics" can be applied to the same data by

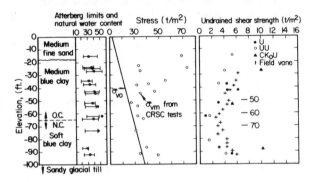

FIG. 16.—Geotechnical Profile at Center of I-95 Test Embankment at Sta. 263, Boston, Mass. (42)

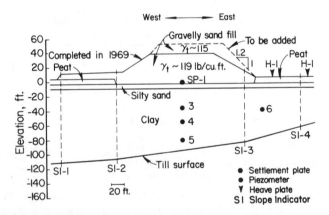

FIG. 17.—X-section of I-95 Test Embankment Before Failure (42)

different investigators with radically different results. On the one hand, an additional correction to the safety factor obtained from a critical circle falling well outside the bounds of the observed displacements was proposed; while on the other hand, a weak seam was postulated to explain the unexpected location of the failure surface. The key point here is that the initial conception of the problem controls the kind of investigation that is made and how the data, thus gathered, will be used, either in design or in back-analysis of failures. Had the possible existence of a weak seam beneath the R-7 embankment been considered, the site investigation would have been different and additional instrumentation would have been installed to define more closely the actual shape of the slip surface.

**I-95 Test Embankment.**—The test embankment was initially constructed in 1967-69 on 1:2 slopes to a height of 35 ft (10.7 m) above the original ground. A geotechnical profile at the center of the embankment (at Sta. 263) is shown in Fig. 16. The clay layer is Boston blue clay—a post-glacial clay deposited in brackish waters. The original purpose was to investigate the reliability of various techniques for predicting settlement, horizontal movements, and pore pressures (16). In 1974, the embankment was widened along the east slope until it was steepened to 1:1.2 and fill was dumped on the crest at a rate of 1 to 2 ft/day (0.3 to 0.6 m/day) to failure. Instrumentation consisted of four piezometers, four slope indicators, a settlement plate, and two heave markers (Fig. 17). A Symposium was organized at the Massachusetts Institute of Technology (42) to predict: (1) Additional horizontal movement at SI-3 and SI-4; (2) additional settlement of SP-1; (3) additional pore pressures at P-3,

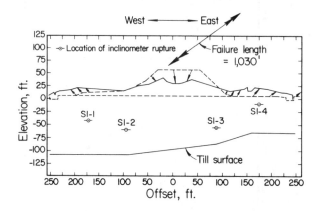

FIG. 18.—X-section of I-95 Test Embankment After Failure (42)

P-4 and P-6, all after 6 ft of fill had been added; and (4) the additional height of fill and the pore pressure at failure. When the height of additional fill reached 18.7 ft, within minutes, simultaneous failures to *both* sides of the embankment caused the crest to drop about 30 ft (9.2 m); a heave of 14 ft (4.3 m) was measured on the east side and 10 ft on the west side. A crack 14 ft (4.3 m) wide at the top and 8 ft (2.4 m) deep opened 140 ft (42.7 m) to the east of the centerline (Fig. 18). The instrumentation gave no clear indications of impending failure.

The approach taken by the predictors ranged from simple hand calculations, to the use of sophisticated computer programs and centrifuge model tests. A comparison between observed and predicted fill height increments at failure is shown in Fig. 19. The prediction error ranged from +44% to −57%. There was no relation between the level of sophistication used in the analysis and the quality of the end result; none of the predictors consistently came the closest in predicting different aspects of behavior; and none predicted the simultaneous failure of the west side of the embankment (where the lateral extent of movement was larger), since a berm remained in place on this side and the slope was much flatter (1:2 compared with 1:1.2). Among the data furnished, the predictors were horizontal displacements measured during the initial construction stage (Fig. 20). These showed that the movements to the west were not only larger but extended a greater distance from the center line than those to the east. None of the predictors took cognizance of this information, apparently because they were "preconditioned" to predict movements to the east. In the post mortem analyses, the most striking aspect of the I-95 test embankment—the fact that

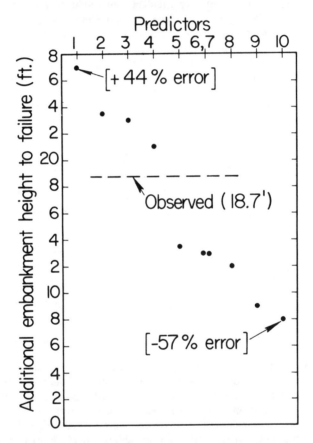

**FIG. 19.—Comparison of Predicted versus Observed Failure Heights I-95 Test Embankment (42)**

failure occurred, and was more extensive, on the side where the net loading conditions were less severe—elicited comment from only one participant who mentioned he thought the vane strengths were somewhat lower on the west side than on the east (this is not borne out by the available test data).

Fig. 21 shows a set of postulated failure surfaces based on the location of inclinometer ruptures and the depths at which the lowest undrained shear strengths were measured (e.g., at 63 ft (19.2 m) in Fig. 16). The selected shape of the slip surfaces was strongly conditioned by the belief that a weak seam existed at the indicated depth, dipping more or less parallel to the till surface. From data such as that shown in Fig. 16, and on the assumption that the tests did not measure the strength of the weakest layer, it was estimated that the shear strength along the planar portions of the slip surfaces could be as low as 1.5 t/m$^2$. Undrained stability analyses gave safety factors ranging from 0.35–1.2, depending on the method used for stability analysis, the strength assigned to the clay in the nonplanar portions of the slip surface, and on whether or not the shear strength of the fill was neglected. While this is not encouraging news for the designer, for a consistent set of assumptions, comparable safety factors were obtained for sliding either to the east or to the west. Thus, although

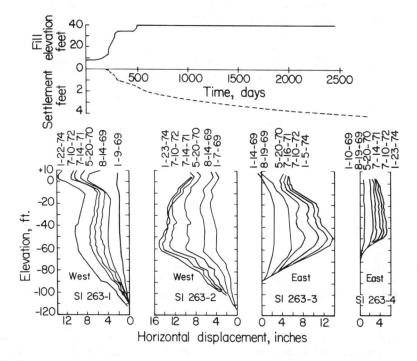

FIG. 20.—Settlement and Horizontal Displacements at Sta. 263, I-95 Test Embankment (42)

no special credence is placed on the magnitude of the safety factors, the results demonstrate that the control exercised by the weak seam on the geometry of the slip surfaces makes failure to either side equally plausible!

**Lessons Learned**

In the past two decades, literally dozens of unexpected failures have occurred in natural slopes or cuts, or in embankments on soft, nearly-normally consolidated clays. Many of these failures have been investigated extensively and reported in the literature. Did all this effort and expense lead to more rational design methods, or to better ways of investigating future failures? My answer is categorically, '*No.*'

The slide on the Kimola canal illustrates the inherent difficulties associated with gaining new knowledge from post-failure investigations of unexpected instabilities: (1) Neither the undrained nor the effective stress strength parameters of relevant soil strata in the slide zone can ever be recovered; (2) although piezometers were read a few days before the slide occurred, it never can be known what the pore-water pressures were on the actual failure surface at the initiation of failure; (3) despite eye-witness accounts, a rapid retrogressive failure cannot be ruled out, thus it is not known whether backcalculations using the full extent of the slide are relevant or misleading; and (4) different failure hypotheses cannot be tested by installing new instrumentation in the old slide.

All these limitations notwithstanding, I contend that more could be learned from investigating failures than has been the case heretofore, and the slides at Kimola were used as an example to support this claim. Although dozens of slides occurred at times after excavation varying from a few months to more than ten years, only the large slide at Sta. 52 + 70 was dealt with in the five publications referring to the Kimola slides. Kankare's (26) outstanding

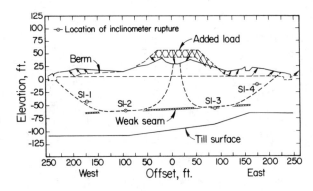

**FIG. 21.—Interpreted Locations of Actual Slip Surfaces, I-95 Test Embankment**

thesis documents the other events, but even in this case no analysis was made to explain the enormous range in time span after construction before delayed failures occurred. In mathematics, a partial solution to a problem often can be considered as a valid contribution to further progress, but a partial explanation of observed failures simply may mean that the real causes were not identified.

At Kimola, a case was made for the existence of a thin seam of weak clay, dipping towards the canal, that controlled the depth, areal extent, and timing of the slides. Three additional case records were also reviewed, at King's Lynn (England), James Bay (Canada) and Interstate Highway, I-95 (U.S.A.). These cases involved embankments on soft clay that were located thousands of miles apart. They were instrumented and loaded to failure to investigate the reliability of the methods used to predict performance. The location and extent of the failures were unexpected, and slip surfaces comprising planar sections were either observed directly (King's Lynn) or were inferred from the measured data. Virtually all of the prefailure stability analyses were made using slip circles. The following question is raised:

Are we so conditioned by repeated slip-circle analyses that planning of geotechnical investigations, including instrumentation to monitor performance, is controlled by the concept that slippage will occur only along a circular arc?

The pattern of noncircular slip surfaces that persists for the cases that were reviewed—and for many others not considered herein—was intended to emphasize how often a circular slip surface, even in soft clays, is a false precept that must be laid to rest. Specifically, a plea is made that in the future at least the following factors be given due consideration:

1. Soft, nearly-normally consolidated deposits of sedimentary clay are likely to contain thin, relatively weak seams dipping parallel to the depositional bedding planes. Within these seams the shear strength is markedly lower than in the abutting clay layers. When their orientation is unfavorable with respect to potential slip surfaces, they can control the occurrence, depth, areal extent, and timing of instabilities. If the weak seams are not identified, the occurrence or the extent of the instability, or both, will be unexpected. A primary aim of the site investigation should be to detect and map these weak zones, and to measure their strength characteristics using appropriate stress paths and strain rates. Fortunately, tools for this purpose are available, and there is much current activity to simplify their operation and to improve their reliability. There is no longer any reason to ignore their use in everyday practice.

2. In planning instrumentation to monitor field performance, consideration should be given to broader ranges of potential behavior than that suggested by the location of critical slip circles. At the very least, zones of major slippage must be identified by direct measurements. For this purpose, inclinometers and slip surface indicators are far superior to piezometers, settlement plates, or surface monuments. Research should be initiated to develop instruments that can detect where slippage begins and how it subsequently propagates; for example, gages that can define displacement or strain vectors within the soil mass.

3. When an instability is investigated, *all* observed events, in proper sequence, must be documented carefully. Initially, every possible hypothesis of failure should be given equal consideration; no potential failure mechanism should be discarded until proved to be very unlikely. A valid analysis must be capable of explaining all observed events, not just selected aspects of the failure. Studies made to reject alternate failure hypotheses should be reported in the same full detail as the one deemed to have caused the failure. If alternate failure hypotheses cannot be rejected on a rational basis, this should be admitted openly.

4. Conventional designs of cuts in, or embankments on, soft clays induce zones of overstressing. Elsewhere along the potential failure surface, stress levels can be sufficiently high that creep strains are important. Naturally sedimented clays possess critical strain levels (which depend on the strain history) at which the clay structure breaks down. The process is accompanied by an increase in compressibility and also may involve a decrease in the effective stress-strength parameters. The breakdown in clay structure generates positive excess pore-water pressures and a progressive undrained failure can occur rapidly even though drained conditions were previously extant for a long period of time. Piezometer measurements made on adjacent, unfailed sections will not

reflect the excess pore-water pressures that existed in the slip zone at failure, and neither will piezometers located within the slide if they are not close to the slip zone and do not have a sufficiently rapid response time. This may explain why, in the past, piezometers generally have been poor indicators of impending failure.

5. Conventional design of cuts and of embankments on soft clay involve local overstressing and creep phenomena which, coupled with nonuniform stiffnesses and ductilities of natural clay strata, induce failures that are generally progressive in nature. At present, the effect only can be accounted for empirically. Thus, the designer must continue to rely heavily on experience within a given geologic region—experience that has been carefully digested in terms of local practice for site investigation, sampling, testing, and analysis. There is nothing to gain—and much to lose—by not openly admitting this fact.

6. Probabilistic approaches are the only way inherent variability in soil properties can be rationally handled. The probability of failure, or the reliability of a structure, is so superior in concept and in utility to that of a safety factor, that its use in routine design must be rapidly assimilated. On the other hand, the properties of soil deposits definitely are *not* random. Probabilistic methods can be successful only if the deterministic model is representative of the real stratigraphy. If not, no amount of statistical manipulation can remedy the error.

7. A new perspective on the roles of theory and field measurements is needed. Repeated calls for more and more field measurements seem to have lulled us into believing that they alone can solve our problems. Field measurements, appropriately conducted, provide the ultimate test of the validity of the theory. However, it is clear that the kinds of measurements made, where they are located, and how they are interpreted are all so dependent on the concepts derived from existing theories, that the tendency persists to "validate" inapplicable procedures. While it may be true that our ability to *compute* is nearly boundless, much of our knowledge of what to input is still rudimentary. Better concepts, and new theories to facilitate their use, are needed just as much today as they were when Terzaghi formulated the theory of consolidation.

## OIL TANK AT FAWLEY, ENGLAND

The following is quoted from the preamble to the only published record describing the failure of a pile foundation supporting an oil tank 260 ft (79.3 m) in diameter and 64 ft (19.5 m) high at Fawley, England (*New Civil Engineer*, 1974):

> On the morning of June 29, 1968, Tank 281 at the Esso refinery, Fawley, Hants—one of the biggest oil storage tanks in Europe—was two-thirds of the way through the water-loading tests that would prove the strength of the welded steel vessel and the stability of its foundations.
>
> The foundations of Tank 281 [had] failed. What went wrong? Unfortunately, after five and half years, some of the most intensive—and expensive—investigations ever applied to a failed foundation and a massive legal action, it is still extremely difficult to say precisely what happened.
>
> The central lesson, whatever the cause of failure, is the need to clearly

define the roles of client engineer and contractor on every contract.

All along the line at Fawley, the decisions were ahead of the data, the warnings after the action. In the story that follows, there are many engineers, experts and, even troubleshooters! But in the hectic rush to get the tanks into service and start collecting savings worth 500,000 [50,000] dollars a month, one character was never written into the script. The consulting engineer or managing contractor who could have prevented disaster.

Other lessons from Fawley have already made their way round the grapevine. Many engineers have come to ask themselves how often they approve piling which contains defects like those found at Fawley and whether they have been testing enough piles in the past.

There has been a general move to specify permanent casing where driven cast in-situ piles are used in soft, wet conditions. The concept of progressive collapse in piled foundations has had to be given more thought.

Wide differences of opinion have been revealed on methods for predicting behaviour of a thin gravel stratum over fissured clay.

The article goes on to sketch out the events that led to the failure and summarizes some of the arguments put forward by the many experts called in to unravel the problem. An abbreviated chronological outline of events is included here; the article in the *New Civil Engineer* is more complete in this respect, and makes interesting reading! However, the NCE editors considered the cause of the failure secondary. They identified the central lesson as "the need to clearly define the role of client engineer and contractor on every contract." Without meaning to challenge this perspective, for the purpose of this paper, the central questions are the following:

1. After the expenditure of so much time, effort, and money, and with so many experts involved, why was it not possible to reach agreement on the cause of failure?
2. Could anything additional have been done to discriminate further among possible failure mechanisms? And, if so, what?
3. Would public records of the many detailed analyses made to investigate the failure contribute much to our store of knowledge? Or to the improvement of future investigations of this nature? Can a mechanism be established that will insure publication of such investigations in the future?

These questions will be explored and answered, but first some basic data is presented.

**Site Conditions and Foundation Design.**—Fawley is located in the estuary of the River Test just south of Southampton, England. Preliminary site investigations were conducted by Soil Mechanics Limited as early as 1955 in connection with the development of a tank farm on the reclaimed marshland site. From borings and soundings made over a large area, a generalized picture of the soil profile was obtained (Fig. 22). The alluvial clay was considered to be unsuitable for sustaining the load imposed by the tanks, but the gravel layer was thought capable of giving a satisfactory set to a driven pile and distributing the load to the Barton clay below. It was predicted that the shear strength of the clay should be adequate to support the transmitted loads. It is noted here how often major decisions are made in the early stages of a project, based on preliminary data and on the designer's experience and judgment. Once a decision is made, subsequent investigations and detailed designs become "locked-in," and most of the design effort then is expended to make the concept work rather than in discriminating between concepts. Better methodologies for making preliminary design decisions are badly needed.

During the period of 1957 to 1963, clayey-gravelly dredged sand fill was placed on the part of Block 51 where two identical tanks (281 and 282) were to be constructed. In 1966, additional granular fill was placed to provide a working platform for construction plant. The total fill thickness eventually ranged from 5–10 ft (1.5–3 m). It was recognized that the underlying alluvium would consolidate and impose negative skin friction on the piles.

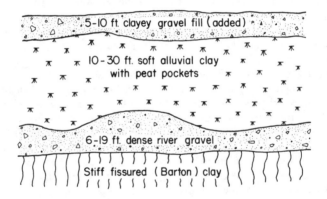

FIG. 22.—Generalized Soil Profile at Fawley Site

The period of September, 1966 to March, 1967 was occupied largely with establishing the tank dimensions. On the basis of operating efficiency, Esso was suggesting diameters of at least 260 ft (79.3 m) and heights up to 80 ft (24.4 m). Soil Mechanics, Ltd. felt that from the information available regarding the strength of the clay, it would be unwise to adopt a height greater than 56 ft (17.1 m), and proposed eight additional borings to determine the shear strength of the Barton clay. Esso's decision, on March 20, 1967, to go to 64 ft (19.5 m) high tanks, was influenced by the fact that engineering for this size tank already had been done. To allow as much time as possible for consolidation of the Barton clay without extending the completion date, water filling of the tank was programmed as tank erection proceeded. This procedure had to be abandoned due to welding failures that developed in the first three horizontal courses.

Early in April, 1967 John Laing and Son was commissioned to design the concrete slab for tank 281 and "to provide design services for the foundation . . ." which was to include the piling layout and specifications for pile load tests. Piling tenders were invited from Frankipile, West Piling, and Alphapile, while

Soil Mechanics, Ltd. continued to carry out additional site investigations: Frankipile offered 100 ton (2,240 lb/t) piles—21 in. diameter. West Piling offered 88 ton piles—20 in. diameter. Alphapile proposed 70 ton piles—17 in. diameter, and were the lowest in price.

Negotiations were initiated with Alphapile, who were given logs of 8 additional borings, 20 sounding records, and results of triaxial compression tests on samples from the Barton clay. The picture that emerged from this new information was gloomy. In some places the upper part of the river gravel was found to be loose or only medium dense, and to contain clayey-gravelly seams. At some locations the dense gravel was only about 5 ft (1.5 m) thick, and in others a medium dense state of compaction was encountered for the entire thickness

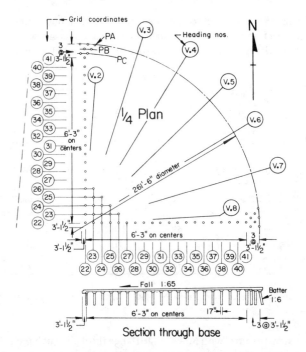

**FIG. 23.—Piling Layout Showing 1/4-Plan of Grid System and Heading Designations**

of gravel. Preliminary results indicated undrained shear strengths for the Barton clay ranging from 930–2,700 psf (44.5–129.3 kPa). Representative strength values near the clay surface were 1,100 psf (57.6 kPa) increasing to 2,000 psf (95.8 kPa) at a depth of 16 ft (4.88 m). Moreover, there were indications that at the interface between the river gravel and the Barton clay, a mixture of clayey gravel existed with a consistency softer than that of either the gravel or the clay. Alphapile proposed enlarging the bases of the piles (to about 2 ft (0.61 m) in diameter) and agreed to control driving to ensure 5 ft (1.5 m) of gravel under the toes of the piles. They also modified their guarantee, which was subject to a limit of £100,000, incorporating the understanding that good firm gravel did in fact overlie the clay, and emphasized that "substantial settlement of the foundation would occur due to the underlying Barton clay, the behavior of which [was] not the responsibility of Alphapile."

On May 17, 1967 Alphapile was notified that they would be getting the contract, but as late as June 2, Soil Mechanics, Ltd. sent a letter report to Esso continuing to express concern about the variability in consistency and thickness of the gravel and of local weak zones at the top of the clay. They also warned about the high intensity of loading that would come to the clay at the perimeter of the tank. The piling layout and slab dimensions, as prepared by John Laing and Son, is shown in Fig. 23. Perimeter piles were spaced more closely and the outer pile ring was battered (inclined). A brief description of how Alphapiles are installed is given:

> The components needed to install Alphapiles are an outer tube, or casing; a conical cast iron shoe filled with hardened concrete and fitted with four anchor stirrups; four reinforcing bars that hook into the anchor stirrups;

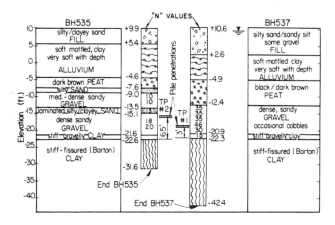

FIG. 24.—Penetrations of Test Piles T.P. No 1 and T.P. No. 2, and Adjacent Boring Logs—Tank 281

a hollow-stem mandrel with holes in the flange through which the reinforcing bars can pass; and a helmet and dolly assembly with slings to support the mandrel, which in turn is supported by slings suspended from the hammer. The casing with its shoe in place is set on the ground with the reinforcing bars hooked into the stirrups. The mandrel is lowered with the bars passing through the holes in its flange and the latter positioned above the pile shoe. The hollow stem of the mandrel and the space between the flange and the pile shoe are filled with concrete. The helmet is positioned on the pile and the casing and mandrel are driven down simultaneously until the desired set is achieved. The mandrel is then raised to allow concrete to pass from the mandrel to the casing, and the mandrel is replenished with concrete. The casing is raised and the mandrel is driven down simultaneously. Concrete is thus forced out to form an enlarged base. The mandrel is raised, replenished with concrete, and then the casing is raised while the mandrel is held in place by the static weight of the hammer. These latter operations are repeated until the casing is fully extracted, leaving a shell-less, cast-in-place concrete pile with an enlarged base in the ground.

Test pile No. 1, 32 ft (9.8 m) long, was driven on June 2, 1967, the driving set being 2 3/4 in./10 blows (80.9 mm/10 blows). Test Pile No. 2, 27.5 ft (8.4 m) long, was driven on June 4 to a set of 2 1/2 in./10 blows (63.5 mm/10 blows). Fig. 24 shows the test pile penetrations in relation to the soil profile at adjacent borings. On June 5, Alphapile began driving production piles without waiting for load test results. Load test No. 1 was completed on June 24; testing was stopped at a load of 132 tons, due to slippage of one of the supporting trestles. Load test No. 2 was completed on July 21. Summaries of the load test results are given in Fig. 25. As the design load was 70 tons, the test results were accepted, and were transmitted to both Soil Mechanics and John Laing.

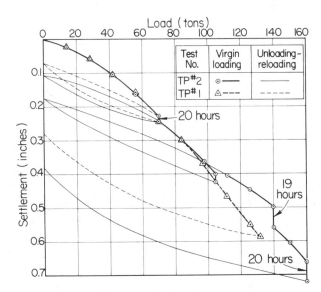

FIG. 25.—Results of Pile Load Tests—Tank 281

Pile driving for tank 281 was completed on August 12, 1967. A total of 1,580 piles were installed in about 10 1/2 weeks. All piles were driven to sets no greater than those recorded on test piles Nos. 1 and 2, with the exception of 14 piles whose driving was stopped when it was felt that an elevation 5 ft above the clay was reached. To compensate for the larger sets the area of their bases was increased an additional 30.

Construction of the reinforced concrete slab commenced on July 26, 1967 and was completed, including the oil sand cushion, on October 10. Thus, completion of pile driving overlapped placing of the slab by about 2 1/2 weeks. (Data to prepare a plot plan showing the portion of the slab placed while pile driving was still continuing, and the location of the piles driven during the overlap period, could not be obtained.)

**Water Filling.**—When construction of the concrete slab was completed, but before erection of the steel tank commenced, a set of levels was taken on the slab and referenced to datum. In addition, on completion of tank erection, levels were taken on the annular projection of the steel base plate and on the adjacent concrete slab at each of 26 locations spaced a distance of 31.4 ft (9.6 m) around the tank perimeter (identified as V.1–V.26 in Fig. 30).

Water filling started on May 5, 1968. Daily check levels were taken but plotting of the recorded data was done only when there was a difference of 0.25 in. (6.4 mm) or more from the previous day's levels. The open circles in the upper

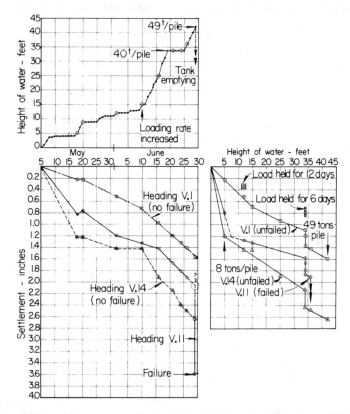

**FIG. 26.—Load versus Time and Load versus Settlement During Filling of Tank 281 with Water**

diagram of Fig. 26 show the water levels at which elevations at headings V.1–V.26 were plotted; the lower diagrams showing settlement versus time and load versus settlement at V.1, V.11, and V.14 were plotted by the writer post-failure. No plots were made for three days before failure (when the rate of filling was 2 ft (0.61 m) of water/day). On Friday, June 28, 1968, due to heavy, continuous rain, level checks were not made but visual inspection revealed no signs of anything untoward happening. On the morning of June 29, during routine level checking, it was observed that the reinforced concrete slab had canted and severe cracking had taken place. Between June 27 and 29, the maximum depression at the tank annular plate had increased from 2 5/8 to 3 5/8 in. (66.7 to 90.1 mm), while the edge of the concrete slab had risen 3 1/8 in. (19.3 mm). Instructions were issued to start emptying the tank immmediately. It may be inferred from Fig. 26 that even if settlements had been plotted daily, it is unlikely that impending rupture would have been perceived soon enough to avert failure.

**Post-failure Investigations.**—Emptying of the tank proceeded as rapidly as possible and was essentially completed in about 10 days. On July 4, with about 11 ft (3.4 m) of water still in the tank, an approximate survey of the tank floor was made by measuring the depth of water through the tank roof supports. The results are shown in Fig. 27. The maximum depressions were 16 3/4 in. (425 mm) and 11 1/2 in. (292 mm) in the SE and SW quadrants, respectively.

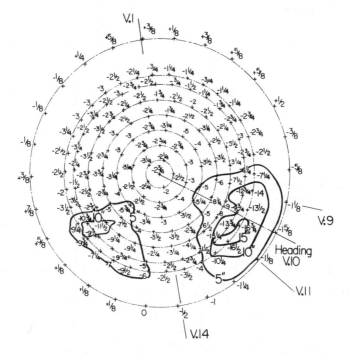

FIG. 27.—Residual Settlement of Floor After Emptying Tank 281

It is noted that even in the "unfailed" areas, the residual settlement of the tank floor was, generally, between 2 and 3 in. (51 and 76 mm).

Esso arranged for an excavation to be made beneath the tank to expose the piles in the vicinity of heading V.10. Although the piezometric level in the gravel layer was only 2 ft (0.61 m) below ground surface, excavation through the alluvium was feasible to a depth of about 12 ft (3.7 m). Table 6 is a summary of the defects that were observed, and Fig. 28 shows photographs of representative defects designated by asterisks in the Table. Interior piles are referenced to an EW-NS grid system 6.25 sq ft (0.56 m$^2$) (Figs. 23 and 30) beginning at the western and southernmost extremities of the tank; perimeter piles are designated PA, PB, PC, beginning with the outer ring of battered piles. In

TABLE 6.—Pile Defects Observed after Exposure by Excavation, Tank 281, Heading V.10

| Pile number (1) | Fractures (2) | Separations (3) | Necking (4) | General (5) |
|---|---|---|---|---|
| PA-41 | — | Crack full-circle at top. Gap in west face at 2 1/2 ft[a] | — | Coarse aggregate segregated at pile surface, top 4 ft |
| PA-42 | At 7 1/2 ft. Shortening $\simeq$ 2 1/2 in. | Crack full circle at top | — | Segregated coarse aggregate above and below fracture |
| PA-43 | At 1 ft. Shortening $\simeq$ 4 in. | — | — | — |
| PB-40 | — | 1 in. gap full-circle at top, Fig. 28(a) | — | — |
| PB-41 | At 3 1/2 ft Shortening $\simeq$ 3 in. | Crack in south face at 1 ft Crack full-circle at 4 ft | — | — |
| N13/E39 | — | Cracks full-circle at 1 ft and 4 ft | — | — |
| N12/E39 | — | Cracks in south face at 1/2 and 1 1/2 ft | — | Segregated coarse aggregate at surface, top 4 ft |
| N13/E38 | — | 1/2 in. and 1 in. gaps in west face at 0 to 1 ft | — | — |
| N11/E38 | At 1 ft and 12 ft. Total Shortening = 14 in. | — | — | — |
| N13/E37 | — | 1 in. gap in north face at 1 ft, cracks at 2 and 3 ft | — | Top 6 ft misshapen into an "ess" in east-west direction |
| N12/E37 | — | 1 in. gap in west face at top | — | Misshapen; sharp projections |
| N11/E37 | — | Crack full-circle at 1 ft | — | Porous matrix at 1 ft |
| N13/E36 | — | 1 in. gap in north face at 1 ft, cracks at 2 and 4 ft | — | — |
| N12/E36 | At 1 1/2 ft. Shortening $\simeq$ 14 in. Fig. 28(b) | — | — | Misshapen; sharp projections |
| N11/E36 | — | 1/2 in. gap in south face at 1 ft | — | Lack of cementation 0 to 1 ft |
| N13/E35 | At 13 ft (No photo) | Crack full-circle at 1/2 ft | — | — |
| N12/E35 | At 12 1/2 ft (No photo) | Cracks in north face at 1/2 and 1 ft | — | — |
| N11/E35 | — | 1 1/2 in. gap full-circle 1 1/2 ft, Fig. 28(d) 1/2 in. at 7 ft and 11 ft | — | — |

**TABLE 6.**—*Continued*

| (1) | (2) | (3) | (4) | (5) |
|---|---|---|---|---|
| N13/E34 | — | 1 in. gaps in north face at 1 ft, in south face at 7 and 8 ft | 12 3/4 in. diameter at 4 ft | Reinforcement exposed at 4 ft |
| N12/E34 | At 1 1/2 ft. Shortening $\simeq$ 15 in. | — | — | — |
| N11/E34 | At 12 1/2 ft (No photo) | 1 in. gap in southwest face at 1 ft | 12 in. diameter at 6 1/2 ft Fig. 28(c) | Reinforcement exposed 6 1/2 ft |
| N13/E33 | At 1 ft. Shortening $\simeq$ 11 in. | Crack full-circle at 2 1/2 ft | 10 1/2 in. diameter at 7 ft | Reinforcement exposed at 7 ft |
| N12/E33 | — | 1 1/2 in. gap in west face at 1 in. | — | — |
| N11/E33 | At 1 1/2 ft Shortening $\simeq$ 12 in. | — | — | Reinforcement badly exposed at 13 ft |

[a]Cracks are separations less than 1/2 in.; gaps are 1/2 in. or greater.

the SE quadrant, the most severe depression in the tank floor is within the grid N9-18/E31-39, with the focal point being at N13/E35.

Nine post-failure soil borings were made in the zones of maximum depression in the tank floor. Particular attention was paid to the details of stratification in the gravel layer, to the interface between the gravel and the Barton clay, and to obtaining good quality samples from the clay. Fig. 29 shows a plan view of the location of these borings in relation to the prefailure soundings and borings. Also shown is the location of piles that were cored to assess the pile condition (cracks, gaps, etc.) the quality of the concrete, and to determine the state of compaction and thickness of the gravel beneath the pile tips. Fig. 29 shows that in only one of the four piles cored in the area excavated beneath the tank did the core reach the tip of the pile, the remaining three cores failing to do this due to misalignments between the pile and the drill hole. The coring operations identified fractures and located gaps in the piles that compared well with direct observations made in the heading. The compressive strength of the cores ranged between 2,200 and 6,600 psi (15,158 and 45,474 kPa) (equivalent cube strengths). In some cases, intact cores were not recovered (pile fractured, or compressive strength less than circa 1,400 psi (9,688 kPa)).

Fig. 30 shows the locations of test piles TP #1 and #2, and of the 14 piles driven to sets larger than 2 3/4 in./10 blows (69.9 mm/10 blows). It is noted that none of these latter piles is associated with the major depressions in the tank floor; in fact, the residual settlement in these areas is as low as those in other locations with pile sets less than 2 3/4 in./10 blows (69.9 mm/10 blows). Fig. 30 also shows the limits of the excavation made beneath the tank and the general condition of the exposed piles.

Table 7 summarizes the thickness and state of compaction of the dense gravel layer (DGL) as revealed by the post-failure investigations. Attention is directed to the following:

(a) Gap full-circle at top

(b) Fracture with shortening ≃ 14″

(c) Necking to 12″ diameter

(d) 1 1/2″ gap full-circle

FIG. 28.—Photographs of Representative Defects in Piles as Disclosed by Excavation Beneath Tank 281 after Failure

1. The elevation of the bottom of the DGL is relatively uniform compared with that of the top.
2. The DGL is covered by a layer of looser clayey gravel (containing relatively soft clay or silt seams) that varies in thickness from 0–8 ft but is generally between 3 and 5 ft (0.9 and 1.5 m) thick.
3. A minimum of 5 ft (1.5 m) of dense gravel exists above the clay in all borings.
4. Borings through the cored piles extending past the pile tips show the minimum thickness of dense gravel below the pile tips to be 4.75 ft (1.4 m).
5. Comparing N-values in borings through the adjacent to cored piles, it appears that the looser portions of the gravel stratum were densified as a result of pile driving.

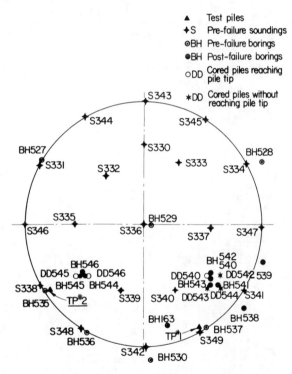

FIG. 29.—Plan Locations of Prefailure Soundings and Boreholes, Pile Load Tests T.P. No. 1 and No. 2, and Post-failure Cored Piles and Adjacent Boreholes—Tank 281

The interface between the DGL and the Barton clay was found to consist of a mixture of clay and gravel, but there were no indications this mixture was weaker than the Barton clay. Fig. 31 shows the shear strength-depth relationship for the Barton clay as obtained from undrained triaxial tests on 1 1/2 in. (38.1 mm) diameter samples tested at a confining pressure of 60 psi (413.4 kPa). The results essentially confirmed those obtained from the prefailure investigations.

Various alternatives were debated for reconstructing the tank foundations.

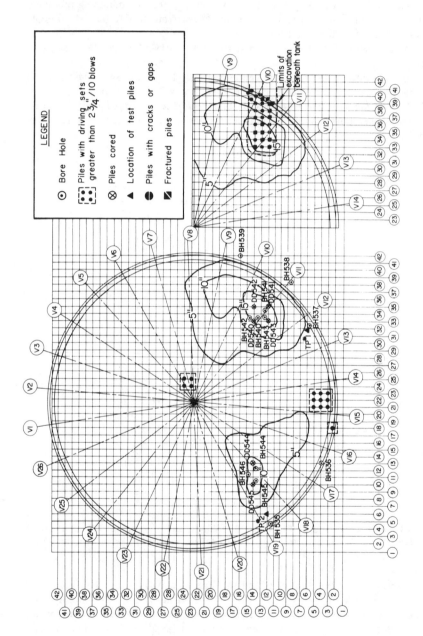

FIG. 30.—Plan of Post-failure Excavation and Pile Condition Surveys—Tank 281

The decision was made to relocate both tanks on new foundations, which involved removal of the marsh deposits by dredging and their replacement with compacted granular fill (31,10), at a cost of more than £1 million. Including losses from two years of potential savings on lower tanker rates, Esso set the bill at about £3 1/2 million and sought to recover this amount by legal action against Alphapile and John Laing and Son. The case against Alphapile alleged that the piles were necked, out of plumb, cracked, discontinuous, and contained concrete that was weak and reinforcement that was out of line. The allegations against Laing included: (1) Bad advice to use driven uncased cast-in-situ piles which were not suitable to the site conditions; (2) failure to discover alleged faults in the piles or prevent them; (3) making insufficient allowance against overstressing of the Barton clay; and (4) designing a base slab that transmitted bending moments to the piles as the slab settled into its expected saucer-shaped profile, which weakened them unduly.

TABLE 7.—Post-Failure Delineation of Dense Gravel Layer (DGL) in Failed Areas

| Location (1) | Method (2) | Thickness, in feet[a] Below pile tip (3) | Thickness, in feet[a] Total (4) | Elevation Top (5) | Elevation Bottom (6) | $N$-values (7) | Notes (8) |
|---|---|---|---|---|---|---|---|
| N13/E13 | DD[a]—544 | 4.75 | — | — | −23.0 | 69–75 | |
| N13/E12.7 | BH[b]—544 | — | 5.0 | −18.0 | −23.0 | 28–33 | 8 ft weaker clayey gravel above DGL |
| N13/E10 | DD—545 | 4.83 | — | — | −22.9 | 65–75 | |
| N13/E11.6 | BH—546 | — | 5.4 | −17.8 | −23.2 | 19–42 | 4–5 ft do |
| N13/E10.4 | BH—545 | — | 6.6 | −16.6 | −23.1 | 58–119[c] | 6 ft do |
| N13/E33 | DD—540 | 5.0 | — | — | −22.5 | 58–71 | |
| N12.8/E33.3 | BH—540 | — | 8.5 | −14.2 | −22.7 | 37–156[c] | 4 ft do |
| N13.3/E33.3 | BH—542 | — | 7.2 | −16.0 | −23.2 | 52–123[c] | 3 ft do |
| N11.3/E33.3 | BH—543 | — | 6.0 | −16.0 | −22.0 | 30–63 | 3 ft do |
| N11.3/E34.7 | BH—541 | — | 6.6 | −16.7 | −23.2 | 34–43 | 0 ft do |
| N7/E38.3 | BH—538 | — | 8.6 | −13.9 | −22.5 | 27–59 | 6 in. clayey seam at Elev. −17.6 |

[a] DD—Core drilling through the pile.
[b] BH—Borehole.
[c] High values due to large flinty particles.
Note: 1 ft = 0.305 m.

Each of the litigants assembled an impressive array of consultants who arrived at widely divergent explanations for the causes of the failure (see New Civil Engineer, 1974 for a summary of these views).

The case was settled out of court after some 3 months of pleadings, with Alphapile and Laing agreeing to pay Esso £500,000 and £200,000, respectively. Although agreeing to a settlement, all parties denied any liability; unfortunately, none of their consultants' reports have been made public. I was one of the consultants to Esso Research and Engineering Company advising on the cause of the failure. A summary of my analysis follows.

**Cause of Failure**

The first task was to identify all possible causes of failure. These are listed below in the order they came to mind.

1. Bearing capacity failure of individual piles, either in the gravel or in the underlying Barton clay.
2. Bearing capacity failure of the pile group in the Barton clay.
3. A combination of pile misalignment and differential settlements of the reinforced concrete slab inducing excessive moments that overstressed the piles and contributed to their failure.
4. Failure of piles due to combinations of excessive necking, inferior concrete, inclusions of foreign matter, and formation of cracks or gaps during construction, or both.

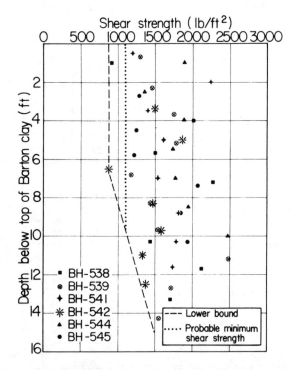

FIG. 31.—Undrained Shear Strength versus Depth for Barton Clay

**Bearing Capacity of Individual Piles.**—The data from the post-failure investigations demonstrated that, in the focal points of the failure areas, the gravel beneath the piles was dense to very dense and contained no weaker clayey seams. As the pressure at the pile tips at failure was only 49/3.14 = 16 tsf (neglecting positive shaft friction and the weight of the pile), and the minimum ultimate bearing capacity of the gravel was

$$q_{\text{ult}}(\text{gravel}) = \gamma'_{\text{ave}} D_f N_q = \frac{35(27)100}{2,240} = 42 \text{ tsf} \quad \dots \dots \dots \dots \dots \dots (1)$$

the possibility of a bearing capacity failure in the gravel was discounted entirely.

In late 1968, at least four papers evaluating routine tests to assess undrained shear strengths of stiff fissured clays were available (60,3,22,41). All these publications demonstrated that tests on 1 1/2 in. (38.1 mm) diameter samples gave mean values of undrained strength that were from 20–50% too high, and that the lower bound of such tests approached those measured using large diameter samples or interpreted from plate bearing tests. More recent tests have confirmed these observations (39,40). Examining Fig. 31, it is considered very unlikely that the operative shear strength of the Barton clay beneath the piles was less than 900 psf (6,401 kPa).

An estimate of the pile load that would induce a bearing capacity failure of the underlying Barton clay can be obtained by distributing the contact pressure at the tip of the pile to the top of the clay (at a slope of 2:1) and comparing this pressure with the ultimate bearing capacity of the clay

$$q_{ult}(\text{clay}) = c\, N_c = \frac{900(9)}{2,240} = 3.6 \text{ tsf} \quad \ldots \ldots \ldots \ldots \ldots \ldots \ldots \ldots (2)$$

The applied pressure at failure, assuming a minimum gravel thickness beneath the piles of 4.75 ft (1.45 m) (Table 7) and a bulb diameter of 2 ft (0.61 m), is

$$q\,(\text{applied}) = \frac{49}{\frac{\pi}{4}(2 + 4.75)^2} = 1.4 \text{ tsf} \quad \ldots \ldots \ldots \ldots \ldots \ldots \ldots \ldots (3)$$

Similar calculations applied to pile load test TP #1 gives an ultimate load of 70 tons, compared with the 132 tons at which the load test was stopped (Fig. 25). Although it is not possible to extrapolate the load test data to obtain the ultimate load (37), it is clearly much greater than 70 tons. Accordingly, it was concluded that the aforementioned calculations are conservative and that failure was not due to exceeding the bearing capacity of individual piles.

**Failure of Pile Group.**—As shown in Fig. 27, the focal points of the depressions are inside the rings of perimeter piles. In these zones the average load/unit area was 1.24 tsf. As the attenuation of pressure is small, this is taken to be the average pressure applied to the Barton clay. The minimum ultimate bearing capacity of the clay (noting that its strength tends to increase with depth) is

$$q_{ult}(\text{clay}) = c \cdot N_c = \frac{900(6)}{2,240} = 2.4 \text{ tsf} \quad \ldots \ldots \ldots \ldots \ldots \ldots \ldots \ldots (4)$$

Accordingly, it was concluded that failure was not caused by group action of the piles exceeding the bearing capacity of the Barton clay.

**Settlement of Base Slab.**—Prediction of the slab settlement pattern during loading is a complex soil-structure interaction problem whose solution, in my view, is beyond the present state-of-the-art. Nevertheless, differential settlements of the slab rotate the pile heads and induce bending moments in the piles that reduce their capacity to sustain axial loads. The effects are magnified if the piles are out of plumb. In their analysis, Laing assumed vertical piles and a parabolic deflection pattern for the slab. As no measurements were made on the base slab during water filling, it is not known how realistic the design assumptions were; thus, the influence of slab settlements on the failure cannot positively be assessed. However, as failure was initiated where the curvature would be relatively small if the pile support characteristics were uniform, slab-pile interaction might have contributed to the failure but it could not have been its primary cause.

A concomitant problem relates to the ability of the slab to distribute axial load to the piles. If a single pile failed, how many adjacent piles would participate in the load transfer? How many piles must fail before failure will progress over a large area? The first question cannot be answered without making assumptions regarding the load-deflection relationship for the pile group as the slab loses support from one or more piles. The nature of this loss in support is important if the pile fractures, as in Fig. 28b, its capacity suddenly reduces to zero, a condition favorable to progressive failure. On the other hand, if the bearing capacity of the ground is exceeded, the pile can continue to carry substantial load and much less load transfer will take place. Thus, a pile fracture is much more detrimental than a local bearing capacity failure from the standpoint of maintaining the integrity of the slab and avoiding progressive failure over a large area.

Using the concept of yield lines, an analysis was made to determine the load at which plastic yielding of the slab would be sufficient to form a mechanism. The slab then acts as a membrane, and the accompanying lateral forces could cause failure to propagate over a large area. The calculations are approximate because of the uncertain influence of the pile capitals and because of the manner in which the high-tensile steel mesh reinforcement was lapped.

The analysis showed that if only one pile is fractured, the applied load of 2,800 psf (19,292 kPa) would not yield the slab sufficiently to form a mechanism, but if two adjacent piles fractured, it is likely that progressive failure would occur. A modest increase in the amount of steel reinforcement and a better mesh and lapping scheme would have maintained the integrity of the slab under the full water load, even if two adjacent piles had fractured, provided the surrounding piles were sound. Thus, the reliability of the structure could have been increased enormously at a very modest increase in cost.

**Defective Piles.**—The combination of compressive strength of the concrete and the necked-down cross-sectional area of the shaft that reduces pile capacity to 50 tons is plotted in Fig. 32, with and without considering the loads that can be carried by the steel reinforcing rods. The minimum (necked) shaft diameter observed (in the 45 exposed piles) was 10.5 in. (267 mm) for tank 281 and 12 in. (305 mm) for tank 282. Assuming a necked diameter as small as 10 in. (254 mm), and all four reinforcing rods exposed for a sufficient length so that they would buckle at low loads, the pile would fracture if the compressive strength of the concrete were less than 1,400 psi (96,863 kPa) (the lower bound to obtain drill cores). In the absence of more definitive data, no quantitative statement can be made concerning the probability of encountering a critical combination of necked-down pile diameter, poor concrete, and exposed reinforcing steel. However, the occurrence of this critical combination at two separated locations (in the SE and SW quadrants), though not ruled out, was considered to be so unlikely that the search for a more likely cause was intensified.

The most common defects observed at Fawley were horizontal cracks and gaps (up to 1 1/2 in. (38.1 mm)) either on one face of full-circle around the piles (Table 6). Were these defects induced by loading the tank with water or were they formed in the construction process? Fig. 33 shows representative defects found in the piles beneath tank 282 *prior* to loading. There can be no doubt that the defects were formed during construction!

The full-circle gaps are particularly striking. How could such sharply-defined gaps more than an inch high be formed during construction? Several mechanisms are possible:

*Ground Displacement Due to Pile Driving.*—If all the movement due to ground displacement were directed vertically upwards, the calculated heave would be 1.25 ft (0.38 m), compared with an observed heave estimated to be about 1 ft (0.3 m). The net upward force/ft length of pile could be hardly more than 800 lb/ft, so that to form a crack at a depth of, say, 2 ft (0.6 m) the tensile

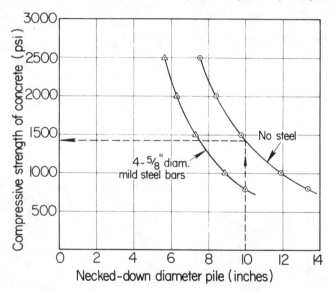

FIG. 32.—Necked-down Pile Diameter versus Compressive Strength of Concrete for Pile Capacity of 50 Tons

strength of the concrete would have to be less than 8 psi (55.1 kPa). There is some question whether a sharply defined gap could form when the concrete is this weak. Moreover, the preponderance of horizontal gaps and cracks are at depths less than 2 ft (0.6 m), and some are near the very top of the pile (Fig. 28a). Accordingly, vertical heaving per se is not the cause of full-circle cracks and gaps.

If heaving occurred after a portion of the slab had already been placed, there would be more than enough force to crack the piles (area ratio of slab to piles = 20:1), and the tendency would be for the separations to be near the top of the pile. I was unable to determine whether or not such heaving might have occurred at Fawley, although the indications are that it was unlikely. If slab construction is commenced prior to completion of pile driving under

(a) Crack at 2' depth

(b) Necking to 12 3/4" diameter

(c) Crack full-circle

(d) 1" gap, full-circle

FIG. 33.—Photographs of Representative Defects in Piles as Disclosed by Excavation Beneath Tank 282 Prior to Loading

conditions similar to those at Fawley, the possibility of cracking the piles due to uplift of the slab should be carefully considered.

Lateral ground displacement could provide sufficient force to damage piles in bending. The tendency would be either to form an S-shaped pile or to crack the pile in the tension zone, i.e. on one face of the pile only. Both of these defects were observed at Fawley (Table 6 and Fig. 33a). Once weakened by a crack, piles would be susceptible to formation of gaps, although an uncracked pile might have been strong enough to resist the induced tensile stresses. Thus, cracking due to lateral displacement of soil could facilitate later formation of gaps, but it could not have caused the gaps themselves.

*Downdrag.*—If heaving in the gravel, and overlying alluvium, raised piles in the interior of the tank more than at the perimeter, and if closer spacing of perimeter piles increased their relative stiffness (as suggested by the initial portions of the load-settlement curves in Fig. 26 and the pattern of residual settlements in Fig. 27), then downdrag due to consolidation of the alluvium later could pull the interior piles away from the slab because it would be restrained from downward movement by the perimeter piles. The downdrag force plus the weight of a 30 ft (9.2 m) long pile could induce a tensile stress of about 160 psi (1,102 kPa) near the top of a 16 in. (406 mm) diameter pile. This would be sufficient to crack the pile if the concrete was of very poor quality, and more than enough to cause a gap in a sound pile if it had already been cracked by lateral ground displacement. The gaps would be full-circle, tend to concentrate near the tops of the piles, and their location in plan would be inside the perimeter piles—all of which is consistent with the observed facts. Thus, differential heaving and subsequent downdrag on piles damaged by lateral soil displacement was judged to be the most likely cause of full-circle gaps. It is also noted that downdrag induces bending in the perimeter ring of battered piles and could damage them even if their bases were firmly seated.

Understanding the causes of observed defects is important because it can help avoid their recurrence in future construction. But regardless of how the gaps were formed, the crucial question at Fawley was 'How much load can a pile with a gap sustain before it fails?' To answer the question, the writer later simulated this condition by casting a concrete column with a preformed gap and testing it to failure in the laboratory.

**Load Test on Column with Gap.**—A 12 × 12 in. (305 × 305 mm) column, reinforced with 45 in. (117 mm) diameter hard steel deformed bars was cast with a 1/2 in. (12.7 mm) gap (Fig. 34a), carefully aligned in a testing machine, and loaded to failure. Concrete of good quality was deliberately chosen; on the morning of the test its average compressive strength was 5,200 psi (35,828 kPa). Fig. 34b shows the condition of the column at a load of 98 kips, at which point the steel yielded and the load dropped off to a small value (the area of the steel rods times the yield stress of the steel corresponds to a load of 92 kips). The two segments of the column were brought into contact, and Fig. 34c shows the condition when the load was again built up to 30 kips. As the load was increased further, additional cracking and spalling was observed. At a load of 430 kips, the concrete suddenly shattered and the load dropped off to zero (Fig. 34d). Recognizing that the testing machine could not follow the column as it shattered, the similarity between the conditions shown in Figs. 28b and 34d is spectacular!

(a) Column with 1/2" preformed gap

(b) Spalling at initial yield (98 kips)

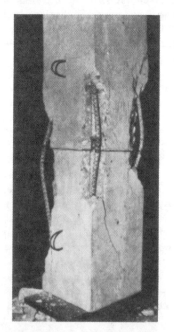

(c) Concrete in contact (30 kips)

(d) Fracture (430 kips)

FIG. 34.—Photographs During Laboratory Testing of Simulated Pile with Gap

In the case of the neatly separated and aligned test column, the effective concrete area at ultimate fracture was approximately 50% of the initial area. For imperfectly aligned piles with irregular gaps, it is reasoned that the effective area could be reduced to about 20%. A plot of the effective area versus the compressive strength of concrete for a pile to fail at a load of 50 tons is shown in Fig. 35. It is seen that failure could occur if a gap existed in an otherwise sound pile. Thus, it is concluded that when the load per pile beneath tank 281 reached 50 tons, piles with gaps failed, and their load capacity was reduced to a small value. The foundation slab transferred their full load to adjacent piles. If one of the adjacent piles was also defective, failure could propagate over a large area. It is noted that piles N14/E38 and N15/E39 beneath tank 282 are immediately adjacent to each other (see grid, Fig. 30), and Figs. 33c and 33d show they both have full-circle defects. It is, therefore, likely that

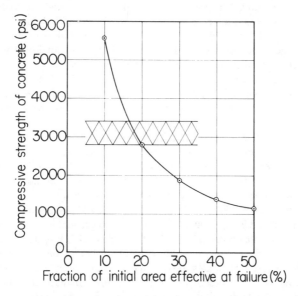

FIG. 35.—Effective X-sectional Area of Pile versus Compressive Strength of Concrete for Pile Capacity of 50 Tons

adjacent piles beneath tank 281 were similarly defective. Accordingly, it is concluded that gaps formed in the piles during construction was the *primary* cause of failure.

### Fawley Pile Foundations in Retrospect

Civil engineering projects generally involve many people in the decision-making process. At Fawley, the client's in-house staff, the geotechnical consultants, the design engineers, and the piling contractor all participated in the process, with the result that responsibility for the decisions taken was diffuse. Recommendations for improving contractual arrangements and management practices for underground projects have since been proposed, and are highly recommended (46,47). However, if no semblance of agreement on the cause of failure emerges, the same mistakes may be repeated again, or expensive measures may be adopted in design or construction, or both, when, in fact, they may not be needed. I have concluded that the primary cause of failure was the existence of piles with gaps and the presence of defects in adjacent piles. Had conditions permitted, post-failure investigations could have been carried out to prove this point, and much other valuable information for future design could have been collected. The nature of these additional studies is outlined below.

### Additional Post-Failure Investigations

**Individual Pile Load Tests.**—Piles would be selected beneath tank 281 within and outside the failure area and carefully cored to establish the condition of the concrete and the thickness and consistency of the underlying gravel. When necessary, they would be grouted to maintain their integrity during loading. By means of tell-tales, the movements of the pile tips and the underlying clay surface, as well as that at the pile heads, could be measured during load testing, which would make possible distinguishing whether failure was in the gravel or in the underlying clay. Results of tests within and outside the failure area would have established beyond doubt whether excessive contact pressures at the pile tips had contributed to the failure of tank 281. They also would have provided a permanent data bank for evaluating present and future bearing capacity theories applicable to piles driven through soft clay, seated at shallow depths in a thin gravel stratum, and underlain by stiff fissured clay.

Similar tests on individual piles beneath tank 282, including some in rows PB and PC, not only would add materially to the data bank, but they could establish whether or not interior piles would settle differentially with respect to perimeter piles. If they did, a necessary condition for downdrag to cause full-circle gaps in the interior piles would be established.

**Filling Tank 282 with Water.**—Before loading tank 282 with water, piles exposed by excavation could be instrumented with strain gages and with load cells inserted at the pile tops. Piles with defects, such as piles N14/E38 and N15/E39 (Fig. 33c and d), would be given special attention. Whether piles with gaps behave in-situ in the manner previously described could have been established positively. Instrumenting the reinforcement in the foundation slab and measuring the deflections of the tank floor would have provided invaluable data for future design of slab-pile systems. Moreover, strain gages on the tank itself would have added useful information on the settlement patterns that large steel tanks can tolerate. Direct evidence of the kinds of pile defects that are critical, and of the influence of slab-pile interaction, could have been obtained.

**Integrity of Cast-in-Situ Piles.**—Excavation beneath all (or a large fraction) of tank 282 prior to filling with water would have documented the number, type, and severity of pile defects that were representative of the job. Thus, an assessment could have been made of the general construction quality, although no formal basis exists to judge whether or not this quality was below normal expectations. There would also have been an unparalleled opportunity to evaluate and improve on available methods for integrity testing of piles.

In their review of problems associated with the construction of cast-in-place concrete piles, Thorburn and Thorburn (56) identified and gave examples of eight commonly encountered types of pile defects. The report clarified the causes of some common problems confronting piling contractors, and made recommendations for improving construction practices that deserve wide-spread support. However, the aim of eliminating all defects on large jobs—a condition generally assumed in design—is unrealistic. The methods available for checking the integrity of piles not only involve considerable time and expense, but they do not always detect significant defects. Moreover, it is not always possible to tell if an observed anomaly identifies a defect sufficiently serious to warrant corrective measures or one so minor it can be ignored (61). What emerges is the need for data on the frequency and level of defects, expressed in terms of the probability of their occurrence, that is associated with good site investigations and construction practices. The design then could strike a conscious and appropriate balance between costs and the probability and consequence of failure. Site inspection would concentrate on maintaining good construction practices instead of attempting to enforce the requirement that all piles be entirely free of defects.

## Lessons Learned

1. The foundations of the oil tanks at Fawley failed due to defective piles. A variety of defects was uncovered: piles that were misaligned or mispositioned, poor concrete, necked-down pile shafts, exposed reinforcement, S-shaped piles, and piles with cracks on one face only. Such defects are endemic to shell-less, cast-in-place piles. Although at Fawley they may have been more pervasive or more severe than usual, or both, pile foundations containing similar defects have served, and continue to serve, their useful purpose. The fatal defect at Fawley was the formation of gaps due to downdrag effects on interior piles, partly damaged by ground displacements, that could settle differentially with respect to those at the perimeter. Good design and construction practices, coupled with diligent and knowledgeable site inspection, can minimize the occurrence of all these defects.

2. There are limitations to post-failure investigations that severely restrict the possibility of pinpointing the principal cause of failure. Often, there is more than one contributing factor, although perceptive and persistent investigation may identify a primary cause. At Fawley, the situation was unique in that a companion tank to the one that failed, constructed under essentially comparable conditions, was available for study. Thus, the as-built condition of the piles could have been established in detail, and its subsequent performance quantitatively monitored under controlled loading conditions. Unfortunately, the means for doing this was lacking.

3. Disaster often places clients, engineers, and contractors in the position of adversaries. Primary efforts are focused on protecting financial interests, a situation that is not always conducive to the determination of the real causes of failure and seldom results in public disclosure of all the lessons learned. To reduce the enormous waste associated with this state of affairs, a National Center for Investigating Failures is proposed.

4. It is unrealistic to presume that large underground works can be constructed without variations from the objectives specified in the contract documents. While it is the obligation of the designer to choose foundation elements that are not highly susceptible to construction defects, variations in support characteristics—even the existence of potential defects—should be accounted for in the design. The only rational way of doing this is to relate the probability of their occurrence

to the over-all reliability of the structure, so that the costs incurred to improve each component of the structure is appropriately reflected in a reduced probability of failure.

At Fawley, the design of the foundation slab was based on the assumption that every pile could support a load of 70 tons, albeit with a nonuniform deformation pattern. Designing the slab to function satisfactorily in the event of the failure of two adjacent piles would have involved increasing the amount of steel reinforcement at a miniscule added cost compared with the cost of the structure, yet the probability of failure would have been reduced enormously. This is not to be construed as encouraging poor construction practices; on the contrary, it is a plea to obtain quantitative data on the variances associated with *good* construction practice and to account for their effects consciously, using a probabilistic approach. In the long run, the profession and the public will reap the benefits of greater productivity, more value for a given expenditure of money, and *fewer unexpected failures*.

## NATIONAL CENTER FOR INVESTIGATING FAILURES

When an unexpected instability occurs, the concerned parties must deal with the problem of liability and with the need for quickly putting the structure back into operation. Seldom does this situation favor complete studies to identify the actual causes of failure and, even more seldom, will efforts be expended solely to advance the state-of-the-art. Legal considerations often preclude publication of reports detailing the bases for the differing views that emerged, and no open dialogue is later developed to reconcile these differences.

Failures are investigated by a small number of engineers who draw largely from personal experience in deciding what to investigate and how to interpret the results. In this task, the interests of the client are of paramount importance. The lack of opportunity to pursue all possibilities with equal vigor, and the absence of a full public account of all the studies that are made, often limits the lessons that can be learned from previous failures. Each successive failure challenges the investigators anew to develop an appropriate approach, and to decide on the methods for interpreting the results. No central library exists where strategies for minimizing bias in making investigations can be evolved. Thus, each new generation of engineers is compelled to relearn the lessons already mastered by its predecessors. There is no continuity and there are no cumulative effects. This state of affairs is especially aggravating in the U.S., where there is no organization specifically charged with the responsibility of investigating a broad range of civil engineering failures. The time has come to remedy this wasteful situation by establishing a central organization whose primary responsibility will be to investigate failures.

### Requirements for Central Organization

To effectively investigate civil engineering failures, a central organization is needed that will necessarily be possessed of at least the following characteristics:

1. No stake whatsoever in the outcome, regardless of the cause or causes of failure.
2. Free access to the site and to all available information.

3. A permanent facility with laboratories, staff, and a technical director.
4. Funds to conduct independent studies, including field measurements, and to engage special consultants when needed.
5. An independent advisory board to review progress and to set general policies.

**Existing Organizations**

Organizations that investigate failures exist in many countries, although this is not their sole function. Some of these will be reviewed briefly. The listing is offered to illustrate different ways that have evolved to cope with the problem; it is not intended to be a complete list, nor to describe the organizations in detail.

**Scandinavian Geotechnical Institutes.**—The Norwegian and Swedish Geotechnical Institutes are sponsored by the government and their main charge is to do research to help solve problems of national interest. Investigation of failures is a part of this task, especially when failure occurs from natural causes that repeatedly involve loss of life. Consulting services are permitted, including jobs outside Scandinavia, and revenue from this work can be used for research and to fund investigations of failures. Results of these studies frequently are published, and have contributed much to the state-of-the-art, although the scope of the problems studied is relatively narrow and confined to a limited range of soil deposits.

**Building Research Organizations.**—The Building Research Division of the National Research Council of Canada, the Building Research Establishment in the United Kingdom, and the Center for Building Technology (CBT) of the National Bureau of Standards, are agencies of their national governments the basic mission of which is to provide research and information services to their respective governments and construction industries. In Canada, the governing board is appointed, while in the United Kingdom, the establishment is a part of a government agency and negotiates directly with other government departments. The CBT obtains a portion of its funding from the Department of Commerce, but the major portion is from other Federal agencies; it is a unit within the National Engineering Laboratory of the Bureau of Standards but maintains an advisory panel of external engineering professionals.

From their inception, these agencies have dealt with building problems involving durability of materials, structural integrity (especially roofs), fire hazards, noise, and vibrations. They also have been active in the investigation of failures, including geotechnical problems such as landslides, construction over peat, and performance of piling, as well as construction accidents and earthquake damage. In Canada, much effort has been expended investigating problems associated with seasonal and permafrost conditions.

None of the organizations has a mandate that permits investigating a particular failure of interest. Those failures to which access is made available generally involve public agencies and can be studied—with some exceptions—only by using funds allocated for "research." Though limited in scope, their contributions to the state-of-the-art are highly valued.

**Belgium's Bureau Seco.**—In Belgium, a designer can obtain insurance against possible failures by submitting his design and calculations to a private office called "Bureau Seco" [*Sécurité dans la Construction*]. If Seco rules favorably, the designer gets the insurance.

An owner can require that construction be supervised by the bureau. For example, a representative of the bureau may participate in pile loading tests and would give an interpretation of the test results.

When a failure occurs, the bureau makes a thorough investigation, with the aim of establishing responsibility and as a lesson for the future. The results generally are not published in the open literature and remain only as documentation within Seco. Bureau Seco covers the entire field of construction activities, and is much appreciated in Belgium.

The Bureau Veritas, based in France, plays a similar role on a much larger scale, being represented in 100 countries around the globe (but not in the United States).

**The Japanese System.**—In Japan, if a publicly owned structure fails, the government agency involved appoints an independent commission and pays for the investigation. If the designer is implicated, the agency accepts liability (having commissioned the work and reviewed the design), but the designer may be barred from future government work for a period of time. If the contractor is implicated, he must repair the damage, and also may be barred from future government work. Either the designer or the contractor may appeal to the courts, but in fact this rarely happens.

If the structure is privately owned, the courts seek expert opinions, decide who is at fault, and assess damages. If the failure is "socially serious," the government also appoints an independent commission and pays for the investigation. Although their report is advisory, the court's decision usually agrees with the commission's conclusions.

In the Japanese system, competent and comprehensive investigations of important failures are carried out by independent commissions. Moreover, there is much incentive for the contractor to monitor his own operations quantitatively, and generally to be alert to potential failures, as those that imply construction defects have a direct impact on his potential for future work.

While it is possible under some circumstances to gain access to the Commissions' reports, they are not published and there is no open debate on their merits.

### Proposed National Center to Investigate Failures

It was the aim of this paper to demonstrate that much more can be learned from the investigation of failures than has been the case in the past. A number of difficulties were identified, and while some of these are inherent in unexpected failures, many of the existing deficiencies can be remedied. Some approaches used in other countries to deal with failures were reviewed briefly; while they all have advantageous aspects, none are free of defects nor can they be adapted readily to the American scene. Accordingly, it is proposed that a national center to investigate civil engineering failures be formed in the United States. To meet the requirements outlined earlier, the following is recommended:

1. The American Society of Civil Engineers sponsor a bill in the Congress to create a National Center for Investigating Civil Engineering Failures. The center would have a mandate to investigate any failure of its choosing, and be afforded access to the site and to all available information. It would be financed by a small tax (say 0.02%) paid by the government on all federally

funded construction contracts (a similar tax to finance building research was proposed by Tschebotarioff (1980)).

2. The center would have a governing board appointed by the civil engineering profession. The board would appoint a technical director, approve proposed budgets, and, at appropriate times, review progress and evaluate performance.

3. The technical director would be responsible for all of the center's operations. He would make staff appointments, recommend which failures to investigate, organize the investigative team (including the appointment of consultants, if necessary), and review the final report. He would supervise the development of laboratories and recommend research programs to be carried out either in-house or by contract.

4. The center's activities would not replace any of the existing mechanisms that now are activated to investigate failures, but would be superimposed upon them. The results of its work would be confidential until all disputes have been settled; it then will be obligated to issue a report on its findings.

The staff at the center would develop expertise and experience in the investigation of failures. They would have access to the best available consultants who would be under no obligation except to help identify the real cause of failure. A comprehensive report including all basic data and a thorough analysis of all possibilities would be prepared and published. Open technical sessions would be organized to debate the merits of the approaches used and the conclusions reached. This would provide opportunities for people with a wide spectrum of knowledge and experience to participate and make contributions, not just those few initially chosen to investigate the failure. As a library of reports on past investigations is built up, the center's staff, or other interested parties, will be able to study trends, identify deficiencies, propose relevant research, and develop strategies to minimize bias in subsequent investigations. The benefits to the profession and to the public at large would be enormous and cumulative. Here is an unparalleled opportunity for civil engineers to demonstrate tangibly that they are, indeed, a people-serving profession.

### Acknowledgment

My deepest appreciation goes to Esko Kankare who, in correspondence extending over a period of six years, painstakingly (and patiently) responded to my probing questions and to continual requests for additional information. Through him I came to know the Kimola, albeit vicariously, as well as any other failure that I have had an opportunity to investigate; never in my experience has the spirit of friendly cooperation been more faithfully exercised.

The portion of the manuscript describing the slides at Kimola and the accompanying embankment failures was reviewed critically by a handful of people. I acknowledge the assistance of P. F. Wilkes who sent me parts of his thesis dealing with the King's Lynn embankment failure. This helped correct some errors in my initial interpretation of his paper to the Purdue Conference (62). The other reviewers are not named to avoid giving the impression they were supportive of my views. Nevertheless, their comments and criticisms challenged me to sharpen my arguments and to present them in a more straightforward manner. They must know how much this exchange of views was appreciated.

I am grateful to the Exxon Research and Engineering Company for the opportunity to investigate the failure at Fawley and for permission to publish the results. John Lyons read a draft of the write-up, corrected some points of fact which could not be known from the limited documents in my possession, and commented on some of the interpretations. However, the views expressed in the paper are my own, and do not necessarily reflect those of Lyons or of the Exxon Company.

Valuable assistance was provided by a number of my colleagues: John Mundell helped make the stability calculations for the R-7 and I-95 test embankments; R. D. Holtz participated in the early analysis of the slide at Kimola; C. D. Sutton and W. Cook cast the column that simulated the defective pile at Fawley, and then tested it in the laboratory; R. H. Lee made the determination that two adjacent piles would have to fracture before the foundation slab would behave as a mechanism; W. L. Dolch and D. N. Winslow added to my insights on the setting properties of concrete; and M. E. Harr and W. R. Judd improved the readability and clarity of the text. I also benefited from discussions with D. L. York and A. Aronowitz concerning downdrag effects on piles.

Special thanks are due Purdue University who, for the past 35 years, nurtured my spirit of inquiry and provided the opportunity to pursue scholarly efforts—including release-time from teaching to prepare this paper. We hope it will stimulate others to contribute views and opinions that further advance the state-of-the-art.

## Appendix.—References

1. Andersen, K. H., "Clay Slides in the Kimola Canal in Finland," Norwegian Geotechnical Institute Publication No. 92, 1972, Oslo, Norway.
2. Bishop, A. W., Discussion on Skempton, A. W., and Bishop, A. W., *Géotechnique*, Vol. 2, No. 2, 1950, pp. 113–116.
3. Bishop, A. W., "The Strength of Soils as Engineering Materials," Sixth Rankine Lecture, *Géotechnique*, Vol. 16, No. 2, 1966, pp. 91–128.
4. Bishop, A. W., and Bjerrum, L., "The Relevance of the Triaxial Test to the Solution of Stability Problems," *Proceedings of the ASCE Research Conference on Shear Strength of Cohesive Soils*, Boulder, Colorado, 1960, pp. 437–501.
5. Bjerrum, L., "Geotechnical Properties of Norwegian Marine Clays," *Géotechnique*, Vol. 4, No. 2, 1954, pp. 49–69.
6. Bjerrum, L., "Engineering Geology of Normally Consolidated Marine Clays as Related to the Settlement of Buildings," Seventh Rankine Lecture, *Géotechnique*, Vol. 17, No. 2, June, 1967, pp. 83–117.
7. Bjerrum, L., "Embankments on Soft Ground," *Proceedings of the ASCE Conference on Performance of Earth and Earth-supported Structures*, Purdue University, Lafayette, Indiana, 1972, Vol. 2, pp. 1–54.
8. Bjerrum, L., "Problems of Soil Mechanics and Construction on Soft Clays and Structurally Unstable Soils," Session 4, *Proceedings of the Eighth International Conference on Soil Mechanics and Foundation Engineering*, Moscow, U.S.S.R., Vol. 3, 1973, pp. 111–159.
9. Bjerrum, L., and Kenney, T. C., "Effects of Structure on Shear Behavior of Normally Consolidated, Quick Clays," *Proceedings of the Geotechnical Conference on Shear Strength of Soils and Rocks*, Oslo, Norway, Vol. 2, 1968, pp. 19–27.
10. Bratchell, G. E., Leggatt, A. J., and Simons, N. E., "The Performance of Two Large Oil Tanks Founded on Compacted Gravel at Fawley, Southhampton, Hampshire," *Proceedings,* Conference on Settlement of Structures held at Cambridge University, Pentech Press, London, England, 1975, pp. 3–9.
11. Brinch Hansen, J., and Gibson, R. E., "Undrained Shear Strength of Anisotropically

Consolidated Clays," *Géotechnique*, Vol. 1, No. 3, 1949, pp. 189–204.
12. Caldenius, C., "Skredet vid Säveån den 18 Januari 1945," *Sveriges geologiska undersökning*, Stockholm, Sweden, Avhandlingar och uppsatser, Ser. c, Vol. 476, 1946, p. 14.
13. Calvert, S. E., and Veevers, I. J., "Minor Structures of Unconsolidated Marine Sediments Revealed by x-Radiography," Sedimentology, Vol. 1, No. 4, 1962, pp. 287–295.
14. Casagrande, A., Discussion of "Symposium on the Panama Canal—The Sea-Level Project," by Binger and Thompson, *Transactions*, ASCE, Vol. 114, 1949, pp. 870–874.
15. Casagrande, A., and Wilson, S. D., "Effects of Rate of Loading on the Strengths of Clays and Shales at Constant Water Content," *Géotechnique*, Vol. 2, No. 3, 1951, pp. 251–263.
16. D'Appolonia, D. J., Lambe, T. W., and Poulos, H. G., "Evaluation of Pore Pressures Beneath an Embankment," *Journal of the Soil Mechanics and Foundations Division*, ASCE, Vol. 97, No. SM6, 1971, pp. 881–897.
17. Dascal, O., Tournier, J. P., Tavenas, F., and Larochelle, P., "Failure of a Test Embankment of Sensitive Clays," *Proceedings of the ASCE Specialty Conference on Performance of Earth and Earth Supported Structures*, Purdue University, Lafayette, Indiana, Vol. 1, 1972, pp. 124–159.
18. Dascal, O., and Tournier, J. P., "Embankments on Soft and Sensitive Clay Foundation," *Journal of the Geotechnical Engineering Division*, ASCE, Vol. 101, No. GT3, 1975, pp. 297–314.
19. Fellenius, B., *Report on the Landslide at Guntorp, Between Nygård and Alvhem near the Railway Line Kil-Gothenburg on April 13th 1953*, Swedish State Railways, Stockholm, Sweden, Geotechnical Department, Vol. 4, 1954, p. 16.
20. Gabilly, M., and Levillain, J. P., "Analyseur Gammamétrique de Carrotes Conteneur," [Gamma-ray Analyser for Soil-Samples within Sampling Tubes], *Bulletin de Liason, Laboratoire Ponts et Chaussées*, Vol. 97, Sept.–Oct., 1978, pp. 113–120.
21. Hamblin, W. K., "X-Ray Radiography in the Study of Structures in Homogeneous Sediments," *Journal of Sedimentary Petrology*, Vol. 32, No. 2, 1962, pp. 201–210.
22. Hooper, J. A., and Butler, F. G., "Some Numerical Results Concerning the Shear Strength of London Clay," *Géotechnique*, Vol. 16, No. 4, 1966, pp. 282–304.
23. Hutchinson, J. N., "A Landslide on a Thin Layer of Quick Clay at Furre, Central Norway," *Géotechnique*, Vol. 11, No. 2, 1961, p. 69–94.
24. Hvorslev, M. J., *Uber die Festigkeitseigenschaften gestörter bindiger Böden*, Ingeniørvidenskabelige skrifter, A45, Copenhagen, Denmark, p. 159.
25. Janbu, N., "Contribution to the State-of-the-Art Report on Slopes and Excavations," *Proceedings of the Ninth International Conference on Soil Mechanics and Foundation Engineering*, Tokyo, Japan, Vol. 2, 1977, pp. 549–566.
26. Kankare, E., "Geotechnical Properties of the Clays at the Kimola Canal Area with Special Reference to the Slope Stability," thesis, presented to the State Institute for Technical Research at Helsinki, in 1969, in partial fulfillment of the requirements for the degree of Doctor of Philosophy.
27. Kankare, E., "Failure at the Kimola Floating Canal in Southern Finland," *Proceedings of the Seventh International Conference on Soil Mechanics and Foundation Engineering*, Mexico, Vol. 2, 1969, pp. 609–616.
28. Kenney, T. C., and Uddin, S., "Critical Period for Stability of an Excavated Slope in Clay Soils," *Canadian Geotechnical Journal*, Vol. 11, No. 4, 1974, pp. 620–623.
29. Krinitzky, E. L., "Effects of Geological Features on Soil Strength," *Miscellaneous Paper No. S-70-25*, Waterways Experiment Station, 1970, Vicksburg, Mississippi.
30. Ladd, C. C., "Stress-Strain Behavior of Anisotropically Consolidated Clays During Undrained Shear," *Proceedings of the Sixth International Conference on Soil Mechanics and Foundation Engineering*, Montreal, Canada, Vol. I, 1965, pp. 282–286.
31. Leggatt, A. J., and Bratchell, G. E., "Submerged Foundations for 100,000 ton Oil Tanks," *Proceedings*, Institution of Civil Engineers, London, England, Part 1, Vol. 54, May, 1973, pp. 291–305 and Nov., 1973, pp. 699–702.
32. Legget, R. F., and Bartley, M. W., "An Engineering Study of Glacial Deposits at Steep Rock Lake, Ontario, Canada," *Economic Geology*, Vol. 48, No. 7, 1953, pp. 513–540.

33. Leonards, G. A., "Behavior of Foundations and Structures," *Proceedings of the Ninth International Conference on Soil Mechanics and Foundation Engineering*, Tokyo, Japan, Vol. 3, 1977, pp. 368–369.
34. Leonards, G. A., Discussion of "Foundation Performance of Tower of Pisa," by J. K. Mitchell, V. Vivitrat, and T. W. Lambe, *Journal of the Geotechnical Engineering Division*, ASCE, Vol. 105, No. GT1, 1979, pp. 95–105.
35. Leonards, G. A., "Special Lecture: Stability of Slopes in Soft Clays," *Proceedings of the Sixth Panamerican Conference on Soil Mechanics and Foundation Engineering*, Lima, Peru, Vol. 1, 1979, pp. 225–274.
36. Leonards, G. A., "Strain Rate Behavior of St. Jean-Vianney Clay," by Y. P. Vaid, P. K. Robertson, and R. G. Campanella, *Canadian Geotechnical Journal*, Vol. 17, No. 3, 1980, pp. 461–462.
37. Leonards, G. A., and Lovell, D., "Interpretation of Load Tests on High-Capacity Driven Piles," Symposium on the Behavior of Deep Foundations, American Society of Testing and Materials Special Technical Publication No. 670, 1979, pp. 388–415.
38. Lo, K. Y., and Morin, J. P., "Strength Anisotropy and Time Effects of Two Sensitive Clays," *Canadian Geotechnical Journal*, Vol. 9, No. 3, 1972, pp. 261–277.
39. Marsland, A., "Large In-Situ Tests to Measure the Properties of Stiff Fissured Clays," *Proceedings*, First Australia-New Zealand Conference on Geomechanics, Vol. 1, 1971, pp. 180–189.
40. Marsland, A., "The Shear Strength of Stiff Fissured Clays," *Proceedings of the Roscoe Memorial Symposium on Stress-Strain Behavior of Soils*, R. H. G. Parry, ed., G. T. Foulis & Co. Ltd., Henley-on-Thames, Oxfordshire, England, 1972, pp. 59–68.
41. Marsland, A., and Butler, M. E., "Strength Measurements on Stiff Fissured Barton Clay from Fawley," *Proceedings of the Geotechnical Conference, Oslo, 1967*, Norwegian Geotechnical Institute, Oslo, Norway, Vol. 2, 1968, pp. 160–161.
42. Massachusetts Institute of Technology *Proceedings*, Foundation Deformation Prediction Symposium, *Report FHWA-RD-75-515*, U.S. Department of Transportation, Cambridge, Mass., 1975.
43. Milligan, V., Discussion of "Embankment on Soft Ground," ASCE Conference on Performance of Earth and Earth-Supported Structures, Purdue University, Lafayette, Indiana, Vol. 3, 1972, pp. 41–48.
44. Milligan, V., Soderman, L. G., and Rutka, A., "Experience with Canadian Varved Clays," *Journal of the Soil Mechanics and Foundation Engineering Division*, ASCE, Vol. 88, No. SM4, Aug., 1962, pp. 31–67.
45. Morgenstern, N., "Contribution to the State-of-the-Art Report on Slopes and Excavations," *Proceedings of the Ninth International Conference on Soil Mechanics and Foundation Engineering*, Tokyo, Japan, Vol. 2, 1977, pp. 567–581.
46. National Academy of Sciences, "Better Contracting for Underground Construction," *NTIS Accession No. PB-236 973*, Standing Committee No. 4, U.S. National Committee on Tunneling Technology, 1974.
47. National Academy of Sciences, "Better Management of Major Underground Construction Projects," *NTIS Report No. NRC/AE-TT-78-1*, Subcommittee on Management of Underground Construction Works, U.S. National Committee on Tunneling Technology.
48. New Civil Engineer, "Esso's Giant Oil Tanks—a Question of More Haste, Less Speed," *New Civil Engineer*, London, England, Feb. 28, 1974, pp. 28–38.
49. Odenstad, S., "The Landslide at Sköttorp on the Lidan River, February 2, 1946," *Proceedings*, Swedish Geotechnical Institute, Stockholm, Sweden, 1951, p. 39.
50. Odenstad, S., *Jordskredet i Göta den 7 juni 1957*, Geologiska föreningen i Stockholm. Förhandlingar, Sweden, Vol. 80, No. 492, 1957, p. 76–86.
51. Quigley, R. M., "Geology, Mineralogy, and Geochemistry of Canadian Soft Soils: A Geotechnical Perspective," *Canadian Geotechnical Journal*, Vol. 17, No. 2, May, 1980, pp. 261–285.
52. Rowe, P. W., "The Stress-Dilatancy Relation for Static Equilibrium of an Assembly of Particles in Contact," *Proceedings*, Royal Society, Vol. 269, 1962, pp. 500–527.
53. Schmertmann, J. H., and Osterberg, J. O., "An Experimental Study of the Development of Cohesion and Friction with Axial Strain in Saturated Cohesive Soils," ASCE Research Conference on Shear Strength of Cohesive Soils, Boulder, Colorado, pp. 643–694

(elaborated in Laurits Bjerrum Memorial Volume, National Geotechnical Institute, Oslo, Norway, 1976, pp. 65–97).
54. Skempton, A. W., "A Slip in the West Bank of the Eau Brink Cut," *Journal of the Institute of Civil Engineers*, London, England, Vol. 24, 1945, pp. 267–287.
55. Sopp, O. I., "X-ray Radiography and Soil Mechanics: Location of Shear Planes in Soil Samples," *Nature*, Vol. 202, No. 4,934, 1964, p. 832.
56. Thorburn, S., and Thorburn, J. Q., "Review of Problems Associated with the Construction of Cast-in-Place Concrete Piles," *Piling Development Group Report PG 2*, Construction Industry Research and Information Association, 6 Storey's Gate, London, England, 1977.
57. Torstensson, B. A., "Pore Pressure Sounding Instrument," *Proceedings*, Conference on Insitu Measurement of Soil Properties," Raleigh, N.C., Vol. 2, 1975, pp. 48–54.
58. Tschebotarioff, G., "Top Foreign-Born Civil Engineers Speak Their Mind," *Civil Engineering*, Oct., 1980, pp. 118–119.
59. Vaid, Y. P., and Campanella, R. G., "Triaxial and Plane Strain Behavior of Natural Clay," *Journal of the Geotechnical Engineering Division*, ASCE, Vol. 100, No. GT3, 1974, pp. 207–224.
60. Ward, W. H., Marsland, A., and Samuels, S. G., "Properties of the London Clay at the Ashford Common Shaft: In-situ and Undrained Strength Tests," *Géotechnique*, Vol. 15, No. 4, 1965, pp. 321–344.
61. Weltman, A. J., "Integrity Testing of Piles: A Review," *Piling Development Group Report PG 4*, Construction Industry Research and Information Association, 6 Storey's Gate, London, England, 1977.
62. Wilkes, P. F., "An Induced Failure of a Trial Embankment at King's Lynn, Norfolk, England," *Proceedings*, ASCE Specialty Conference on Performance of Earth and Earth-Supported Structures, Purdue University, Lafayette, Indiana, Vol. I, Part 1, 1972, pp. 29–63.
63. Wissa, A. E. Z., Martin, R. T., and Garlanger, J. E., "The Piezometer Probe," *Proceedings*, ASCE Specialty Conference on Insitu Measurement of Soil Properties," Raleigh, N.C., Vol. 1, 1975, pp. 536–545.

# INVESTIGATION OF FAILURES[a]

## Discussion by A. Isnard[2]

The writer would like to point out the real role devoted to the so-called French "Bureaux de Contrôle Technique" compared to the author's suggestions concerning a center for investigating failures. Bureau Veritas and others (SOCOTEC, CEP, etc. . . .) are, in France, only in charge of safety inspection of design and construction or erection operations on the site. They are commissioned by the building owners to watch over the reliability of the design and the quality and safety of the Contractors and suppliers services on the site.

Of course, the final aim should be to reach a very significant level of safety and reliability of the construction which has to last and remain in good running order far beyond the guarantee period (the French "assurance décennale" system); this period being mainly considered as a test period for the major part of building and equipments.

A corollary aim is to help the owners to get an appreciable discount of their compulsory insurance prime rates.

Consequently, the "Bureaux de Contrôle Technique" are often involved when some failure and proceedings occur. Then it becomes very difficult for them to keep out of legal consideration and to let free access to their informations. That information is, of course, processed and used by the "Bureaux de Contrôle Technique" to improve their own knowledge and their efficiency.

The French Government decided recently to create an agency for construction safety and for the improvment of the Building rules of Art. This agency should comply with the author's proposals for a center for investigating failures. If so, this Sixteenth Terzaghi Lecture is the best introduction to such an initiative and entitles the author to our congratulations.

## Discussion by T. J. Pilecki, P. E.[3]

The author must be congratulated for his colossal effort in gathering information on soil failures and, at the same time, his search for answers as to the causes of these failures. The author also calls for new ideas, new concepts and new methods pertaining to soil investigations as well as mathematical solutions.

This writer feels that the causes of such failures are manifold and therefore the solutions would be complex. In fact, Soil Mechanics reminds this writer of the pattern that the history of "Western Civilization" went through. First, there were the "Ancient Times" when bearing capacities, settlements, etc., could only be evaluated from recorded localized experiences.

---

[a]February, 1982, by Gerald A. Leonards (Paper 16864).
[2]Chf. of Soil and Foundations Dept., Bureau Veritas, France.
[3]Consulting Soil Engr., Berkeley, Calif.

Then came the "Middle Ages" with Coulomb, Rankin, Fellenius and Terzaghi—Soil Mechanics had commenced. Soil Mechanics is purely technical and as such, is based on the empirical philosophy of science. It is rightly called Geotechnics which, at the present time, consists of "Confused Aggregates," based on observation of failures and laboratory test results, empirically interconnected in a system of "Crude Rules." "Intelligible Principles," leading to the creation of a modern science, are still to come.

Terzaghi and others introduced into Soil Mechanics two negative concepts. One is the strict division of the soil mass into cohesive or noncohesive materials. This naturally is false. All "real" soils are both cohesive and noncohesive, at the same time. Pure granular materials, composed of minerals that are absolutely neutral, will still have some cohesion when moist, which is generally the case in the field, due to the bipolar effects of water molecules. On the other hand, what appears to be purely cohesive is often found to have electrically charged particles, not exceeding 10%–15% of the overall volume.

The second negative concept introduced years ago is the classification of soil slides into two groups, one with planar and the other with a circular surface. Such a classification is erroneous as, de facto, both surfaces are one and the same. The difference between them is only Geometric as a circular surface has a limited radius of curvature, whereas a planar surface has an infinitely large radius of curvature.

In this "second period" of Soil Mechanics, many different types of testing apparatus were introduced. However, it is rarely comprehended that test results depict the properties of the actual soil together with those of the apparatus. The author very clearly indicated that, depending on the tests, the values of internal friction of the same soil varied between 0.25 through 0.33–0.42 [Anderson (1)]. At this stage of Geotechnics, the effects of soil testing has not been adequately explored. The future Geomechanical Engineer must learn to interpret test results in order to find the existing parameters of the real soil, and also forecast the expected future parameters of the same soil under physical and mechanical changes.

The author calls for new ideas, concepts, and principles. For this he has to wait for the arrival of the "Modern Ages" of Soil Mechanics when "Intelligible Principles" of a true science will be born. The new science called Geomechanics will be not only based on empirical observations and rules, but also on such theories as Newton's Laws of Motion and Gravity, as well as on Newton's concepts of Forces, Inertia, Mass, etc.

The author describes the Kimola slide together with a few others. The other slides cannot be discussed here because of the "short cut" methods of soil testing based on the Vane Test where data on independent but highly important parameters, such as cohesion and angle of internal friction, are not obtained. Judging from the description of these slides, one can conclude that the failures, in all instances, were composite. They represented events that closely resemble liquefaction rather than regular soil slips.

It appears that the Kimola slide, as elaborated upon by this writer and shown in Fig. 36, consists of one single slip surface, having the radius of curvature equal to 286.6 m. The two short radius curves located at

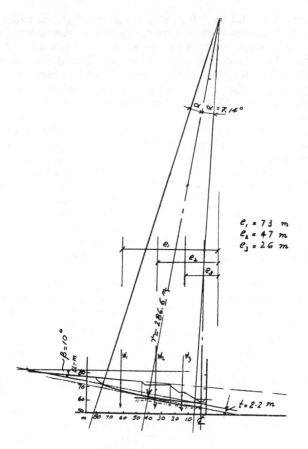

FIG. 36.—Kimola Slide

the two extremities, are mostly of a secondary nature and have no practical mathematical effects upon the overall stability analysis. Judging from the large size of the main radius, one has to conclude that the cohesive parameter was extremely small. As a matter of fact, the calculations indicate that the cohesion was much smaller than the one quoted in the author's article.

The main issue here is whether the failure could have been predicted. In the case of the Kimola slide, this appears to be the case. Here are some of the reasons:

1. Finnish surface soils are mostly post-glacial and therefore not pre-consolidated—thus capable of sliding.
2. The construction of the canal passes through some gentle swales where the loose (soft) soil layer is thickest, and, at the same time, the flow of underground moisture is more concentrated as it comes from more than one direction. The actual major slides occurred in two such areas.

3. The main canal was located in a longitudinal valley and here, the thickness of the upper loose soil layer is expected to be the greatest. At the same time, this layer will have the highest saturation with rain waters, often reaching the ground surface. This implies that at such a location, water may be of paramount importance in connection with the slides, due to high seepage pressures.

4. The climate of Finland in many ways resembles an oceanic climate due to the high frequency of medium and low magnitude rainfalls. This implies that soils are predominantly saturated and dessication is negligible. Thus, an increase in cohesive strength, equivalent to preconsolidation, cannot be expected.

5. The Test results were performed with different laboratory devices, but adequate adjustments of the "Apparatus Effects" were not elaborated upon. The test results, shown in Table 3, seem to be rather too high.

6. By taking Fig. 8 into account, where the moisture content is very high and thus the soils have a rather low density, this writer evaluated that the angle of internal friction of the soils should range somewhere between 12° and 15° which is, of course, very low and most banks of the Kimola type would easily break down.

7. Another indication of danger, is the fact that the soil mass predominantly consists of quite uniform and rather small particles, additionally implying that low angles of internal friction are present. Such soils may even liquefy.

8. The criteria used in computations of the stability were based on "the best fit straight line." This is wrong. Such a procedure may apply to calculations of consolidation, bearing capacity, etc. In the case of failures, only the smallest parameters should be used.

9. As indicated in Fig. 4, the natural soil slopes down at an angle of 5°, and some of the banks were formed at a much steeper slope. This automatically implies that the final stability will be in doubt in unconsolidated soils.

10. The soil in the vicinity of the canal was removed up to a depth of some 10 m which implies that a very large "rebound effect" is to be expected, which may imply that the cohesive strength will be destroyed and, at the same time, frictional granular resistance will also be reduced.

11. Since the end of glaciation, the subsurface waters in the Kimola Valley flowed down through the full thickness of the upper loose soil layer equal to about 15 m. The proposed cut reduced the final cross section to about 5 m–6 m. Such an event automatically indicates that conditions equivalent to "springs" may occur. This implies that, with time, numerous "boiling" or "piping" events will take place and small soil slips will occur. This writer's computations revealed that in Kimola, a "piping" caused slide of 4 m diam may occur, and the thickness of the soil of the sliding layer will be 1.25 m whereas, for a larger diam, the thickness of the "piping" slide would not exceed 1.4 m. These slides were not only predictable, but they also could easily be calculated.

12. The calculations further revealed that by using a reduced angle of internal friction of 12°, and taking the slope of the slide as equal to $\beta = 10°$, the safety factor drops below 0.60. Another calculation revealed that by taking the cohesion of 0.070 ($k/c^2$) and the angle of internal friction

of 20.5°, which are the smallest values at Station 52 + 70; the safety factor is equal to 0.90.

13. These types of adjustments of structural parameters could easily be performed prior to the construction, and thus the formation of the destructive Kimola slides could be predicted and eliminated. This could be achieved by placing suitable rip-rap along the banks and thus eliminating the formation of the "piping" slides, and by constructing an adequate system of subdrains, extending a reasonable distance up the slope, particularly in the swale area, so that the water pressure could also be eliminated or at least considerably reduced. Furthermore, an additional hazard caused by the vibrations created by logging trucks could be reduced by shifting the road away from the danger spots.

With reference to the other event described by the author, that is, the collapse of the pile foundation under the oil tank at Fawley, England, the issue is caused by faulty foundation engineering. Several authors, such as Terzaghi and Jumikis, quoted that a foundation engineer must be a person educated and experienced in most of the civil engineering subjects, especially Soil Mechanics, Structural Engineering and Hydraulics. It seems that, on an individual basis, the foundation was perfectly designed on perfectly submitted soil data. However, a true Foundation Engineer should be able to analyze the following issues:

1. A 30 ft–40 ft thick, soft, alluvial layer of clay, mixed with pockets of peat, is squeezed between a layer of gravelly fill, 5 ft–10 ft thick, located at the top of the finished grade, and a dense river gravel, some 6 ft–19 ft thick, located below and resting on a stiff clay. Such conditions automatically indicate that any pile driven through the upper or lower granular material will tend to densify this material. Simultaneously, it will cause heaving of the soft alluvial layer, due to the low permeability of this stratum and thus a "swelling" phenomenon, due to the increase of the water pressure.

2. Taking into account the diameter of the piles and their spacing, the soil mass will have to heave up (swell) by 4.0%, that is, the upward movement must be equal to a maximum of 22 in. This is the amount of lifting of the top of the piles driven at the beginning. This amount of uplift decreases with time and the last group of piles will suffer nearly a zero amount of uplift.

3. The above implies that a thorough record of the elevation of each pile should be kept. These elevations should be checked after the completion of the driving of all of the piles, so that the amount of heaving, caused by the forced uplifting, will be established.

4. The piles must then be driven back into the subsoil, sometimes on several occasions until the earlier elevations are reestablished for all of the piles. However, selected piles, poured in situ, cannot be redriven. As a result, a settlement of the piles under the influence of the superimposed loads will take place in the magnitude of up to 22 in.

5. From the structural point of view, the piles failed because the longitudinal reinforcing bars were not strengthened according to the theory of Reinforced Concrete, which states that the slenderness ratio of each bar between the stirrups must be at least equal to the slenderness ratio

of the columns. In the case of the Fawley foundation, the concrete piles must be considered as columns, at least within the distance equal to the thickness of the soft clay material through which they penetrated. The structural failure of piles was basically caused by the lack of placement of suitable stirrups. This implies that a wrong selection of the types of piles was made for the type of soil that was described by the Soil Mechanics Ltd.

6. The structural slab resting on top of the driven piles can be designed for the actual span between the piles, because this procedure is always used and can be considered safe as long as the above-noted process of designing, checking and controlling is followed. In the case of the Fawley tank foundation, the uplifted piles could not be driven back to their original elevations, and thus the idea of increasing the span by skipping one and even, as suggested, two piles, will not work because there may be uplift on a line of a much larger number of piles, say 10 or more. By designing and building a slab on the basis of skipping one or two piles, the bending moment, and therefore the cost, will increase approximately according to the square of the selected span. It will be four times larger when skipping one pile and nine times larger when skipping two consecutive piles. Such a procedure would, however, not eliminate the settlements of all of the earlier driven piles.

Among the other points raised by the author in his article, the call for the formation of a National Central Organization for gathering failure data seems to be unwarranted. Such an organization could become an undemocratic and bureaucratic entity, which could only become an inefficient burden on the taxpayer. It seems that the current practice of engineers reporting failures in the appropriate journals and magazines of the ASCE and others are quite adequate. The authors or the discussers should elaborate upon the causes and effects of failures, as they see them, and, wherever possible, they should try to find a pattern of events and relate these to a general principle or to mathematical theory. Only then will the future Geomechanical Engineering as a theoretical science be created and Soil Mechanics enter into the family of Theoretical Sciences, such as Elasticity, Hydraulics, Structures, Reinforced Concrete, etc.

## Discussion by G. E. Bratchell[4]

NCL Consulting Engineers, of London, England, was appointed by Esso Petroleum, Ltd., in 1968 to investigate the failure of the tank foundations which were the subject of part of the sixteenth Terzaghi lecture. NCL subsequently presented evidence to the English High Court in a case which lasted several weeks—the outcome being that Esso received compensation from both defendants.

NCL was also involved in the technical papers which are the subject of Refs. 10 and 31. It should be noted that other technical references on

---

[4]NCL Consulting Engrs., 192–198 Vauxhall Bridge Road, London SW1 V DX, England.

the subject are papers by Penman (66), and Davies and Dunn (65).

The author points to the lack of published data on the analysis of the causes of such failures. A great deal of technical information on this failure is available in this firm, and the kernel of it was presented in the Court, so that it was available together with the transcripts. The writer should mention in answer to one point, that NCL's Terms of Reference were to report on all the probable causes of failure regardless of whose responsibility, and this NCL believes they succeeded in doing.

After the Legal Settlements (as distinct from the foundation ones!), NCL cooperated with one of the defendants, namely the raft designer, and wrote a paper on the matters investigated and the technical lessons established. These are listed at the end of this contribution. The paper was submitted to the two leading engineering institutions in England, but was not published, partly due to legal implications.

The writer was interested to note that the author concluded that defects in the piles were the primary cause of the foundation failure. NCL found very strong evidence that this was so, but we spent a great deal of effort in examining all likely contributory causes. Unfortunately, NCL's reasoning differs somewhat from that of the author's, because there are a number of important pieces of evidence which he has not mentioned. For example, one such piece to be found in Court transcript is that piles had suffered direct compression failure, and in fact, one pile was found fully intact and continuous but on the point of compression failure at less than the working load. This pile was found in an area of the foundation unaffected by the general failure zones and had no gaps. Other examples are given later.

Referring to the recommendations made in respect of the second ('unloaded') tank 282, examination of 80 of the piles beneath this tank was in fact done. Incidentally when NCL floated that tank off its slab, NCL discovered a depression in the slab some six in. deep and about seventy ft diam, involving about a hundred piles, all of whose temporary steel casings had been driven to a set equivalent to seventy tonnes working load! The writer is confident that the author can be persuaded that foundations failed because there were general problems with the piles involved, not just gaps.

The author is no doubt aware that 'the New Civil Engineer' is produced by journalists, as distinct from the Proceedings of the Institution which publish papers by members and which are subject to scrutiny by the Institution. NCL did not contribute to the article from which the author quotes extensively, because NCL believed that such an article ought to be based on the evidence presented to the Court, and was anxious that all material used in the article should be capable of being checked against the Court records. All in vain, I fear, since others were much less reticent, and NCL certainly does not agree with many of the views expressed in the article.

As far as Fawley is concerned NCL would claim that an independent investigation has already been done and the expertise of our NCL firm is available to assist in clarifying any aspects of this case.

The writer would now like to comment on particular items in the author's paper, which are located by page numbers.

**Page 216.**—Although some interface problems should have been avoided, there is little evidence that a managing contractor or consulting engineer would have prevented the failure, which bad though it was, it was not the disaster it could have been. A number of tests on working piles could have revealed bad piles and heaved piles, but this was not dependent on there being one manager.

**Page 217.**—Concerning decisions "locked in": There were several alternatives considered, but the one chosen was cheapest by far, and each piece of data from site was reviewed by Esso and treated in a proper engineering fashion before continuing.

Concerning "decision was influenced by the fact that engineering for the bigger tanks had been done": The influence is that Esso had no access to designs for smaller tanks, which is clearly not so.

**Page 218.**—It should be noted that, regarding the bearing capacity of the gravel, two piling firms with a world-wide reputation offered piles with much heavier loads per pile on the gravel. Also, all such piles were to be driven to a set in the gravel. Subsequently, Soil Mechanics Ltd. calculated that the factor of safety against an Alphapile punching through gravel was 3.3 at the time of failure.

**Page 222.**—When tank 281 was floated off, the depressions were as shown in Fig. 37. Humps were found which belied a general soil failure. Fig. 38 shows one of these humps (hump A) at the edge of one of the

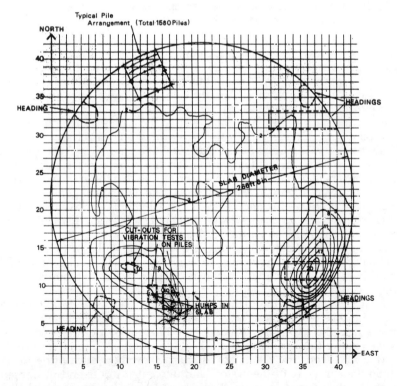

FIG. 37.—Tank 281-Hump A Showing Sound Pile beneath Slab in Area of Cratered Depression; Note Shear Failure in Slab

FIG. 38.—Tank 281 Slab after Failure, Showing Contours of Settlement of Slab, (In Inches); also Showing Headings and Humps

craters of depression, and the punching of the twelve in. R.C. slab which occurred. Calculations on the slab showed that the punching load was probably about 140 tonnes, so that the piles under the humps were carrying their ultimate bearing capacity with settlements comparable with those in unfailed slab areas. In addition, the downward movement of the cratered part of the slab related well to the crumpled lengths of certain piles beneath. Vibration tests were carried out on certain piles and in December 1974 (ICE Proceedings Part 2, Volume 57 and Discussion) Davies and Dunn presented a paper on this vibration testing, giving their assessment of the tests at Fawley which was that the expected lengths were confirmed in only 11 piles out of 43. The writer is in general agreement with these results and believes that vibration tests are valuable if assessed with a sound knowledge of foundation work, particularly in comparing similar piles on the same site.

All the aforementioned evidence is quite incompatible with a general soil failure, or of piles punching through the gravel. On the subject of raker piles, NCL concluded that these would have a detrimental effect when significant ground settlement occurred.

With vertical piles, the dished profile of the slab tends to follow the dished profile of the gravel layer as the clay beneath consolidates. With rakers near the edge, their toes are in a different part of the settlement curve relative to their heads and this tends to impose a different profile on the slab. Thus the rakers act as bracing members attempting to restrain the curvature of the system and unwanted forces and unwanted stresses are induced in the soil, piles, and slab. It was accepted that these raking piles would have effectively improved the edge factor of safety in the 'level support' condition, but the settlement profile, as it developed, would cause increased load on the rakers. The magnitude of these increases was not resolved but it can be concluded that where differential settlement is significant, raking piles should not be used unless satisfactory behavior can be confidently demonstrated.

**Page 226.**—Comparison between pre- and post-failure borings showed that the horizon between gravel and Barton Clay had not deformed in a pattern similar to the cratering of the aforementioned slab.

**Page 228.**—Concerning: "consultants who arrived at widely divergent explanations for the causes": There were at least four plausible causation factors (and four parties involved!) namely bad concept, bad ground, bad piling, and bad design of slab and slab/pile joints.

NCL examined each in great depth, and none was perfect. NCL discovered under tank 282, piles which, although undamaged by loading, were so bad that the other causes paled into insignificance. The Esso evidence which was given in court by Esso Research and Engineering, NCL Consulting Engineers, Soil Mechanics Ltd., and others was subjected to intense cross examination. None of the Defendants' experts gave evidence in court and hence were not exposed to examination, much to the writer's personal disappointment. Incidentally the writer was not aware until now, that the author was involved.

**Page 230.**—NCL attempted an analysis of the behavior of the slab and the slab/pile influences. In view of the uniform vertical loading, the piles were expected to act only as simple columns. They were therefore, expected to carry only nominal moments and shear forces associated with a column of that kind. The piles were reinforced, and were tied into the slab through conical heads. There was no horizontal binding reinforcement linking the vertical bars.

The piling contractor was aware that the load would cause substantial differential settlement in the Barton Clay. The adoption of the piled foundation had little significant effect on the settlement pattern in this clay.

The differential settlement results in a dished slab which tends to rotate the tops of the piles. Their behavior depends upon the degree of rotation, the stiffness of the piles, their load, and their fixity in the ground, the amount, and position of pile reinforcement and the connection with the slab.

The predicted settlement had been 18 in.–24 in. at the center and 9 in.–12 in. at the edge. On a diameter of 260 ft a differential of 9 in. represents 1/350 of the 'span.' The settlement profile was not predicted, but was assumed to be a parabolic shape for the purposes of the slab design. During the trial it was argued that in practice, the more likely shapes would be substantially flat over the central area and steeper at the edges. The effects of the shape assumed and of an alternative are compared:

|  | Assumed Parabola | Alternative Profile |
|---|---|---|
| Minimum Radius of Curvature (affecting design of slab) | 10,000 ft (at center) | 1,000 ft (near perimeter) |

Maximum Slope   1/115 (at perimeter)   1/35 (at perimeter)
(influencing
pile rotation)

Calculations based on limiting concrete strain for a 17 in. diameter pile show that a rotation of about 1:31 is achieved before a compressive strain of 0.0035 is exceeded. Half scale model piles were made and tested, by one defendant, with various degrees of ground density and demonstrated that rotations of this magnitude could be accepted by the piles without total failure. Even so, it is undesirable to apply significant rotation to piles of this type unless the head detail is carefully considered together with the effect of possible lateral ground movement and the effect on the slab.

For the design of a slab, the empirical flat slab method from BSCP114 was employed, and pile bending moments were assumed to be nominal.

The short term performance of the slab was satisfactory and it was accepted by NCL that the slab was not a significant contributor to the failure.

The design did not investigate the effect of individual pile failure on the slab, nor of random differential settlement, but this was studied later. It was shown by yield line theory that the load factor with any one pile ineffective was about 1.7 with a stress in the bottom steel approaching 40,000 lb/sq in. at working load. The question of whether the slab would fail in shear at a lower load factor was posed, but not argued to a conclusion.

In considering the effect of long term settlement on slab stresses the parabolic profile had been assumed. A more damaging profile was calculated to cause a maximum total stress in the bottom reinforcement of 45,000 lb–50,000 lb/sq in. at the most critical position under full working load. The steel reinforcement used was high tensile wire fabric with a minimum 0.2% proof stress of 70,000 lb/sq in.

The question of whether local steel stresses of about 45,000 lb/sq in. with calculated surface crack widths of 0.012 in. would be acceptable in such a structure was not argued to a positive conclusion.

It is very important to note that the slab and piles must be investigated as an enforced strain case, and not designed simply on a bending moment basis derived from the anticipated curvatures. When settlement occurs, slabs suffer enforced curvature, and no practical thickness or amount of reinforcement will significantly reduce it. This suggests the need for a flexible slab in order to reduce enforced stresses but in the case of a suspended slab the bending and shear stresses, which will be additive to the enforced stress, will dictate the minimum thickness. It should be noted that if the curvature exceeds a certain amount, then a reinforced concrete slab may not be a viable solution. With regard to durability, the relationship between cracking and corrosion is not fully understood. The conclusion which NCL drew at the time was that cracked sections are more susceptible to corrosive attack than uncracked ones and that code limitations on crack widths were the best guide. Recent research in connection with offshore concrete structures may provide more guidance on this matter.

Incidentally NCL has also made detailed settlement measurements of the tanks on their new foundations, which are also available in our offices.

**Page 231.**—NCL agrees that a single pile failure would not have precipitated progressive failure of the slab. Two adjacent piles failing is probably a border line case. NCL found three very defective piles adjacent to each other in tank 282. It should be noted that there were no links in the piles, so sudden failure was more likely than in a column provided with links. NCL accepts that the sensitivity of the structure could have been decreased, and reliability increased, at a modest cost. It would probably have been done if it had been appreciated at the time. It is worth warning that since these structures are being strained as they undergo long term settlement, such increase in reliability needs to come from plasticity, rather than from increase of strength. It would have been much worse if the base slab had failed under long term strains when full of oil. Of course, exposed steel bars in piles will corrode causing eventual loss of support.

Fig. 32 appears to ignore any bending effects from either the slab or from eccentric defects, of which we found many. Bending moments increase the compressive stresses comsiderably of course, and it is easy to demonstrate that pile head rotations cause high stresses.

**Page 232.**—The writer is of the opinion that the gaps were formed by a combination of two factors: (1) Arching of the concrete in the temporary tube during its withdrawal, causing preformed planes of weakness with water and soil inclusions; and (2) Heaving due to driving immediately adjacent piles. It should be noted that the overlying gravel was compacted by the act of driving piles through it, and this would cause adjacent pile tops to be strongly gripped and lifted when the alluvium is displaced by the ensuing driving.

It was apparent that the concrete had slid up the steel bars. This could only have occurred within a few days after concreting and precluded downdrag effects.

**Page 234 et seq.**—NCL notes the tests with great interest. In general the cracks full circle were found in full dimension piles with flat surfaces at the crack. NCL investigated the behavior of such piles by analytical methods and came to a similar conclusion as reached in the tests, as to their probable behavior but that yield of the bars would occur at less than 30 tonnes. Had this been the cause of failure, NCL would have expected to find only one or two in. deflection in some areas compared with generally over the slab, since once the separated faces are in contact again, the load capacity is again very high. The 12 in. by 12 in. columns tested carried about 200 tonnes, and it could be expected that a 17 in. diameter pile of the same strength would carry about 300 tonnes. Even if the concrete strength were only half, say 2,600 psi, then the load capacity would be 150 tonnes. Thus only if gaps exist in conjunction with very low strength concrete is the settlement under 30 or even 50 tonnes load likely to exceed 1 or 2 in. more than that expected. In fact the strength has to be so low, that the gaps become almost incidental, although they would certainly be the triggers in any combination of those two defects.

**Page 237.**—NCL would like to have used tank 282 for some loading tests, but Esso wanted it for storing oil, so we had to move it to a new foundation. NCL examined 80 of the piles beneath 282, but excavating for, and inspecting 'a large fraction' of over 1,700 piles implies nearly a

1,000 involving huge expense. In the event, a small sample sufficed. The writer was rather keen to extract a few of the piles which had showed short responses on the ultrasonic testing, but it did not warrant the expense. Investigation costs have to be fully justified, even more than in construction, because when one side capitulates, they then challenge the costs!

**Page 239.**—The writer strongly endorses the plea for a center for investigation. In England, the Building Research Establishment (BRE) and the Construction Industries Research and Investigation Association (CIRIA) carried out a model study on failure investigation (64) but the main stage has not got very far yet. CIRIA are also considering the subject of 'Gross Errors' in civil engineering. There are hundreds of case histories available; it needs only the funds to be provided to catalogue them in detail and establish general causative factors. However, the writer believes that it would then be established that design faults arise because of a lack of understanding of the behavior of the structures in their environment, especially the total foundation system, and that construction faults are due to lack of craft skills and the proper control of them.

The writer lists the lessons established on this investigation as follows:

1. The organization of design and construction should leave no party in doubt of his responsibilities and authority. It should ensure that there is no discontinuity within and between the organizations of the client, professions and contractors.

2. A decision has to be made on the appropriate amount of site investigation and its interpretation. Timely information is required by each party, and if not provided, the risks must be explained to, and be accepted by the client.

3. Appropriate attention must be paid to the prediction of possible soil settlement patterns under loading, and assessment of soil-structure interaction. (Highly significant in this case.)

4. The meaning of the results of pile testing must be carefully considered. Since the completed pile cannot be seen, better and cheaper methods of pile testing need to be developed so that greater numbers of piles may be checked for integrity and bearing capacity.

5. The possibility of failure or excessive settlement of piles under immediate elastic and long term conditions must be assessed and, if considered significant, studied in depth. Progressive collapse must not result from the failure or settlement of single piles or groups of pile designated as "acceptable failures" by the designer. In addition, the detection of structural distress in service and the difficulties in executing remedial work should be taken into account.

6. The choice of pile must be made in relation to the sub soil, installation conditions, feasibility of supervision and testing, and the effect of failure. It was shown that cast in situ piles are liable to suffer major defects.

7. Contractor's control should take into account the gravity of the consequences arising from defective work, and if these are not apparent they should be spelled out in the contract documents. Inspection should aim to ensure that the design concept is maintained, as well as sound construction.

8. The effects of negative friction can be more far reaching than merely increasing pile loading. Consideration should be given to changes of the overburden pressure on the bearing stratum and the consequences of inadequate penetration.

9. The use of raking piles in conditions of significant settlement can result in substantial undesirable forces.

10. The sensitivity of the chosen solution for both design and construction must be appreciated in relation to the behavior of soil, pile, and structure.

11. Documents should be explicit.

12. Piled foundations can sometimes introduce more problems than they solve. They have been a useful method with a reasonable success record, but they can be sensitive to certain conditions and it is easy to get things wrong. Settlements are not eliminated by piling.

## REPLACEMENT FOUNDATIONS

A number of solutions were considered, but the one chosen gave most reliability at reasonable cost, and it had the added benefit that it left the failed foundations available for investigation. Since it was the 30 ft depth of alluvium which posed most of engineering problems, it was decided to remove it and replace it with gravel fill dug straight from an adjacent quarry. The excavation was done by a mini-dredger, and a full account is given in Ref. 31. The new foundations were close by, and the tanks were floated across from their failed bases to the new ones in just a few feet of water, utilizing the bund walls as dams.

For the record, the settlements recorded under the hydrotests on the new positions confirmed that there was no unexpected behavior in the existing soils beneath. As far as 'bad ground' theories are concerned, it is worth mentioning that many engineers have regarded piling as a standard solution to 'bad ground' problems.

## APPENDIX.—REFERENCES

64. Bre and Ciria, "Survey of Building Failures," *Symposium on Structural Failures in Buildings*, Institution of Structural Engineers, United Kingdom, London, Apr., 1980.
65. Davies and Dunn, "Vibration Tests on In-Situ Piles," Institution of Civil Engineers, London, United Kingdom, Proceedings Part II, Vol. 57, 1974.
66. Penman, A. D. M., "Soil Structure Interaction and Deformation Problems with Large Tanks," *Proceedings of the International Symposium of Soil-Structure Interaction*, Roorkee Univ., Vol. 1, 1977, pp. 521–536 and BRE CP14/78.

### Discussion by H. Cambefort[5]

The writer is very pleased to note than an approved authority did not hesitate to reveal the starch way of thinking in the soil mechanics field.

[5]Honorary Prof. of Soil Mechanics, Ecole Nationale des Travaux Publics, Paris, France.

Using Andersen's computations, the author should have pointed out that the Kimola slip surface was very close to the zone of maximum simple shear in the soil. If such a coincidence is observed for other slides, the criterion applicable to some soils failures does not correspond to maximum values of stress but to maximum values of strain or simple shear. Such a criterion was already proposed by Reiner (68) for the solid of Kelvin. The writer thinks that it might be so and pointed it out in his introduction to the 12th session of the Xe ICSMFE held in Stockholm in 1981.

The failure of the foundations of the Oil Tank at Fawley seems to have been induced by the very poor quality of the piles. The writer may relate that, on a site where similar piles were driven, he noticed uplift and lateral displacements of the top of the piles reaching 2 or 3 cm during the driving of the other ones. The concreting of such piles is usually very tricky particularly when reinforced. On this site, the reinforcement was restricted to some starter bars driven into the concrete just after placing. The further inspection of the piles revealed shaft failures at a depth of about 2 m. To avoid such failures, the contractor had to drill the piles through the corresponding stratas.

Down drag effects may be dramatic but hardly to be expected during the short duration of a loading test since the raft induces the same shortening to the soil and to the piles. The down drag force may have only appeared during the few months after the completion of the piles.

The author's project for creating a Center to Investigate Failures has to be encouraged since the most we have to learn comes from investigations that have failed. Had such a center existed in France, the dramatic failure of the arch dam of Malpasset, occurred more than twenty years ago when the water level reached the first time 55 m upstream could have been explained differently than pointing out the effects of water over pressures in the left bank rock abutment.

The writer briefly exposed in 1963 (67) how much more satisfactory might be the consideration of the yielding of the right bank rock of foundation to match the related observations before and after failure. Before this failure on the right bank, the foot of the vertical cantilevers situated close to the haunches of the arch moved down stream at the same rate as the crowing. This movement, 5 mm measured when the water upstream was 13 m lower than the maximum design level, reached 15 mm six months after (i.e., 4 months before failure) for an extra upstream water elevation of 6 m whereas, during this period, the little measured movement close to the left bank abutment, initially downstream, reversed upstream. Unfortunately, these measurements were available only after the failure. Then the reservoir was filled up to the maximum designed level very quickly: the last 4–5 m in three days.

In addition, some water leaks appeared through the right bank, 20 m downstream, ten or twelve days before failure. Few days later, these leaks became real outflows. No leakage was observed on the left bank. The right bank rock mass apparently moved. Assuming the failure of the left bank, the cantilevers located in the center part of the dam should have rocked without any drag of their foundations. But this drag reached 60 cm and was 45° inclined on the axis of the river (85 cm for the extremity still in place, approximately located just in front of the left bank haunches of the arch). The right bank part of the arch has turned and dragged the rock mass of foundation. The result was the huge vertical crack following the upstream face of the arch down to several meters depth in the rock and gaping up to 50 cm at the surface of the ground.

A further investigation after failure revealed that the gneiss structure supporting the dam foundations was a locally decomposed rock (thin inclusions of clay or somewhere else, a mixing of clay and gneiss rock particles). During several years, these weak inclusions in the rock did not affect the stability of the arch but might have increased, undoubtedly, the progressive yielding of the rock mass concerned by the foundation. Unfortunately, the investigation boreholes have apparently been located in order to comply with a preconceived pattern of failure and the situation right underneath the foundations was not really investigated.

This case will justify the author's project for a Center for Investigating Failures. But we may imagine all the problems such a center would have to face when investigating the dramatic collapse of the near shore embankment designed for the new harbor of Nice, France! Here, the legal enquiries have not been achieved and the experts conclusions are not yet known. How can we imagine the free action of the center officers in charge with such a failure as far as new facts and new investigations may involve the reconsideration of the trial? On the other hand, the restriction of this center's activities to the only cases where legal proceedings are not engaged would drastically reduce its efficiency in our knowledge improvement. Here remains the main difficulty to be cleared up before the creation of a center attractive and useful for all the engineers. However that may be, this conference presents the greatest interest and entitles the author to our congratulations.

## Appendix.—References

67. Cambefort, H., "Etanchement et Consolidation des Roches," *Felsmechanik und Ingenieurgeologie*, Vol. 1/2, 1963.
68. Reiner, M., "Twelve Lectures on Theoretical Rheology," *Rheologie Theorique*, (authorized translations and additions by North Holland Publishing Company, Amsterdam, the Netherlands), Dunod, Paris, France, 1955.

### Discussion by Claudio A. Mascardi,[6] M. ASCE

The author deserves gratitude for the topic of his Terzaghi lecture, and the intelligent and patient work he has done to the benefit of the profession. This contribution will be limited to the Fawley case.

**Causes of the Faults in the Piles.**—Piles such as those executed at Fawley have cross-sectional dimensions depending on the pressures exerted by fresh concrete against surrounding soil and by the stress-strain behavior of the latter.

In soft soils (such as the alluvial clay at Fawley) it can be necessary to decrease the pressure of the mandrel on the fresh concrete in order to avoid excessive extra size of the pile stem, which would in turn imply

[6]Sr. Engr., Studio Geotecnico Italiano, Milano, Italy.

undesirable consequences: increase in uplift and downdrag forces exerted by the soil on the piles and additional cost of cast in situ concrete. Therefore, the mandrel is lifted by the winch cable and the sensibility of the operator is sometimes assisted by a spring connecting the cable to the mandrel. If the workability of the concrete and the smoothness of the internal surface of the temporary casing are not constantly kept into an optimum condition, the task of the operator can become too difficult, and he can incidentally give too little weight on the fresh concrete, not balancing the skin friction of the concrete against the casing as it is withdrawn.

If the concrete slump is too low, necking or even circular fissures can occur. Friction against the reinforcing bars and the soil can reduce the tendency to close up the fissures in the fresh concrete. Thornburn and Thornburn (56) quote cases in which gaps were formed by such effects.

If the concrete slump is high and the operator overconservative, the diameter of the piles will be significantly larger (10%–15% or more) than the nominal one. It would be of interest to know whether this source of pile faults has been considered, the reasons that possibly led to its exclusion and which was the ratio of the actual pile diameter to the nominal one in the upper zone.

**Measures that can be Taken to Prevent Similar Failures in the Future.**—The author has pointed out that the structural design of the slab did not take sufficiently into account the possible nonuniform behavior of the piles; maybe a less common design, with a layer of compacted gravel above the top of the piles instead of an r.c. slab, could have better performances in such cases, where a flexible superstructure (the tank bottom) has to match a rigid deep foundation. The main point is however that the soundness of each pile can be tested with little cost by the vibrations method, which in the writer's experience is most effective in detecting gaps within a depth of 20 diam below the pile top. This method (69) was at its early applications in 1967 when the piles at Fawley were driven, but is now fairly often used and accepted in Europe.

**APPENDIX.—REFERENCE**

69. Paquet, J., "Étude Vibratoire des Pieux en Béton, Réponse Harmonique et Impulsionnelle, Application au Controle," Annales I.T.B., No. 245, May, 1967.

Discussion by Farrokh N. Screwvala,[7] M. ASCE

The author has done the profession a service by pointing out some of the reasons why past investigations of instabilities may not necessarily have contributed insights into the reasons for such instabilities. A review of the case studies presented in the paper suggests that there may be another reason beyond the five cited at the outset of the paper for this lack of insight. This additional reason may be the inadequacy of the analytical technique used to evaluate the failure. Every analytical technique

---

[7]Pres., Farrokh N. Screwvala Inc., Cleveland, Ohio.

represents a simplification of nature. In order to determine whether a given analytical procedure results in a reasonable explanation of natural phenomena, it is necessary to apply the analytical procedure to a real situation where the assumptions which are part of the analytical procedure are more or less satisfied. Under such circumstances the failure of the analytical technique to accurately explain natural phenomena may then properly be assigned to either an incorrect model of nature or possibly to inaccurate input data for a more or less realistic model.

The slip circle method of analysis, both total stress and effective stress, was used to evaluate the instability of the Kimola Canal in Finland, the King's Lynn embankment in England, the James Bay embankment in Canada, and the I-95 embankment in U.S.A. The author correctly states that the behavior of these slopes differs considerably from the expected behavior based on slip circle analyses. In a slip circle analysis a presumed critical surface, along which rupture is most likely to occur, is determined. The behavior of the slope after the development of the critical surface or after the onset of motion is not, however, considered. Under such circumstances the expectation that a slip circle analysis can satisfactorily explain the configuration of a slope or embankment after failure merits further discussion.

The presence of a weak seam at Kimola is suggested by the author as a possible cause of the slide. The failure surface at Kimola, as shown in Fig. 4, was about 80 m long and varied in elevation about 17 m. The expectation that a single layer of soft clay was deposited along this surface seems to suggest unusual geologic conditions. With some exceptions, it is widely believed that sedimentary layers are nearly level when formed. Does the local geology at Kimola suggest the likelihood of inclined strata? An explanation for the nine month delay between the completion of the excavation and the occurrence of the slide between stations 51 + 80 and 54 + 20 would also be helpful.

The paper points out clearly the differences in the present methods of estimating the strength of soils. Expressed another way, one could say that our understanding of the strength of soils is too limited to allow us to explain the behavior of slopes or embankments with the precision that we would like. Despite these limitations, it is important to identify conditions where failure is likely. One common thread connecting all of the case histories appears to be that all of the failures occurred in very weak layers which were stressed at levels which might be regarded as approaching their ultimate bearing capacity. Could the possibility of some form of bearing capacity failure be a credible alternative to the possibility of failure due to the presence of thin weak seams?

### Closure by G. A. Leonards[8]

Many thanks to Bratchell, Cambefort, Isnard, Mascardi, Pilecki, and Screwvala for their contributions to the Discussion. A number of important points were raised, and it was difficult to determine how the

[8]Prof., School of Civ. Engrg., Purdue Univ., W. Lafayette, Ind. 47907.

response should be organized. In the end, it was decided to treat the concerns of each discusser separately.

The slip surface for the main slide on the Kimola canal was sloped 1 vertical to 8 horizontal. Screwvala states that "the expectation that a single layer of soft clay was deposited along this surface seems to suggest unusual geologic conditions. With some exceptions, it is widely believed that sedimentary layers are nearly level when formed." Screwvala then asks if the local geology at Kimola suggests the likelihood of inclined strata? The answer is yes. The inclination of depositional bedding planes is approximately parallel to the slope of the surface on which deposition is initiated. This is clearly evident not only in Fig. 4 at Kimola, but also for King's Lynn (Fig. 12) and for I-95 (Fig. 21) trial embankments. For the latter case, the selected shape of the slip surfaces was strongly conditioned by the belief that a weak seam existed at the indicated depth, *dipping more or less parallel to the till surface.* Such geologic features are not at all uncommon. As noted in the paper, determining the location and properties of thin weak seams should be a primary goal of site investigation. Identifying the dip of the depositional bedding planes is a useful guide in easing the effort involved in this task.

Screwvala also asks for an explanation of the 9-month delay between the completion of the excavation and the occurrence of the main Kimola slide. Both Kankare (26) and Kenney and Uddin (28) attributed this to an increase in pore pressure that developed with time. A number of arguments were presented in the paper to refute this claim of which the most telling are: (1) Numerous slides that took place in the slopes of the Kimola canal and were "delayed" for periods of time ranging from a few days to decades; (2) and 90,000 $m^3$ of soil that slid down and blocked more than 200 m of the canal in less than half a minute. The explanation given in the paper (item 4, p. 214) is that drained creep strains, which can occur at constant levels of effective stress, reach a critical strain level at which the clay structure breaks down. The increase in compressibility accompanying the breakdown in clay structure induces large excess pore water pressures and a rapid, progressive, undrained failure occurs. I hold strongly to this explanation of "delayed" failures at Kimola; the different times to failure are due to different rates of creep at various locations along the canal.

Yes, Mr. Pilecki, the main issue is whether or not the failure at Kimola (or elsewhere, for that matter) could have been predicted. However, to be useful, the prediction must be sufficiently quantitative to allow economic design of stable slopes. Of the 13 reasons cited in support of your contention that the Kimola slide could have been predicted, twelve are qualitative in nature. It does not help at all to state that "Finnish surface soils are mostly post-glacial and therefore not preconsolidated, thus capable of sliding." The one quantitative reason given involves assuming an angle of internal friction equal to 12°–15°, in which case you calculated a safety factor equal to 0.6. An effective angle of shearing resistance of 12°–15° would correspond, approximately, to the residual strength of the Kimola clay. As there is no evidence of sliding prior to construction of the canal, the residual strength is not applicable. In fact, if such friction angles had been used in the design, slopes of about 1 on 10 would have been proposed. Such a recommendation may have precluded construction of a canal that has served its useful purpose already for two decades.

At Fawley much attention in the design stage was paid to evaluate the properties and thickness of the river gravel, to the possibility of overstressing the underlying clay, and to establishing criteria for pile driving to insure adequate (but not excessive) pile penetrations into the river gravels. I agree with Pilecki that insufficient attention may have been given to insuring the integrity of the pile shafts during construction. However, I disagree, strongly with his contention that, ". . . the piles failed because the longitudinal reinforcing bars were not strengthened according to the theory of reinforced concrete, which states that the slenderness ratio of each bar between the stirrups must be at least equal to the slenderness ratio of the columns." It has long been established that piles, even when embedded in very weak soils, behave structurally as short compression members. If further specific evidence is desired, one need only note that test pile No. 2 sustained a load of 160 tons (Fig. 25) with no evidence whatsoever of structural damage. This is more than three times the load/pile at which the foundation failed during proof testing. The evidence is unequivocable that the failure was due to *defective piles*. The specific nature of these defects will be reviewed later.

Mascardi points out some common causes of pile defects associated with the operations of raising the mandrel and with using too high or too low concrete slumps. He asks what was the ratio of the actual pile diameters to the nominal one in the upper zone and the reasons why these sources of pile defects were excluded from consideration.

According to the information available to me, twenty-seven piles were exposed in a heading beneath tank 281 whose foundation failed during proof loading. The condition of all but two of these piles is described in Table 6 with representative photos in Fig. 28. Twelve piles failed structurally. The nominal shaft diameter was 17 in.; three piles exhibited sufficient necking (to 12, 12-3/4 and 10 in. diam) to be recorded. The diameter of 10 piles was measured. These generally ranged between 15-1/2 and 16-1/2 in., with two piles having diameters of 17-1/4 in. Eighteen piles were exposed in a heading beneath essentially unloaded tank 282. The diameter of fourteen of these piles was measured and they fell within a narrow range of 15-1/4–16-1/4 in. Two piles exhibited necking to 12-3/4 in. diameter, and one pile had bulged to 23-1/4 in. Significantly, five piles had full circle cracks, and one pile [Fig. 33(*d*)] had a 1 in. full circle gap 2 ft below its top. Notations for two piles whose diameters were not recorded were: "Top 1 ft: no strength through entire cross section. Disintegrates with probing" and "Top 1 ft: no strength in partial cross section. Disintegrates with probing." Thus, it is likely that all of the mechanisms described by Mascardi contributed to the formation of defective piles at Fawley. These defects were not excluded from consideration; in fact, along with the effects of ground displacement, they were considered very carefully. For reasons given in the paper, it was concluded that only piles with gaps, or those with a full cross section consisting of incredibly poor concrete would have had low enough capacities to fail under the proof loading. I will deal with these pile defects subsequently. The question to be answered now is: how did the full-circle gaps develop? Three possibilities were postulated:

1. "Ground heaving pulled the piles apart after the concrete had taken its initial set." As all full-circle gaps were observed to occur within the top two ft of the pile, and two were found at the very top of the pile [e.g., Fig. 28(a)], there is not enough shaft friction to overcome the tensile resistance even of poor concrete. If the slab had been in place over a portion of the foundation while adjacent piles were driven, uplift on the slab could easily separate the piles, even if the concrete was of good quality. At the time the paper was written I was unable to obtain evidence either to confirm or reject this possibility. Since that time, information obtained through the courtesy of NCL Engineers—the engineers of record investigating the failure—shows that the last piles for tank 281 were driven on August 12, while the contractor's proposal for concreting the slab was dated August 18, 1967. Accordingly, ground heaving is rejected as the cause of full-circle gaps.

2. "The concrete slump was so low that circular fissures developed by arching, and friction against the reinforcing bars and the soil reduced the tendency to close up the fissures in the fresh concrete." It is inconceivable (to me) that this mechanism could produce horizontal full-circle gaps of constant width and with matching jagged edges characteristic of tensile fracture [Fig. 33(d)]. Thus, this mechanism is also rejected.

3. Differential downdrag on the piles after the slab had been poured was shown to be sufficient to fracture the pile, especially if a defect was already extant. The critical locations would be towards the top of the pile; the displacements would be symmetric, and the irregular edges would be matching. For these reasons, downdrag is judged to be the only viable mechanism for the formation of full-circle gaps of constant width and matching serrated edges.

Given the circumstances at Fawley, the pile designer could choose between three alternatives:

1. Insist on rigorous construction control, integrity testing of every pile (about 3,000 piles), and remedial grouting or replacement of every suspect pile.
2. Insure that reasonable care is taken during construction and design the slab to sustain the load in the event one or two adjacent piles are defective.
3. Select a pile type whose integrity is unaffected by the site and construction conditions.

The choice is a matter of economics and of time for construction. In the paper I expressed the opinion that the second option would have been appropriate at Fawley.

Mascardi also suggested that a layer of compacted gravel above the top of the piles instead of an r.c. slab could perform better in cases where a flexible superstructure (the tank bottom) has to match a rigid deep foundation. According to Roberts (74), granular pile caps have been used successfully for over 70 years. Roberts describes a "typical design" using 4-1/2 ft of compacted crushed rock and 7 ft of "compacted fill" over the piles; however, no rational methods for designing granular pile caps or documented records of their performance appears to be available, which may be a substantial deterrent to their use. In any case, the behavior of granular pile caps at Fawley, where settlement of the alluvial clay under the weight of the gravel fill is expected to continue during the service life of the tanks, is open to serious question.

Bratchell's contribution to the foundation failure at Fawley is valuable and much appreciated. The extensive investigations conducted by his firm were exemplary and no criticism of this effort was intended in the paper. However, there are limitations of time and resources surrounding every failure investigation by an interested party. Moreover, regardless of the terms of reference, there are restrictions against reaching open consensus on the causes of failure and of the lessons to be learned therefrom. These restrictions were emphasized in the paper, as it was my experience with the investigations at Fawley that led me to perceive the need for a National Center for Investigating Failures.

There are substantial areas of agreement between Bratchell and myself regarding the cause of failure. We agree that it was not due to a general shear failure in the underlying Barton clay or to individual piles punching through the river gravel. We agree that the sole cause was defective piles, and that at least two adjacent piles must have been seriously defective before progressive failure of the foundation slab could occur. We did not agree on the nature of the pile defects primarily responsible for the failure (full-circle gaps versus poor concrete), on the mechanism responsible for the formation of full-circle gaps (separation >1/2 in.), or on the load-carrying capacity of piles with gaps. These differences merit further discussion.

To be correct, an explanation for the formation of gaps must be capable of dealing with the following three features: (1) Symmetry of the applied tensile forces (horizontal gaps of constant width); (2) failure after the concrete had taken its initial set [matching jagged edges, Fig. 32($d$)]; and (3) separation at the head of the pile [Fig. 28($a$)]. Downdrag forces after the slab had been placed is the only explanation that satisfies all three requirements; I remain convinced it is the only plausible explanation for the formation of gaps!

Bratchell is of the opinion that piles with gaps could sustain loads on the order of 150 tonnes before fracturing (and hence large deflections) would occur. He bases this conclusion largely on the results of the test on a 12 × 12 in. column with a preformed gap described in the paper. The purpose of this single test was not to estimate the capacity of piles with gaps but to confirm the mechanism of failure from which relevant strengths could be reasoned; i.e., yielding and buckling of the reinforcing steel, spalling of concrete at the pile periphery, contact of the two separated pile segments, and fracture of the concrete in compression. The 12 × 12 column was constructed of uniform, high quality concrete (5,200 psi compressive strength), the reinforcing rods were accurately placed, the preformed gap was bounded by two aligned plane surfaces, and a concentric vertical load was applied to the vertical column. Under these conditions the effective concrete area at ultimate fracture was found to be 50% of the nominal area of the column. What effective concrete area would Bratchell consider appropriate for imperfectly aligned piles, separated by gaps with jagged faces and loaded unsymmetrically? The cross-hatched area in Fig. 34 (p. 236) shows that I believe 10%–30% to be a plausible range in effective area; hence, piles with concrete strengths in the range of 2,000–3,000 psi would readily fracture at the applied water load of 50 tons. An important point emphasized in the paper (p. 237) is that whether or not piles with gaps would behave in the manner previously described could have been established positively by load testing such piles after tank 282 had been floated off. This would certainly have been done if an independent party with resources and access to the site had been available.

In support of his assertion that failure was due to poor concrete in intact piles Bratchell states, "We discovered under tank 282, piles which, although undamaged by loading, were so bad that other causes paled into insignificance." How bad is 'so bad'? An intact pile with a shaft diameter 15-1/4 in. (minimum recorded in the field notes—barring necking) with four 5/8 in. mild steel reinforcing rods could sustain the test load of 50 tons in direct compression if the compressive strength of the concrete were only 360 psi! What evidence is there that concrete in adjacent piles was defective to this extent? Construction Services, Ltd. (CSL) were commissioned by Esso to investigate the quality of piling concrete. Up to the time the paper was written I was unable to obtain a copy of their report. However, through the courtesy of Mr. Bratchell, I have since received copy of a report dealing with tank 281.

CSL tested 26 three-in. diam cores taken from piles beneath tank 281; seventeen of these cores were previously obtained by Soil Mechanics, Ltd. and the remainder by CSL. All tests were on samples obtained from the poorest looking piles. The average estimated 28-day compressive strength was 3,060 psi; the range was 1,690–4,300 psi, and the standard deviation 697 psi. Ten of the 26 cores had compressive strengths below the design value of 3,000 psi. The estimated cube strengths on the day of failure averaged 4,400 psi with the lowest value being 2,450 psi. CSL also tested six 4-in. cube samples sawed from piles that visually lacked fines, and appeared to be poorly compacted. Four of the 6 cubes had estimated 28-day compressive strengths below 3,000 psi; however, on the day of failure, all but one of the estimated strengths were well above 3,000 psi, with the one lowest value being 2,010. I have no reports of tests on samples taken from piles beneath tank 282. Incidentally, sulphate resisting cement was used in the piling concrete.

It should be mentioned in connection with the test results just cited that, in general, concrete was most poorly compacted around the perimeter of the pile, but improved considerably towards the center. Usually, cores or cubes could not be obtained within 3 in. of the outer face; hence the test results reflect concrete quality in the interior 10 in. of the shaft. In one pile (P.A. 42), it was possible to saw a 4-in. cube only from the very central portion of the shaft.

I acknowledge that Bratchell had the advantage of onsite inspection while I had to be content with the examination of photographs. Nevertheless, although there is no doubt that a considerable amount of concrete failed to meet design strength, the physical evidence that many piles contained concrete with average compressive strengths at any section less than 360 psi is simply not available. I continue to hold the view that piles with gaps played an important role in the failure at Fawley.

Bratchell's discussion of slab/pile behavior, of the criteria adopted in the design of the slab at Fawley, and of the post-failure studies examining the ability of the slab to sustain failure of a single pile, is of interest. On the one hand, the shears and bending moments in the slab must be in equilibrium with the applied load and the pile reactions, and they must be compatible with the strains induced by the deflected shape of the slab and the stress-strain relations for reinforced concrete. On the other hand, the pile reaction and head deflection must be compatible with the load/settlement relation of each pile within a very large pile group. Due to local variations in soil support, in pile stiffness, and in "seating" of the pile tip, this latter relation can vary within wide limits. Where the stiffness is higher the pile will draw more load and the maximum bending moment may be negative; where it is lower the pile reaction will be lower, and the maximum bending moment may be positive. It is important to recognize that while two adjacent piles may have the same head deflection, their reactions against the slab are by no means the same. As stated in the paper, and reiterated here, the prediction of local pile/soil stiffnesses to calculate the "enforced strains" in the slab with sufficient precision for design is beyond the present state-of-the-art. Accordingly, arbitrary assumptions of critical curvature and maximum slope of the slab must be made to design the slab/pile system. Unfortunately, the experience at Fawley shed no new light on the validity of the assumptions that were made. This is a problem where field measurements of actual system response would be very valuable.

I agree with Bratchell that, within practical limits of slab thickness, its stiffness has only a modest influence on the settlement pattern and that to avoid excessive rotations of the pile heads the slab should be as flexible as possible. This suggests the use of a minimum thickness necessary to effect shear transfer and an excess of reinforcement to provide reserve resistance against the formation of a mechanism in the event some of the piles are defective. This was the approach advocated in the paper for the conditions extant at Fawley.

Bratchell lists 12 lessons established from the Fawley investigation. Six of these (nos. 3, 4, 5, 6, 9 and 10) were cited in the paper. I also support five others (nos. 1, 2, 7, 8 and 11) although these were not cited because they were beyond the scope of the paper. Regarding piled foundations being "sensitive to certain conditions and it is easy to get things wrong. Settlements are not eliminated by piling" (no. 12); I am concerned that this generalization may be misinterpreted. A shallow foundation applies load to the soil at the level of the excavation dictated by the function of the structure being supported. If the bearing capacity is deficient or the expected settlements are excessive, a variety of measures can be considered for improvement; piling is one such measure. Piles are no more "sensitive to certain conditions," nor is it any easier "to get things wrong," than is the case for many other methods of ground improvement—including grouting, dynamic compaction, overexcavation, preloading with or without artificial drainage, etc. When selected, designed and installed with due consideration of the relevant factors involved, pile foundations do indeed perform successfully.

The problem at Fawley was not due to any mysterious behavior of the piles but to their lack of quality and integrity. From the viewpoint of design, the most important lesson from the experience at Fawley concerned the approach to be taken in dealing with this problem. I have already listed the applicable alternatives in my response to Mascardi's question and expressed the opinion that more care during construction, and designing the slab to sustain the load in the event two adjacent piles were defective, would have been appropriate at Fawley. This view was generalized in the paper (lesson 4, pp. 238–39): it states, basically, that if the *design concept* accommodates the probability that foundation construction will not conform exactly to the design documents, the result will be *fewer unexpected failures*.

Four of the six discussers commented on my proposal to create a National Center for Investigating Failures; three were strongly supportive and one was opposed. Isnard commented briefly on the French "Bureau de Controle Technique," which is an agency concerned with the reliability of performance in the long-term. The agency often becomes involved when failures occur, but due to legal considerations the information thus garnered is processed and used only internally. Of special interest is the fact that the French Ministry of Urbanism and Housing recently created an "Agency for Construction Safety and for the Improvement of the Building Rules of Art." The primary aim of this agency is to improve building quality and reduce failure risks by making known to designers and builders the "pathogenerating causes" of failures. This Agency has some of the attributes proposed for the National Center, including free access to failure investigations; moreover, funds to supplement ongoing failure investigations have been requested from the Ministry of Scientific Research and Development. What is not yet clear is how the legal problem will be bypassed. A precedent for dealing with legal questions exists in the law establishing the U.S. National Transportation Safety Board. Features of this law applicable to the proposed National Center will be discussed subsequently.

Cambefort presents an alternate explanation for the cause of the Malpasset Dam failure and suggests that if a national center such as the one proposed in the paper had existed in France, the cause of the failure—and the lessons learned therefrom—would have been different from the prevailing view. Malpasset dam failed on December 2, 1959. Over 400 people lost their lives and a good part of the nearby town of Fréjus was destroyed. It was a national disaster for France. It was also the first total failure in the history of arch dams. Of some six hundred dams of this type that had been constructed throughout the world, there had been only a few incidents that gave rise to serious concern, and in these cases the vital parts of the dam remained in place. This time, almost the entire arch structure was destroyed in an instant. An explanation of the actual causes of this unique event is of great importance in order to determine how to recognize and control the same potential dangers elsewhere.

It is not feasible to present herein all the relevant data on the design, construction, and postfailure investigations of the Dam. An English translation of the report by the official Commission investigating the failure is available (71); it did not clearly explain the cause. Mary (73) presented a digest of the displacements measured July 1958–July 1959, at which time the reservoir level was 5 m below that at which failure occurred. The dam had a double movement of rotation; one around its right abutment and another simultaneously about the crest. Significantly, in the central third of the dam, the foundation displaced downstream much more than had been expected but this information was not communicated to the engineer until after the failure. A number of famous designers of dams published explanations of the cause of failure. Some of their views were artfully summarized by Jaeger (72). The causes ascribed to the failure differed, and none were consistent with all the observed facts. More recently, a consistent and plausible explanation of the Malpasset failure was published by Jean Bellier and Pierre Londe (70). Basically, they attributed the failure to instability of a rock wedge bounded by foliation features that dipped downstream and a downstream fault plunging beneath the structure—both at the left abutment—on which high pore pressures developed. This explanation has resulted in special attention to foundation drainage measures for arch dams, which previously had been considered to be of little importance. It is this explanation to which Cambefort objects, attributing the failure to foundation movements due to locally decomposed rock on the right bank.

It is of vital importance to establish whether or not the foundation movements played a role in initiating the failure because movements in excess of the 14 mm measured in July 1959 at Malpasset have been recorded on other arch dams and are not presently considered as a cause for alarm. According to Cambefort, the foundation movements caused a redistribution of stress that resulted in bulging of the arch and cracking of concrete at the left abutment. It is not clear why these facts were not reported promptly to the engineer. In fact, the reservoir level was allowed to rise 5 m in only three days immediately prior to failure. It will never be known how much the foundation moved prior to rupture of the arch; hence, Malpasset cannot contribute to our knowledge on how much foundation movement is tolerable. However, the massive, well-constructed thrust block on the left bank very likely was more resistant to movement than the base of the arch. Load concentrations would develop in the concrete near the left bank and foundation contact pressures would be simultaneously reduced, thus permitting more slippage and more load concentration, etc. Actually, the reduction in total stress on the downstream fault may have had as much to do with sliding of the rock wedge as the increase in pore pressure previously emphasized. Is it possible that the relatively unyielding thrust block was an important contributor to the demise of the arch? Is there a lesson to be learned here?

I cannot say that the existence in France of a national center to investigate failures would have resulted in a different explanation of the Malpasset failure than the prevailing one, as suggested by Cambefort, but I do believe that the Center could have served two important functions: (1) As a focal point for intensive and sustained research on the potential mechanisms of failure, it would have the opportunity to engage the best talent available to assist with the investigations; and (2) it could hold periodic conferences—especially after the results of new research had been published—with ample opportunity for experts to debate their respective explanations for the failure. The debates would lead to new research to resolve differences in points of view and, perhaps, to a consensus on the lessons learned. These, then, would be published for the benefit of the profession at large. Such a conference on Malpasset would have relevance even today.

The response to my proposal for a National Center has been overwhelmingly favorable although some engineers—like Pilecki—fear that it could become "an undemocratic and bureaucratic entity, which would only become an inefficient burden on the taxpayers." It seems appropriate to reemphasize the key points in the proposal so that it may be judged on its true merits and not on the basis of vague charges, like "just another bureaucratic federal agency." These key points are:

1. It is necessary that the Center be established by an act of Congress; otherwise it will not have authority for free access to the site of failures and to all available relevant information. Neither will it have authority to conduct on-site tests, or to obtain specimens for detailed laboratory studies. There must be a source of revenue that is both sustaining and reasonably predictable so that the Center can plan the scope of its operations in terms of staffing and facilities, the number and size of failures that can be dealt with effectively each year, the outside consultants that can be engaged, and the kind of long-term research to be initiated either in-house or by outside contracts. A levy of 0.02 percent on federally sponsored construction—currently estimated to yield about $8 million annually—would hardly be noticed in the federal budget but it could be an enormous boon to the profession and the public alike.

2. Federal funding is *not* synonymous with federal control. Surely, no one would accuse the National Center for Disease (NCD) or the National Transportation Safety Board (NTSB) of being federally controlled. The proposed National Center for Investigating Failures (NCIF) would be even further removed from federal control than the aforementioned federal agencies. Its Board of Directors would be appointed by the engineering profession which, in turn, would appoint the Center's Director. The Director would be responsible for staffing, selecting failures to be investigated, publication of reports, holding seminars and symposia, and initiating needed research. The Board would conduct annual reviews, arrange for audits of expenditures, and study long-term trends. There is no possibility for political intervention within this framework, and adequate controls to minimize bureaucratic action are provided.

3. Contrary to prevailing opinion, there need be no legal deterrents to the establishment of NCIF. The Act establishing NTSB could serve as a model in this respect (75). A few relevant quotes from this Act are:

> a) "The Board shall . . . report in writing on the facts, conditions, and circumstances of each accident investigated . . . and cause such reports to be made available to the public at reasonable cost, and to cause notice of issuance to be published in the Federal Register."
>
> b) The Board will . . . "conduct special investigations on matters pertaining to safety in transportation . . ."
>
> "assess and reasses techniques and methods of accident investigation and to prepare and publish from time to time recommended procedures for accident investigations"
>
> "evaluate the adequacy of safeguards and procedures concern-

ing the transportation of hazardous materials . . ."

c) The Board may . . . "hold such hearings . . . and require by subpoena the attendance and testimony of such witnesses and the production of such evidence as the Board may deem advisable"

d) The Board is . . . "authorized to enter any property . . . and do all things therein necessary for a proper investigation . . . and is authorized to employ experts and consultants . . ."

e) *"No part of any report of the Board, relating to any accident or the investigation thereof, shall be admitted as evidence or used in any suit or action for damages growing out of any matter mentioned in such report or reports."*

4. NCIF will *not* encroach on the investigation of failures by engineers in private practice because it will investigate only a small fraction of the total and, in any case, its findings will be inadmissible in a court of law. On the contrary, more people will be involved than is now the case and better studies will be forthcoming from all concerned parties.

The Board of Direction of ASCE has appointed a special committee—responsible directly to the Board—to advise on the development of Policy concerning appropriate responses after failures have taken place and, especially, on measures to adopt that would reduce the frequency of their occurrence. In my opinion, nothing that ASCE has done in its 130-yr history will be of greater benefit to the public, or bring more credit to the profession, than the establishment of a National Center to Investigate Civil Engineering Failures.

## APPENDIX.—REFERENCES

70. Bellier, J., and Londe, P., "The Malpasset Dam," *Proceedings of the Engineering Foundation Conference on Evaluation of Dam Safety*, Pacific Grove, Calif., ASCE, 1976, (original by Jean Bellier and published by Travaux, Paris, France, July, 1967).
71. "Final Report of the Investigating Committee of the Malpasset Dam," Office of Technical Services, U.S. Department of Commerce, Washington, D.C., 1960.
72. Jaeger, C., *Rock Mechanics and Engineering*, Cambridge University Press, London & New York, 1972.
73. Mary, M., "Barrages-voûtes, Historique," *Accidents et Incidents*, Paris, France, 1968.
74. Roberts, D. V., "Foundations for Cylindrical Storage Tanks," *Proceedings of the 5th International Conference on Soil Mechanics and Foundation Engineering*, Vol. I, Paris, France, 1961, pp. 785–788.
75. *United States Code Annotated, Title 49 Transportation, Chapter 28 National Transportation Safety Board*, West Publishing Co., St. Paul, Minn., pp. 131–140.

# THE SEVENTEENTH TERZAGHI LECTURE

Presented at the American Society of Civil Engineers 1981 Convention and Exposition

October 29, 1981

ROBERT V. WHITMAN

# INTRODUCTION OF THE SEVENTEENTH TERZAGHI LECTURE

## By Robert L. Schuster

Robert V. Whitman was educated in the public schools of Edgewood, Pennsylvania. His undergraduate education was at Swarthmore College, where he recieved his B.S. degree in 1948. Graduate study followed at MIT, where he received his S.M. degree in 1949 and his Sc.D. in 1951. He joined the engineering research staff and then the Civil Engineering faculty at MIT. Although his field of specialization for his doctoral studies had been structural engineering, he soon became deeply involved in the study of soil dynamics. As one of the nation's first workers in this field, he served as a consultant to several goverment agencies studying the effects of nuclear explosions.

Beginning in 1964, Bob Whitman's attention turned to earthquake engineering, a field in which he has excelled. He has been involved in the aseismic design of nuclear power plants, earth dams, and oil refineries, and has held important advisory posts to numerous United States and international agencies.

Bob has authored or co-authored more than 60 technical publications, especially on soil dynamics, earthquake engineering, and seismic risks. In his own estimate, with which thousands of past and present students of geotechnical engineering agree, his most important contribution has been the book *Soil Mechanics*, co-authored with T. William Lambe.

He is a Fellow of the American Society of Civil Engineers and a member of the Earthquake Engineering Research Institute and the Seismological Society of America. In 1962, he received the Walter L. Huber Civil Engineering Research Prize from the American Society of Civil Engineers. Since then he has received the Structural Section Prize, the Desmond Fitzgerald Award, and the Ralph W. Home Fund Award from the Boston Society of Civil Engineers. In 1975, he was elected to the National Academy of Engineering.

For several years Bob was Chairman of the Publications Committee of the Soil Mechanics and Foundation Engineering Division of ASCE, and was heavily involved in the Awards and Soil Dynamics Committees. In addition, he was a founding member of the Executive Committee for the ASCE Technical Council on Lifeline Engineering.

# EVALUATING CALCULATED RISK IN GEOTECHNICAL ENGINEERING

## By Robert V. Whitman,[1] F. ASCE

**ABSTRACT:** Recent years have seen rapidly growing research into applied probability and increased interest in applications to geotechnical engineering practice. Unfortunately, probability still remains a mystery to many engineers, partly because of a language barrier and partly from lack of examples showing how the methodology can be used in the decision-making process. The following types of applications are described and illustrated in general terms: separating systematic and random errors when evaluating uncertainty in the stability of slopes; safety factors in connection with analysis of liquefaction; optimizing the design of an embankment in the face of uncertainty concerning stability; risk evaluation for an industrial facility built over potentially liquefiable soils; and risk evaluation for earth dams. Even when a precise quantification of probability of failure is not possible, systematic formulation of an analysis aids greatly in understanding the major sources of risk and thus points the way for cost-effective remedial measures. Analyses of reliability and risk are potentially most valuable during the early stages of a project in guiding the decision as to whether or not to proceed and in helping to establish design criteria.

## INTRODUCTION

The title of this paper was deliberately selected to evoke memories of Arthur Casagrande's Terzaghi Lecture of 1964, "Role of the 'Calculated Risk' in Earthwork and Foundation Engineering" (11). The main points made by Casagrande in that lecture were that risks are inherent in any project, that the existence of such risks should be recognized, and that steps—representing a balance between economy and safety—should be systematically taken to deal with these risks. The notion that risks exist was hardly new in 1964. However, there was then a widespread reluctance on the part of many owners to recognize such risks explicitly and to deal with them openly.

Casagrande went to considerable length to make clear what he meant by the words "calculated risk." To him this phrase implied the process of recognizing and dealing with risks, in two steps:

"(a) The use of imperfect knowledge, guided by judgment and experience, to estimate the probable ranges for all pertinent quantities that enter into the solution of the problem.

(b) The decision on an appropriate margin of safety, or degree of risk, taking into consideration economic factors and the magnitude of losses that would result from failure."

Casagrande noted a dictionary definition of "calculated," when used as an adjective, as meaning "estimated." "Therefore," he said, "the oft-

---

[1]Prof. of Civ. Engrg., Massachusetts Inst. of Technology, Cambridge, Mass. 02139.

Note.—Discussion open until July 1, 1984. To extend the closing date one month, a written request must be filed with the ASCE Manager of Technical and Professional Publications. The manuscript for this paper was submitted for review and possible publication on July 25, 1983. This paper is part of the *Journal of Geotechnical Engineering*, Vol. 110, No. 2, February, 1984. ©ASCE, ISSN 0733-9410/84/0002-0145/$01.00. Paper No. 18569.

heard joke remark that a calculated risk is the type of risk that nobody knows how to calculate, is really a play on words on the ambiguity of the adjective 'calculated.' " Casagrande used phrases such as "grave risks" and "great uncertainties," but he did not use numbers to quantify these adjectives. Indeed, he characterized the problems about which he spoke as ones ". . . which at present defy quantitative analysis."

The 18 years since Casagrande's lecture have been a time of great interest and activity in the areas of probability theory and risk analysis. The Geotechnical Engineering Division of ASCE has had, for the better part of a decade, a Committee on Reliability and Probabilistic Concepts in Geotechnical Engineering Design. Several special workshops and meetings have been held (e.g., Ref. 4). A steady and increasing number of papers concerning probabilistic approaches have been published in the *Journal of the Geotechnical Engineering Division*, and sessions concerning probability and geotechnical engineering have been a part of numerous national and international conferences (e.g., Ref. 3). There have been three International Conferences on Applications of Statistics and Probability to Soil and Structural Engineering (2), and a fourth is scheduled for 1983. Furthermore, risk assessment has appeared in the practice of several geotechnical consulting firms.

There has been even greater interest in probability and risk in the broader areas of civil engineering, engineering as a whole, and in society at large. A few developments of note are as follows:

1. A basic subject in probability theory is now required in many engineering curricula.
2. Recently proposed building codes are using load factors established by careful analysis of the relative frequencies and effects of different types of loadings (17).
3. Risk has been quantified and publicly discussed in connection with projects capturing the public attention, e.g., space missions.
4. There have been man-made and natural catastrophes—e.g., Teton Dam and the 1971 San Fernando earthquake—that have driven home to all the fact that risk is ever present.

Indeed, perhaps the most important developments during these 18 years have been the recognition that all risks cannot be eliminated and a growing willingness to face such questions in an open manner.

What use can now be made of these developments? Two questions arise in particular:

1. How can advances in geotechnical probability be applied in practice?
2. Is it now possible to evaluate risk?

These are questions I wish to explore in this lecture, and thus the title, "Evaluating Calculated Risk." The word "evaluate" means, according to one dictionary, "to work out the value of or to find a numerical expression for." I fear Casagrande might have objected somewhat to this choice of title, since, to him, "calculated risk" implied a process of dealing with risk rather than just the evaluation of projects. I trust, however, that he would have agreed that it is timely to discuss once again the important questions raised in his lecture.

## BRIEF OVERVIEW

The word "probability" means different things to different people; indeed, probability is a large and diverse discipline in its own right. There is, of course, a fundamental mathematical theory, some of it quite simple and some very sophisticated. However, there are many different applications of this underlying theory, and each tends to take on a character of its own. Among the applications of interest to geotechnical engineering are the following:

1. Optimized search, exploration, and testing. For example, theory may be used to design a network of borings so as to minimize the chance of missing a significant weak zone within a soil or rock mass.
2. Reliability theory. This theory provides a way to evaluate the safety of components, substructures, or entire facilities in a logical and consistent manner.
3. Optimization of design in the face of uncertainty. In this application, uncertain knowledge concerning loads and behavior is taken into account in arriving at an optimal design.
4. Risk evaluation. This involves a set of concepts and procedures, such as fault trees and event trees, for studying structures or facilities with many components and different modes of failure.

This listing gives only one way in which potential applications may be viewed, and the listed areas are indistinct and overlap. For example, reliability theory and risk evaluation have much in common.

Then there is *statistics*, which provides a systematic set of procedures for extracting information of value from a set of quantitative measurements (30,52). Statistical procedures are an inherent part of each of the applications described above. In addition, often an engineer uses statistics for purposes other than a probability-based assessment of safety. A simple example is fitting a regression line to a set of data.

## LANGUAGE OF PROBABILITY

It is safe to say that today (1982) probability theory is regarded with doubt and even suspicion by the majority of geotechnical engineers. One reason is a language barrier. As in every specialized discipline, probabilists have developed a language of their own, replete with words and phrases that carry little meaning to the uninitiated. Many phrases that have a general meaning to the engineer have been given a very restricted meaning by the specialist. For the benefit of those unfamiliar with the language, a few words and phrases are mentioned here. Detailed treatment of these concepts may be found in any textbook on elementary probability, e.g., Ref. 10.

**Random Variables.**—When the magnitude of some quantity is not exactly fixed, but rather the quantity may assume any of a number of values and we cannot know just what value it will take, we say that this quantity is a random variable.

Almost every factor with which we deal in engineering analysis is truly a random variable, although the values of some are much less uncertain than those of others. With a proper appreciation of uncertainty, we can, of course, do good engineering without explicitly taking this randomness into account. Probability theory helps us, as engineers, to evaluate the relative importance of various uncertainties and to decide upon appropriate levels of conservatism.

**Probability Distribution.**—A probability density function (PDF) describes the *relative likelihood* that a random variable will assume a particular value in contrast to taking on other values. A typical PDF is sketched in Fig. 1(a) for a common case in which the random variable is continuously distributed (i.e., can take on all possible values). The area under a PDF is always unity.

Another way to describe the same information is with the cumulative distribution function [see Fig. 1(b)], which gives the probability that the variable will have a value less than or equal to any selected value. This new function (CDF) is the integral of the corresponding probability density function, i.e., the ordinate at $x_1$ on the cumulative distribution function is the area under the probability density function to the left of $x_1$. Note that $f_x$ is used for the ordinate of a PDF and $F_x$ for a CDF.

Other types of probability distribution functions are also encountered. One such form is often used to assemble results obtained by sampling some variables, e.g., measuring many values for the shear strength of some soil deposit. Here the observed values are grouped by intervals,

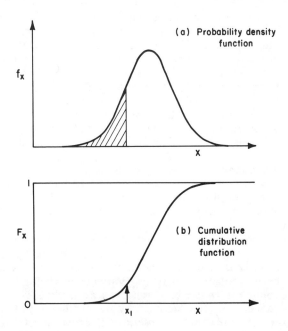

**FIG. 1.—Probability Density (PDF) and Cumulative Distribution (CDF) Functions**

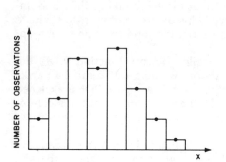

**FIG. 2.—Histogram Depicting Number of Observations of Random Variable Falling within Different Intervals**

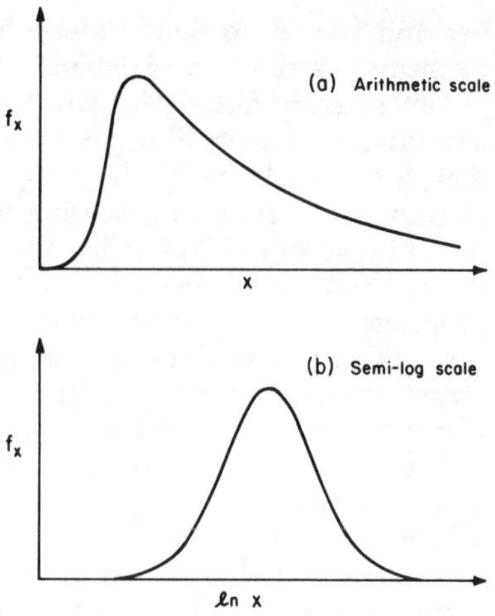

**FIG. 3.—Lognormal Distribution**

and the decimal fraction of all observations falling within an interval is plotted above that interval. In this case, the heights of all the rectangles sum to unity. Often the number of observations within each interval is plotted instead of the fraction of the total observations. This type of presentation (see Fig. 2) is known as a *histogram*.

A very common distribution is the so-called *normal*, or Gaussian distribution. Its PDF is the bell-shaped curve shown in Fig. 1(*a*). This PDF can be expressed analytically, and there is a mathematically provable reason why many actual random variables should conform to this distribution. Another common form is the *lognormal* distribution, the PDF of which is shown in Fig. 3(*a*). If the random variable, *x*, is lognormally distributed, then the variable, ln *x*, has normal distribution (Fig. 3).

**Moments.**—For many applications, it is not necessary to have or to obtain all of the information contained in a distribution function. Quantities summarizing only the dominant features of the distribution may suffice.

The most common such description is the *mean*, or *expected value*. The mean indicates the center of gravity of a probability distribution. For a normal distribution, the mean falls at the peak of the distribution. With other distributions, including lognormal distribution, this is not necessarily the case.

There are three quantities that are often used as measures of the scatter or dispersion of a random variable. *Variance* is defined as the weighted average of squared deviations from the mean, where each value is weighted by its probability density. (Variance is analogous to moment of inertia about the center of gravity.) *Standard deviation* is the square root of variance. It is often denoted by the symbol $\sigma$ and has the same units as the random variable. *Coefficient of variation* (COV or *V*) is the ratio of standard variation to the mean: COV is dimensionless, and is a particularly useful measure of the uncertainty. COV = 0.05 would typically represent small uncertainty, while COV = 0.5 indicates considerable uncertainty.

The mean and variance are referred to as the first and second *moments*, respectively, of a random variable. These quantities, as well as σ and $V$, may be computed for any distribution.

**Correlation and Independence.**—In most engineering analyses, there are several quantities—often many—that may vary randomly. If the value taken on by one variable has no influence upon the value assumed by another variable, then these variables are said to be *independent*. Independence of random variables greatly simplifies the representation and analysis of uncertainty, and often independence is assumed even where it is not really true.

A complete probabilistic treatment of dependent random variables requires joint probability distributions, which for two variables may be depicted as a surface. Again, simple descriptors suffice in place of full probability distributions for many applications; the descriptors *covariance* and *correlation coefficient*, which will not be defined or used in this paper, indicate the degree of dependence among the variables.

A special case of correlation arises when a quantity is a random function of space, time, or both. Such a quantity is said to be a *stochastic process*. Obviously, the value assumed by the variable at one point in space (or instant in time) will be associated to some degree with the value of a nearby point (or shortly later time). The degree of influence is expressed by an autocorrelation function, the essential features of which may be summarized by an *autocorrelation distance* or *scale of fluctuation* (45,47). At distances (or times) larger than the scale of fluctuation, probabilistic dependence is small.

**Probabilities and Conditional Probabilities.**—One important application of probabilistic concepts is to express the risk that some event—perhaps an unwanted event, such as a failure—will occur. In symbols, if $A$ is the event of interest, then this probability is denoted as $P[A]$. $P[A]$ ranges from zero to one. For example, the event might be just that the undrained shear strength, $S_u$, of a soil is more than some limiting value, $s$; this would be written as $P[S_u > s]$. Or the event might be a failure of some component, $C$: $P[C$ fails$]$.

In many problems, especially those involving natural hazards, the probability of a failure depends upon the occurrence of the hazard. Here we speak of *conditional* probabilities. For example, $P[C|A = a]$ would indicate the probability that component $C$ fails *given that* an earthquake with a peak acceleration $A = a$ occurs. To obtain the overall probability of an earthquake-induced failure, it is necessary to combine conditional failure probabilities for all possible levels of shaking with another set of probabilities for the occurrence of each level of shaking.

## SAFETY FACTORS AND RELIABILITY

All engineers are aware of the difficulties surrounding the phrase "safety factor." This concept is simple and useful: safety factor is the ratio of the allowable value of some quantity to the calculated or (in some instances) measured value of that quantity. The allowable value is the capacity, $C$, while the calculated value is the demand, $D$. Thus

$$FS = \frac{C}{D} \tag{1}$$

A common situation involves allowable and calculated stresses. Having a safety factor greater than unity guards against the possibility that actual stresses might exceed calculated stresses (because of approximations in the calculation or loads not considered in the calculation) and that failure might occur at a stress less than the allowable stress (because of especially faulty material or causes of failure not considered when selecting the allowable stress).

If there are carefully prescribed procedures for selecting capacity, for defining applicable loads, and for accomplishing the calculations, then the resulting safety factor has an unambiguous meaning—although the number itself tells little as to the possibility that a failure may actually occur. However, often the steps whereby either the capacity is selected or the demand is determined are not well defined nor followed uniformly by all engineers. For example, when selecting allowable stress (strength) for a slope in soil, some engineers will use a mean of measured strengths while others will assume the most conservative of the measured strengths. Moreover, the same engineer may adopt a different approach for different jobs. Thus, one slope with a reported safety factor of 1.5 may actually have little margin of safety, while another with the same reported safety factor may be virtually proof against failure.

One branch of probability theory, known as reliability theory, provides a rational framework for accounting for the uncertainties in both capacity and demand. Reliability theory also offers the prospect of a systematic method for selecting the safety factor appropriate for some particular application or class of applications. The following subsections review the basic elements of reliability theory.

**Fundamentals of Reliability Theory.**—The basic idea behind reliability theory (1,20,55) is shown in Fig. 4($a$), which contains probability density plots for both actual capacity and actual demand. If the actual demand is $x$, then there is some probability—indicated by the shaded area in the figure—that the actual capacity is less than $x$, implying a failure. If $C$ and $D$ are independent, the overall contribution to the probability of failure from the event "$D$ between $x$ and $x + dx$" is, in symbols:

$$P[x \leq D \leq x + dx] \cdot P[C < x] = f_D(x)\, P[C < x]\, dx \tag{2}$$

Integrating over all possible levels of demand gives the total probability of failure, $P[F]$:

$$P[F] = \int f_D(x) \cdot P[C < x]\, dx \tag{3}$$

The first function in the integrand is the probability density function for demand, while the second function is the cumulative distribution function of capacity. $P[F]$ is thus related to the degree to which the distribution curves for demand and capacity overlap (but is *not* equal to area

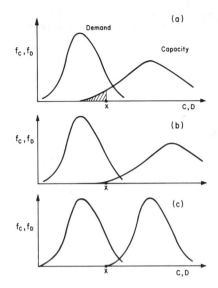
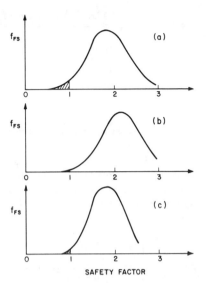

**FIG. 4.—Overlapping Distribution Curves for Capacity and Demand. Probability of Failure Is Decreased by: (b) Increasing Difference In Mean; or (c) Decreasing Uncertainty In Capacity**

**FIG. 5.—Distribution Curve for Safety Factor, with Probability of Failure Decreased: (b) by Increasing Mean; or (c) by Decreasing Variation**

of the overlap). Any action that reduces the overlap decreases the probability of failure, such as increasing the separation between the means of the two distributions or decreasing the spread in either or both of the distributions [see Figs. 4(b) and 4(c)].

The information contained in the distribution curves for demand and capacity may also be used to compute a distribution curve [see Fig. 5(a)] for actual safety factor. Now the probability of failure, $P[F]$, is the area under the new probability density curve to the left of $FS = 1$. When viewed in this alternative way, we see that the probability of failure is decreased if the mean safety factor increases [Fig. 5(b)] or if the spread of the distribution decreases [Fig. 5(c)].

If the capacity and demand both follow normal distributions, then the integral in Eq. 3 may be evaluated as

$$P[F] = 1 - \Phi(\beta) \quad \quad (4)$$

$$\beta = \frac{\bar{C} - \bar{D}}{\sqrt{\sigma_C^2 + \sigma_D^2}} \quad \quad (5)$$

in which $\bar{C}$ and $\bar{D}$ = the mean values of capacity and demand; and $\sigma_C$ and $\sigma_D$ = the standard deviations for these quantities. $\Phi$ denotes the standard normal cumulative distribution function, i.e., the CDF of a normal variable with zero mean and $\sigma = 1$. Tables of $\Phi$ are readily available.

**TABLE 1.—Values of $P[F]$ as Function of $\beta$**

| $\beta$ (1) | $P[F]$ (2) |
|---|---|
| 1 | 0.16 |
| 1.5 | 0.067 |
| 2 | 0.023 |
| 3 | 0.0014 |
| 4 | 0.000032 |

A few values of $P[F]$ as a function of $\beta$ are shown in Table 1. The parameter $\beta$ is commonly called the *reliability index*.

If capacity and demand are both lognormally distributed, then Eq. 4 is still valid, but $\beta$ becomes

$$\beta = \frac{\ln\left[\frac{\bar{C}}{\bar{D}}\sqrt{\frac{1+V_D^2}{1+V_C^2}}\right]}{\sqrt{\ln[(1+V_D^2)(1+V_D^2)]}} \quad \ldots \ldots (6)$$

in which $V_D$ and $V_C$ = the coefficients of variation for the demand and capacity, respectively. Actually, even though the capacity and demand are lognormally distributed, it still is possible to obtain a very good estimate for $P[F]$ by using Eqs. 4 and 5, provided that $\beta < 2.5$ and that $V_D$ and $V_C$ are not very large. For $\beta$ within this range, the relation between $\beta$ and $P[F]$ is not very sensitive to the distribution function. Thus the reliability index, $\beta$, is a very useful parameter for characterizing the degree of safety.

For larger $\beta$, $P[F]$ is quite sensitive to the exact shape of the distribution. So long as either capacity or demand is normally or lognormally distributed, the probability of failure is never zero, no matter what the safety factor, although it may be very small. However, if there is an upper limit to the demand that is less than a lower limit for the capacity, then $P[F]$ is really zero; however, this situation almost never occurs.

**First-Order Analysis.**—In most engineering problems, there are a number of random variables contributing to uncertainty in capacity. The same generally is true for demand. Fortunately, there are simple rules that can be used in many cases to find, at least approximately, the overall mean and coefficient of variation of capacity (or demand), given the same parameters for the variables contributing to the capacity (or demand). Thus, simple results such as Eqs. 4 and 5 or 6 may still be used.

The simplest situation occurs when several independent random variables, $X_1, X_2, X_3$, etc., contribute directly to a dependent random variable, $Z$, e.g., if $Z = X_1 \cdot X_2 \cdot X_3 \ldots$, then

$$\bar{Z} \approx \bar{X}_1 \cdot \bar{X}_2 \cdot \bar{X}_3 \ldots \quad \ldots \ldots (7)$$

$$V_z^2 \approx V_{x_1}^2 + V_{x_2}^2 + V_{x_3}^2 + \ldots \quad \ldots \ldots (8)$$

in which the bars = means; and $V_{x_1}$, etc. = the COV's for the several variables. Eqs. 7 and 8 result from using the first two terms of a series expansion for the exact values of $\bar{Z}$ and $V_z$.

Analyses using Eq. 5 together with some version of Eq. 8 are known as first-order second-moment (FOSM) analyses. These analyses, which are identical to the "error propagation" methods used in undergraduate subjects in experimental physics, are neither new nor exotic. They are, however, very powerful tools. In the simplest cases involving independent random variables, they make it possible for an engineer with only a modest knowledge of probability to evaluate reliability index and, thus, with certain further assumptions, probability of failure. While more complex problems with correlated variables may require further skills, the calculations are nonetheless simple and straightforward. Baecher (7) has given an up-to-date status report concerning the use of first-order reliability analysis in geotechnical engineering. Ref. 14 is a recent advanced treatment of the general subject of reliability. Of course, as will be noted subsequently, judgment must be exercised in the use of results of these calculations.

## APPLICATIONS OF RELIABILITY THEORY

There are several requirements for any formal treatment of reliability:

1. Clear delineation of the criteria for success or failure.
2. Selection of a deterministic model relating the basic variables to the criteria for success or failure.
3. Identification of the uncertainties concerning the basic variables.
4. Evaluation of the distribution functions or moments of the basic variables.

Once these steps have been carried out, an analysis may then be made. The following examples illustrate the evaluation of reliability in connection with two common engineering problems.

**Stability of Slopes.**—One very obvious application is the analysis of the static stability of slopes. The deterministic model for this problem involves the delineation of a soil profile and a method for computation of the safety factor. While different methods are used for different problems, they are reasonably straightforward and generally accepted. In many cases, capacity and demand are expressed by resisting moment and driving moment, and the uncertainty in the driving moment can be ignored relative to that in the resisting moment. Under such conditions, Eq. 5 becomes

$$\beta = \frac{\overline{FS} - 1}{\overline{FS} \cdot V'_C} = \frac{1 - \frac{1}{\overline{FS}}}{V'_C} \quad \dots \dots \dots \dots \dots \dots \dots \dots \dots \dots \dots \dots \dots \dots \dots \dots (9)$$

in which $\overline{FS}$ = the mean safety factor ($\overline{FS} = \bar{C}/D$); and $V'_C$ = the coefficient of variation for the average shearing resistance along the failure surface.

Numerous papers (e.g., Refs. 12, 31, 48, 57, and 59) have presented the application of reliability theory to slopes and have discussed typical values for the parameters that affect $V'_C$. A first step is to examine the coefficient of variation, $V_c$, for measured values of strength. The findings depend very much upon whether the analysis is made using total stresses or effective stresses, i.e., whether undrained or drained failure is being analyzed:

1. For shear strength in terms of effective stress, the coefficient of variation usually is quite small, about 0.1. (Larger COVs apply where cohesion is significant, e.g., in very shallow slides or slopes in cemented materials.) If the general ground-water regime is well known, e.g., in the steady-state flow through an earth dam, techniques for quantifying uncertainty in effective stress have been developed (49). The overall $V_c$ might typically rise to 0.15. However, as will be considered subsequently, in many problems there is much greater uncertainty concerning pore pressures and, thus, effective stresses.

2. Considering scatter in undrained strength as observed in vane shear tests in situ, and comparisons among shear strengths as determined at a given site by various methods, $V_c$ for total stress analysis is typically evaluated as about 0.3.

The next step is to examine the nature of the uncertainties that lead to values of $V_c$. Several sources are present:

1. A "statistical" error in evaluation of mean strength because a finite number of samples are taken.
2. Measurement bias, e.g., from sampling disturbance or from errors associated with in situ testing, with the attendent uncertainties in corrections introduced to compensate for these effects.
3. Spatial variability in the strength of the soil. This is a random variability, as distinguished from identifiable trends that can be incorporated into the deterministic model.
4. Measurement "noise" resulting from random testing errors.

The first two of these errors are *systematic errors*. They do not average out over a large volume of soil. Their contribution to uncertainty in strength propagates unchanged through the analysis, i.e., a COV of 0.3 caused by systematic errors leads to a COV of 0.3 in safety factor. On the other hand, *spatial variability* and *measurement noise* average over the volume of soil involved in a problem. Thus, the contribution of spatial variability to predicted safety factor decreases as the affected volume of soil increases.

Suppose, for example, that the scatter in measured (and, if appropriate, corrected) strengths from a set of tests is $V_c = 0.3$. Using Eq. 9 and assuming $V_c' = V_c$, this implies a low reliability index, $\beta = 1.1$, for a nominal $\overline{FS} = 1.5$. If, however, the COV associated with spatial variability is 0.2, and if the scale of fluctuation is small compared to the length of a possible failure surface (see Fig. 6), then only the systematic error remains and by Eq. 8 it is

$$V_c' = V_{\text{systematic}} = \sqrt{0.3^2 - 0.2^2} = 0.22 \quad\quad\quad\quad\quad\quad\quad\quad\quad (10)$$

so that the reliability index increases to about 1.5.

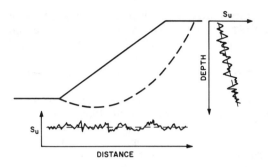

**FIG. 6.—Scatter of Strength with Small Scale of Fluctuation, Unimportant to Overall Uncertainty in Safety Factor**

In other words, if shear strengths from different depths and different borings are combined in a single plot, the observed scatter may overemphasize the uncertainty in the average shear strength along a failure surface. A correction for this effect is potentially a matter of considerable practical importance. Making such a correction requires: (1) Evaluating the autocorrelation distance or scale of fluctuation; and (2) determining how much of observed scatter results from spatial variability and measurement noise, and how much from systematic error. Neither step is easy to accomplish at this time. The classical tools for determining autocorrelation distance can give misleading results, often reflecting the spacing of bore holes more than the actual spatial correlation. New and more powerful tools are being developed and applied (6,50).

Several interesting tabulations may be prepared from Eq. 9. If the mean safety factor = 1.5, then $\beta$ is related to $V'_C$ as shown in Table 2(a). On the other hand, if $\beta$ is fixed, then the required mean safety factors may be determined as shown in Table 2(b). The value of $\beta = 2.0$ was selected because, in my judgment, it is typical of the reliability implied by much commonly accepted geotechnical engineering practice. However, examples can be found where $\beta$ is both larger and smaller than this value.

These results lead to two obvious conclusions: (1) Use of $FS = 1.5$ for all slope stability problems implies wide differences in reliability; and (2) if $V'_c$ for undrained resistance really were as large as 0.3, then a much larger mean safety factor—about 2.5—would be required to achieve a target reliability index of 2.0. The latter observation seems rather out of line with experience. One possible explanation is that engineers tend to adopt conservative values for undrained strength for design purposes,

**TABLE 2.—Relation of Mean Safety Factor, $\beta$, and $V'_C$ from Eq. 9**

| (a) If Mean Safety Factor = 1.5 | | (b) If $\beta$ = 2.0 (Fixed) | |
|---|---|---|---|
| $V'_C$ (1) | $\beta$ (2) | $V'_C$ (3) | $\overline{FS}$ (4) |
| 0.1 | 3.3 | 0.1 | 1.25 |
| 0.2 | 1.7 | 0.2 | 1.7 |
| 0.3 | 1.1 | 0.3 | 2.5 |

and hence the mean safety factor is actually greater than the safety factor they calculate. Another explanation, as already discussed, is that $V'_c$ is less than scatter in measured data. This emphasizes, again, the importance of identifying the sources of scatter in measured values and, where appropriate, reducing or eliminating those sources that tend to average out over distance.

*Use of Reliability Theory in Practice.*—The spectrum of slope stability problems is quite broad. Toward one extreme are situations for which there is already extensive practical experience and for which there is considerable agreement as to an appropriate safety factor to use for design. On the other hand, there are problems involving soils for which there is very little direct experience. In this case, there may be considerable doubt as to the proper method for computing safety, and the probability of failure is rather high. Some of the problems of stability in very soft clays described in the previous Terzaghi Lecture by Leonards (29) fall in this category, as do many slopes in residual soils. In both these situations, reliability theory is of little use in the selection of an appropriate safety factor. Many problems fall into one or the other of these categories, which may explain why there are few, if any, papers in the literature describing the actual application of reliability theory in design decisions concerning a slope.

However, there are problems lying between these extremes for which reliability theory potentially can be of considerable use. These include situations where the behavior of the soil is reasonably well understood, but there is no precedent for the design of slopes in that soil. Reliability theory may be used to select a safety factor consistent with the reliability index (say 2.0) implied by conventional practice. This suggestion is very much in line with the major current application of this theory to structures, which is in the development of appropriate load factors for building codes (17). A suitable value for the reliability index, $\beta$, is found by analyzing buildings where design or actual performance is judged to be satisfactory. Then this value of $\beta$ is used in other analyses to determine the load factors—for dead, live, wind, and other loads—that will provide approximate consistency in $\beta$ over a wide range of conditions. (Incidentally, a $\beta$ of 3 or higher is implied by conventional structural practice. As we already knew, there is often more risk in geotechnical engineering than in structural engineering.)

Another potential application is in helping to achieve consistency of safety where many slopes of different height and length, and possibly involving different soils, are involved in a single project. One interesting aspect of this problem has been studied by Vanmarcke (48) and is associated with the so-called "end-effect problem." A very short section of slope may fall at a location where the strength is randomly low, but will be stabilized by the resistance at its ends. On the other hand, when a failing mass is long enough so that the end effects are negligible, any low strengths along part of the length will have been averaged out by high strengths along another part. Not only does the theory predict the most likely length, but it shows that the reliability of a particular portion of a slope is significantly greater than the reliability predicted by considering a cross section without end effects. Then there is the question of the effect of the total length of a slope or embankment upon the probability that a slope failure will occur somewhere along the length. Vanmarcke shows that, for long slopes and a low probability of failure, the risk of failure increases approximately linearly with the total length of slope.

An interesting application has been described by Duncan and Houston (15) in a study of levees around the many islands composing the Delta Project of California. Here there have been numerous failures, and the problem was to develop a plan for remedial works, balancing probabilities of failure, costs, and realities of scheduling. The following steps were adopted:

1. By studying past failures, it was found that failure was most frequent where the depth of peat underlying the levee was greatest.
2. Thus, levees were grouped according to depth of peat, and an empirical probability of failure was established for each group.
3. A stability analysis was made for the critical section of each island. Using estimates for strength, and accounting for uncertainty in the various parameters, the corresponding $P_f$ was computed and corrected for the total length of levee for that island.
4. The computed $P_f$ were summed over all islands having the same depth of peat. Each individual $P_f$ within the group was then adjusted up or down in the same ratio, so that their sum equalled the empirical $P_f$.

In this way, a probability of failure was established for each island, and this list provided a basis for the planning and scheduling of remedial works. This is an excellent example of using reliability theory to guide the interpretation and use of a historical record of failures.

**Probability of Liquefaction.**—The second example deals with the liquefaction of estimated level ground during an earthquake. Several probabilistic analyses of this case have been published (19,22). Without going into great detail, I shall review the steps in the analysis by Haldar and Tang and discuss the implications of the results.

The deterministic model used for computation of safety factor against liquefaction is, in this case, the Seed-Idriss simplified method for analysis of liquefaction (43). This model asserts that the occurrence or nonoccurrence of liquefaction is determined by the ratio $\tau_C/\tau_D$, in which $\tau_D$ = the dynamic cyclic shear stress caused by an earthquake; and $\tau_C$ = the corresponding stress required to cause liquefaction, both evaluated at the same depth. In the simplified method, equations are given for evaluating $\tau_C$ and $\tau_D$ as functions of such variables as the peak acceleration, $a_{max}$, at ground surface; the number of uniform cycles equivalent to the random cycles in actual earthquake motions; the resistance of a standard soil as evaluated in cyclic triaxial tests; and corrections to be applied to this resistance to account for the grain size and relative density of the actual soil. The relative density is evaluated from penetration resistance measured in situ.

The correct values of the variables to use in a particular situation are uncertain, as are the functional relations between basic and dependent variables. Haldar and Tang examined the many factors and concluded that the basic variables were independent of each other. They employed results from various tests and studies to establish COVs for the several variables, and then combined this information—using previously mentioned techniques—to evaluate the COVs for $\tau_C$ and $\tau_D$. While the results varied somewhat, depending upon the details of the case studied, a typical value for the overall COV for $\tau_C$ was $V_C = 0.35$. Most of this uncertainty arose from uncertainties in the resistance of the standard sand and in the relative density as evaluated by means of penetration resistance. If peak acceleration is known or specified, then the COV for $\tau_D$ is very small compared to that for $\tau_C$. On the other hand, if only the magnitude and epicentral distance of an earthquake are known, $V_D$ may be at least as large as $V_C$.

Haldar and Tang also concluded—partly from logical reasoning and perhaps partly from intuition and experience—that $\tau_C$ and $\tau_D$ are, approximately at least, lognormally distributed. Thus Eq. 6 may be used to calculate the reliability index and, thus, the probability of liquefaction. Using 30 published case records of liquefaction or nonliquefaction during actual earthquakes, they showed that the probabilities of liquefaction as predicted by their model were reasonably consistent with actual experience.

Having thus established the reasonableness of the several models and of the values for the coefficients of variation, one may now proceed to use Eq. 6 to study the relationship between safety factor and probability-of-liquefaction. The following interesting results are given in Table 3:

1. If the safety factor is 1.5, if the peak surface acceleration is certain, and if the coefficient of variation is as indicated (0.33) by the Haldar-Tang analysis of uncertainty, then the predicted probability of liquefaction is 0.14. However, if steps could be taken to reduce the uncertainty in resistance to $\Omega_C = 0.15$, the probability-of-liquefaction can be reduced dramatically.

2. If only the magnitude and epicentral distance of the earthquake are known, then—assuming about the lowest reasonable value for the coefficient of variation of $a_{max}$—the probability-of-liquefaction cannot be reduced greatly by reducing uncertainty of $\tau_C$.

**TABLE 3.—Probability-of-Liquefaction versus Safety Factor, Based on Seed-Idriss Simplified Procedure and Haldar-Tang Analysis of Uncertainty**

| $a_{max}$ Certain ($V_D = 0$) | | $a_{max}$ Uncertain ($V_D = 0.33$) | |
|---|---|---|---|
| (a) If Mean Safety Factor Is 1.5 | | | |
| $V_C$ (1) | $P[L]$ (2) | $V_C$ (3) | $P[L]$ (4) |
| 0.33 | 0.14 | 0.33 | 0.18 |
| 0.15 | 0.005 | 0.15 | 0.10 |
| (b) To Have $P[L] = 0.1$ | | | |
| $V_C$ (5) | $\overline{FS}$ (6) | $V_C$ (7) | $\overline{FS}$ (8) |
| 0.33 | 1.6 | 0.33 | 1.8 |
| 0.15 | 1.25 | 0.15 | 1.5 |

Table 3($b$) gives values of safety factor required to have the probability-of-liquefaction be less than 0.1. Once the statistical parameters have been derived from available data, calculations for various desired $P[L]$ can readily be made by the engineer. All of these results assume use of mean resistance (and mean $a_{max}$ when $a_{max}$ is uncertain) to compute safety factor. Other definitions of safety factor—such as that corresponding to the lower range of measured resistance values—may also be used in the analysis.

This example has illustrated in some detail the steps necessary to conduct a reliability analysis. The importance of a sound deterministic model for safety, to use as a starting point, and of a careful statistical analysis of data to evaluate uncertainties cannot be overemphasized.

The example has also shown that there is considerable uncertainty in the analysis of liquefaction by the Seed-Idriss simplified procedure. If $a_{max}$ is certain, and if we were to seek a probability of liquefaction comparable to that used in structural problems—about 0.001, corresponding to $\beta \approx 3$—it would be necessary to have a nominal mean safety factor of 2.8! (It should be noted that the model deals with the occurrence of initial liquefaction, and not necessarily of catastrophic failure.) The result emphasizes the desirability of having a better method for evaluating liquefaction. By undisturbed sampling and testing, plus the use of sound judgment, it might well be possible to reduce $V_C$ to 0.15. In that case, the nominal safety factor needed for $P[L] = 0.001$ reduces to 1.6, if $a_{max}$ is known or specified.

As more case records of liquefaction and nonliquefaction during earthquakes have been developed, quite a different approach to the analysis of liquefaction has evolved: use of empirical charts relating standard penetration resistance to $\tau_A$ (see Fig. 7). Solid dots represent combinations of dynamic shear stress, $\tau$ (normalized by the vertical effective stress, $\sigma'_v$), and penetration resistance for which liquefaction has been observed. Open dots correspond to no observed liquefaction. The solid curve represents the lower limit to combinations of dynamic stress and resistance for which liquefaction would be expected. Since these charts are based upon actual experience, and the manner in which the chart is constructed is based upon principles developed during basic research in the laboratory, they represent the soundest approach to assessing the possibility of liquefaction. However, the problem of selecting a suitable margin of safety still remains. Note, for example, that a few solid dots lie below the curve in Fig. 7. Yegian and Whitman (58) have outlined a technique that can be used to evaluate probability-of-liquefaction for a new situation, based upon the location that a point representing that situation plots on such an empirical chart. Much work along these lines would be valuable.

**Other Applications of Reliability Theory.**—There have been a number of other studies dealing with the problem of choosing a safety factor and the use of reliability theory in this regard. To mention but a few: the interpretation of pile load tests (25), design of a retaining wall (24), design of foundations (32,56), and the usefulness of partial safety factors in the design of slopes (41). Many other potential applications remain to be explored.

**Final Thoughts Concerning Reliability Analysis.**—It has often been

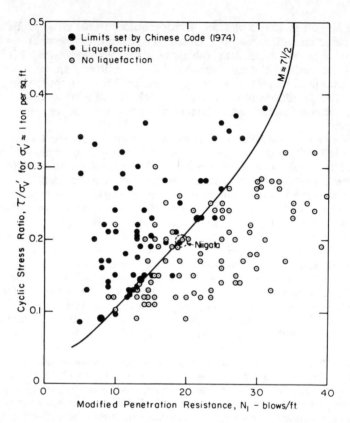

**FIG. 7.—Empirical Chart Relating Dynamic Shear Stress for Initial Liquefaction to Penetration Resistance (43)**

observed that the frequency with which actual structures fail exceeds that predicted by a reliability analysis that accounts only for uncertainties in loads and material properties. This has recently proved to be the case for structures in the North Sea, for instance. Presumably, the higher rate of actual failure is the result of failure modes not taken into account in the reliability analysis and of human errors in design and construction. A corresponding example in geotechnical engineering would be the unexpected failure of slopes because of unsuspected and undetected weak strata within the soil or because the ground-water regime was totally different from that envisioned during analysis. As deMello said in his Rankine lecture (13), safety rests ". . . not upon the accuracy of our calculations, but upon the adequacy of our hypotheses."

Then the question arises: Why do a reliability analysis if it is unable to account for the most important causes of failure? The answer is that an engineer always wants to, and should, ensure that the factors he understands and can control do not contribute significantly to the overall risk of failure.

The results of any analysis, obviously, will only be as good as the input. A final step is an interpretation of the results, including an assessment of their validity. As is the case with many analyses made in geotechnical engineering practice, the true value of the analysis often lies in the insights and understandings that come from careful formulation of the problem. Thus an analysis may be useful even if the numerical results are not used directly in a decision-making process.

## OPTIMIZATION IN FACE OF UNCERTAINTY

**Great Salt Lake Crossing Revisited.**—When using the methods described in the previous section, an engineer is employing historical precedent to select a suitable reliability index and, thus, safety factor. However, for a large and unusual undertaking, it may be appropriate to choose for design a factor of safety selected with an eye toward the costs and risks associated with that project. This approach may be illustrated by an example based upon an actual construction project: the railroad embankment built across the northern end of the Great Salt Lake during the 1950s.

This project, which is one of the case studies discussed in Casagrande's Terzaghi Lecture, involved construction of an embankment for railroad tracks across 12 miles of normally consolidated clay, partially reinforced by a layer of salt. The general location is shown in Fig. 8. The embankment was to replace an existing wooden trestle, which had been constructed many years earlier following an unsuccessful attempt to construct an earthen fill out from the edge of the lake.

The general plan called for dredging of the softest of the clay, and dumping sand/gravel plus rock to build up the embankment. It was recognized that flat side slopes or berms, or both, would be necessary, and that it would be very difficult to predict the stability of the embankment. It was decided to proceed based upon: (1) The construction of several test fills; (2) a safety factor near one, based upon strengths derived from observed failures of test fills; and (3) a program for mon-

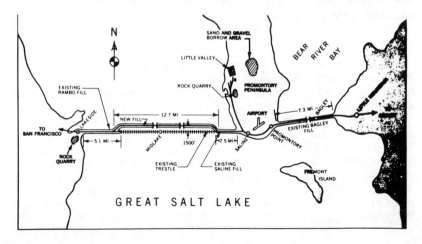

**FIG. 8.—Location of New Fill for Railroad Crossing of Great Salt Lake (Courtesy of Dirk Casagrande)**

itoring settlement and heave, with contingency plans for dumping sand/gravel, as needed, at sections where observations might indicate incipient failure.

> "The decision to design the embankment for the construction stage using a safety factor near one was based on economic considerations. The calculated risk taken by the engineers and consultants weighed the costs of failure repairs against the cost of a conservative design. This one decision, made in the early planning stage, contributed more than any other to the economic completion of the job." (44)

As described by Casagrande, several unexpected failures did occur despite the observational program. These failures were corrected without undue difficulty by constructing berms, and even parallel embankments, to hold down the toes of the slide zones. The project was completed a year ahead of schedule at a cost of \$50,500,000 (1950 dollars)—almost exactly the limit (\$50,000,000) established by the railroad for the economical feasibility of the project.

This project was apparently undertaken without any formal probabilistic assessment of the chances of success; it was an excellent and successful case of a calculated risk based on sound engineering judgment. It is interesting to see how the problem might have been formulated in probabilistic terms, and what might have been concluded from such a study. For this purpose, a simplified (and hence hypothetical) version of the Great Salt Lake Crossing problem has been defined, which is believed to contain the basic elements of the actual project.

Assume that the test sections have been completed and interpreted, and that the only question to remain is, What safety factor should be used to design the embankment sections? To aid in this decision, we wish to evaluate the possible total cost of the project as a function of the factor of safety, considering the uncertainties involved in the actual behavior of the embankment.

The total cost is the sum of the two types of costs:

1. $C_d$: The cost of the embankment "as-designed," which is a function of the safety factor, $F_d$, used for design. This cost may be most conveniently expressed as $C_1 + C_2 F_d$ in which $C_1$ includes mobilization costs and other costs that are independent of the actual design.

2. $C_c$: The cost of the additional fill placed to stabilize any section where observations detect incipient failures or to rebuild a section that has failed. $C_c$ is a function of the safety factor assumed for design, of the probability distribution of the actual safety factor at each section, and—since rebuilding is more costly than stabilization before failure—of the probability that the observational program will fail to detect an incipient failure. $C_c$ includes any costs of extending the time required for construction.

In general, $C_d$ and $C_c$ are both uncertain; but, for simplicity, $C_d$ will be assumed to be known accurately.

A basic assumption in the analysis is that the embankment is composed of 30 segments, each about 1,000 ft in length, each of which behaves independently of its neighbors. The basic random variable is the actual safety factor, $F$, of each section: If $F$ is less than unity, the section will fail unless the incipient failure is detected and corrective action taken. The actual safety factor, $F$, has a random value at each section as a result of uncertainties in the properties of the soil and the behavior of the embankment at that section. $F$ also varies randomly from section to section, and is assumed to have normal distribution. Thus, if the mean value of $F$ is unity, there is a 50% chance that each section of the embankment might fail, and incipient failure is expected to occur along 50% of the total length of the embankment.

A few details of the analysis are presented in Appendix I. For an actual project, the various costs would be evaluated by means of experience and preliminary design studies; in this example, they have been chosen so as to give results consistent with the actual case. The results of test fills provide a basis for estimating the mean and standard deviation for the actual safety factor; the mean has been taken equal to the design safety factor, and there would be reason to hope that the standard deviation might be about 0.15. The probability that the observational program will fail to give a warning of an incipient failure can only be estimated from the experience with such projects; Casagrande and Smith both discuss the difficulties of such programs, including human fallibility. In hindsight, it would appear that this probability was about 1/6 for the actual crossing project.

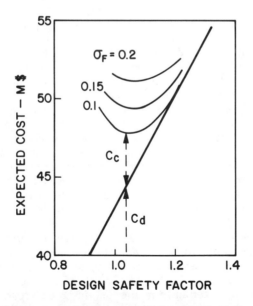

FIG. 9.—Total Expected Cost for Hypothetical Great Salt Lake Crossing, as Function of Safety Factor Used for Design and Standard Deviation of Actual Local Safety Factor

Fig. 9 shows one common form for the output from a probabilistic optimization analysis: curves of expected cost versus $F_d$. "Expected" cost is the most likely value of cost; there is a chance that the cost might be greater than the expected cost, and a chance that it might be less. Curves are presented for three values of the standard deviation on the actual safety factor. The main conclusion is that designing with $F_d = 1.05$ gives the minimum cost. (In principle one might decide to design with $F_d < 1$, so as to take advantage of the probability that some sections will be stronger than average and will not fail. If the cost of adding full to stabilize an observed incipient failure is only a little more than placing this fill in the first place, there might be some advantage to such an approach. In practice, however, it is unlikely that designing for $F_d < 1$ would be adopted.)

It may also be desirable to know the risk that the expected cost might be substantially exceeded because the number of either incipient failures or undetected incipient failures is higher than average. The probability (for the $\sigma_F = 0.15$ case) that actual cost might exceed, say, $51,000,000 is about 27% for $F_d = 1$; 15% for $F_d = 1.05$; and only about 11% for $F_d = 1.1$. Given this type of result, an owner might choose to use a safety factor slightly higher than the optimum $F_d$ on the basis of expected cost alone.

The Great Salt Lake Crossing project entailed substantial financial risk, but very little risk to life, and involved private investors. Such a project is a fertile field for probabilistic optimization of risks. This example has, of course, only confirmed what Casagrande and other engineers know from judgment and intuition: Given the results from the test fills, it made good sense to adopt the observational approach and develop as-designed sections with a safety factor of nearly unity. Moreover, the apparent accuracy of the example is possible only with the benefit of hindsight. Yet, I am convinced that such an analysis would be worthwhile for a similar project undertaken today, to aid the client and the engineer in optimizing the decisions that still remain after the important and difficult work of interpreting the test fills has been done. Engineers use analyses to help sharpen their judgment and to study the effect of parameter variations, and we have been speaking here of one more such tool. The same approach can also be applied at an early stage of project planning, before test fills are constructed; the uncertainty in the probable behavior of the embankment would then be much greater, and considerable "guesstimating" would be necessary to establish the probabilistic model, but the analysis could provide quantitative estimates of the risk of soaring expenditures. The same approach can also be used to optimize the scope of the observational program.

Whatever the aim of the analysis, an experienced probabilist can be of great help by improving the oversimplified assumptions made in the numerical example presented here, and by drawing upon experience to estimate parameters such as the coefficients of variation. It is a simple matter to include uncertainty in the "as designed" cost, $C_d$; and the correlation of actual safety factor along the length of the embankment—i.e., choosing the number of segments and accounting for correlations between segments—may be handled in a systematic manner (50).

**Other Uses of Optimization.**—In principle, this same approach may be used as an aid in selecting design criteria for a wide variety of projects. As an example, this methodology has been used to analyze the optimum seismic design criteria for various parts of the United States (54). Usually, there are risks other than financial loss involved in a project: loss of life, property damage, temporary suspension of operation, etc. The existence of multiple risks complicates the analysis greatly, unless all risks can be quantified on the same basis, e.g., in dollars. There are, of course, many philosophical difficulties when one attempts to place a dollar value on life. Techniques have been developed for dealing with multiple risks without having to explicitly place a dollar value upon emotionally-associated losses (26), but use of these approaches is still very controversial.

## Risk Evaluation

The three preceding examples all dealt with risk, but involved simple situations in which there was a single meaning to failure. In the discussion of slope stability there was a single element, the slope, that might fail, and the consequences of this failure were not explored. The example concerning liquefaction dealt only with behavior at a point. While the analysis of the Great Salt Lake railroad crossing considered an actual project, the treatment was highly simplified and did not concern itself with interactions among the parts of the project.

In contrast, most engineering projects are quite complex. Numerous components interact with one another, and generally there is more than one way in which a project might fail. Thus, a project may be thought of as a system, and some of the techniques of systems analysis must be employed when evaluating the risk of failure for the project as a whole.

Two examples of such analyses are presented.

**Seismically-Induced Off-Site Oil Spill.**—The tank farm that is the subject of this example lies along the fringes of a bay (27). It is built upon hydraulic fill with a depth of about 14 m. Underlying the fill is a considerable depth of moderately soft clay, parts of which are still consolidating under the effects of the fill. There has been concern about the effect of a major earthquake upon this installation. Failure of any tank would constitute an economic loss and a potential fire hazard. The major concern, however, has been about a significant off-site spill of oil into the bay.

The major elements of the tank farm are indicated in schematic form in Fig. 10. There are a number of very large tanks, which are grouped together into patios. Each patio is surrounded by a fire wall, capable of retaining the contents of at least one of the tanks within the patio. In some cases adjacent patios are interconnected, to provide additional storage capacity to retain spillage from tanks. The entire site—i.e., all the several patios—is enclosed by another fire wall. Along three edges of the tank farm there is a sea wall (or revetment); at several locations, this wall is very close to the outer fire wall, so that a large seismically-induced outward movement of the sea wall would breach this wall.

From the foregoing, it is clear that there are several lines of defense against off-site spillage of oil. If only one tank ruptures, then both the

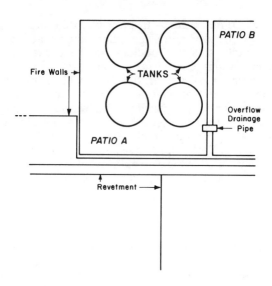

**FIG. 10.—Typical Arrangement for Tanks and Fire Walls in Oil Tank Farm**

fire wall around that patio and the outer fire wall must be breached if there is to be a spill. Simultaneous failure of several tanks in a patio might cause overtopping of the enclosing fire wall, but the outer fire wall might still retain the spill. On the other side of the coin, there is the possibility that tanks and fire walls might be simultaneously ruptured by a deep-seated slide through the not-yet-fully-consolidated clay. In addition, it is necessary to consider that some or all of the tanks will be only partially full at the time of a major earthquake. An *event tree* (Fig. 11) is used in such a case to indicate the various sequences of events that might lead to an off-site spill. Development of an appropriate event tree is a major step in risk evaluation, and by itself leads to a much greater understanding of the nature and degree of risk involved in a project.

*Performance of Components.*—At each nodal point of the event tree, it is necessary to assign probabilities to the branches that represent failure or nonfailure of a component of the overall defense against a spill. Doing so requires a detailed analysis of the performance of the component during earthquake shaking. This was done for four intensities of ground shaking, which bracketed the intensities that might occur in the future.

The *soil* that supports the structures is of a type that is potentially susceptible to liquefaction. Indeed, liquefaction-caused settlement of tanks and fire walls is much more likely to cause rupture and breaching than ground-shaking by itself. Results from numerous standard penetration tests, and from laboratory tests on both undisturbed and reconstituted samples, were analyzed to evaluate, for each intensity, the expected liquefaction potential index (LPI)—the reciprocal of safety factor against initial liquefaction—at a typical point within the hydraulic fill. The coefficient of variation for LPI at a point was established, considering both uncertainty in blow count, possible errors in the model relating blow count to the build-up of excess pore pressures, and the differences in

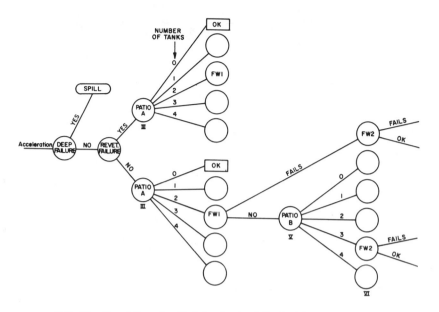

FIG. 11.—Event Tree for Risk Analysis of Typical Tank Farm Patio

ground shakings that would be assigned the same intensity. LPI became the basic random variable for the subsequent analysis.

Several different studies were combined to establish the probability that a *tank*, or the associated piping, might rupture. Finite element analysis was used to predict foundation settlement as a function of LPI. Bearing capacity analyses were performed, considering the excess pore pressures caused by shaking, and were calibrated to settlements observed during model tests upon a shaking table. Experience from actual tank farms was used to relate settlement to the probability of rupture. Sensitivity studies were made to account for uncertainties in the methods of analysis used and in the values for the parameters inserted in these analyses. The end point of these several efforts was an estimate for the probability of rupture as a function of LPI, as illustrated in Fig. 12. [In today's parlance, this graph is called a *fragility curve*. Where more than one definition of failure is considered, a *damage probability matrix* with several failure states is substituted for this one-dimensional graph (54).]

Failure of the *revetment* was defined as enough lateral movement to cause breaching of the outer fire wall. The probability of such an event was established from data for the movement of similar sea walls during actual earthquakes.

As mentioned earlier, there is some possibility of an earthquake-induced *deep-seated slide* through the clay below the hydraulic fill and revetment. Here, failure was defined as enough movement to rupture tanks and breach fire walls. A sliding-block type study (33) together with stability analyses were used to evaluate the probability of such an event.

It proved most difficult to determine the probability that the *fire walls* might be breached as the result of liquefaction-induced settlement. In

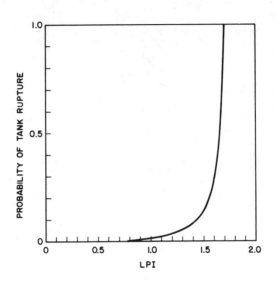

FIG. 12.—Fragility Curve for Tank

the end, two assumptions were made. One was that all fire walls would remain intact unless affected by deep-seated slides or movement of the revetment. The other assumption led to a conservative estimate for the probability of breaching. In connection with this latter assumption, it was necessary to take into account the length of the walls, especially that of the outer wall. The greater the length, the more likely it is that a breach might occur at some point.

The final result for risk of the overall system can be no better than the performance evaluations for the several components. The foregoing paragraphs have merely sketched the steps involved in such evaluations, omitting many details. As will be true for many one-of-a-kind civil engineering projects, the data base for these results was far from adequate, and it was necessary to rely heavily upon judgment and experience.

*Interactions and Correlations*—For this tank farm, there are, potentially, many ways in which the performance of several components might be correlated. For example, in some cases tanks were located quite close to a fire wall, and a bearing-capacity failure of the tank might well induce severe movements of the fire wall. Where two fire walls are in close proximity, settlements—and thus distortions—of the walls will be strongly correlated.

Of greatest importance to this example is possible correlations among the performance of tanks in a given patio. Such correlations can arise for two reasons: (1) Correlation of subsoil conditions, i.e., existence of a loose sand beneath one tank will tend to imply the presence of a similar condition under a nearby tank; and (2) errors in the model used to predict performance of a tank, i.e., any error will apply in the same degree to all tanks. These factors are especially significant because simultaneous failure of several tanks proved to contribute the greatest portion of the total risk of an off-site spill.

Accounting for these interactions and correlations was a major contribution to this risk-evaluation study. The correlations related to the possible lateral extent of loose soils were studied using the concept of scale-of-fluctuation and data from the penetration tests. Probability theory was used to guide systematic evaluation of other correlation effects.

*Analysis of Risk*—With the event tree established and probabilities of failure or nonfailure assigned to each of the branches, computation of the probability of an off-site spill becomes a straightforward exercise in multiplication and addition. Because the computation was repeated for each intensity of shaking, and also for different assumptions about some of the parameters, a simple computer program was written for this purpose.

Accounting for the contributions from all of the tanks and patios at the site was much more complex. A rather complex simulation model was developed, involving additional simplifying assumptions. This model

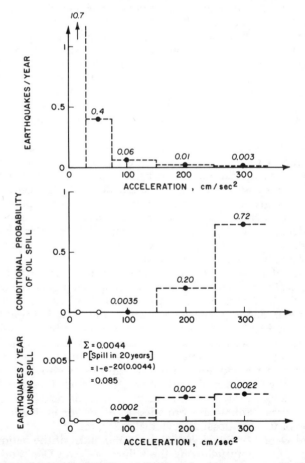

**FIG. 13.—Combining Probability of Earthquake Occurrence with Conditional Probability of Off-Site Spill**

was first applied to one patio and adjusted to give essentially the same results as from the more accurate event-tree analysis. Analyzing the results for the site as a whole, it was found, eventually, that a much simpler model could be used to give much the same risks; and this calculation procedure was used for sensitivity studies and to analyze the possible reductions in risk that might be achieved from various proposed remedial measures.

Once risk of a spill is determined for each intensity of shaking, it then is necessary to consider the likelihood that the various intensities will actually occur. The procedure is shown in Fig. 13, and in equation form is

$$P[S] = \Sigma P[S|I] \times P[I] \quad \dots\dots\dots\dots\dots\dots\dots\dots\dots\dots\dots\dots\dots\dots\dots (11)$$

in which $P[S]$ = the overall probability of a spill; $P[S|I]$ = the conditional probability of a spill given that intensity, $I$, occurs; and $P[I]$ = the probability that a shaking of intensity, $I$, actually occurs. (Strictly speaking, $P[S]$ and $P[I]$ are rates of failure per unit of time.) $P[S|I]$ increases with $I$, while $P[I]$ decreases. The summation is over all possible $I$.

The simple concept embodied in Eq. 11 is especially important when dealing with risk from extreme environmental loadings. It can be quite misleading to consider what might happen, for example, only during a single intensity of ground shaking. This equation brings together the expertise of the seismologist, who is best equipped to evaluate $P[I]$, and that of the engineer, whose expertise lies in determination of $P[S|I]$. It is vital that these two groups of professionals work together closely in any such study.

*Results.*—I have deliberately devoted considerable space to this example to make clear the scope of a study of this type. One might well wonder how reliable the result can be, considering all of the assumptions and approximations that have been made. Such a concern has been expressed strongly when this same approach has been applied, usually with more thoroughness, to controversial projects such as nuclear power plants. Indeed, one can have great confidence in the resulting numerical value for risk only if there is an adequate data base for statistical analysis of the probability of failure of the many components.

In the case of the tank farm, it would be fair to characterize the study as a sophisticated probability analysis based upon a highly variable quality of input data. However, this type of study has been entirely appropriate for the intended purpose. The careful and detailed formulation of the problem led to a much clearer understanding of the overall performance of the site. Sensitivity studies were used to explore the implication of the doubts about some of the information used as input. The study confirmed beyond any doubt that, despite the double protection provided by the fire walls, the risk of an off-site spill is too great. Moreover, the analysis pinpointed the major source of risk—simultaneous failure of several tanks—and provided a means for evaluating the effectiveness of various possible remedial measures.

**Safety of Dams.**—Several years ago, in the midst of the national concern about dam safety following the failure of Teton Dam and several other smaller dams, it was proposed (9) that the risk of failure be explicitly taken into account in cost-benefit studies for new projects. Furthermore, it was suggested that a *default value* for annual risk of $10^{-4}$ be used in the absence of other information about the safety of a particular project. This value of $10^{-4}$ was simply the average annual rate of failure for actual dams, as reported by several investigators.

Needless to say, this proposal stirred considerable discussion and controversy. Each major dam-designing organization was sure that its projects were safer than the average dam, but understandably was unsure how it could ever demonstrate that fact. There was fear—justified in our regulation-minded society—that it might become necessary to *prove* that one's dam was safer than the criterion of $10^{-4}$ annual probability-of-failure, or some even more stringent criterion.

As a consultant to the U.S. Army Engineer Waterways Experiment Station, Casagrande was actively involved in this controversy, and I recall discussing it with him during the spring of 1979. While sharing the doubts of many regarding the numerical evaluation of risk, he recognized that some future dams would be riskier than others, despite the best efforts of the engineers involved, because of more difficult conditions in the foundations and abutments. Further, he agreed that it was certainly appropriate to take this fact into consideration when deciding whether or not the benefits of a proposed dam exceeded the possible costs. He envisioned that it might be possible to develop a subjective rating system for dams, but worried whether there were enough experienced engineers who could be expected to apply such a system in a reliable manner.

Based partly upon that discussion with Casagrande, I have suggested—in a very preliminary form—a rating system cast in the format of risk-of-failure. This approach has two facets:

1. An event tree to give structure to the rating process.
2. A set of criteria to guide choice of the probabilities at each branch of the event tree.

I include it in this lecture, not as the answer to the very real problem of recognizing the riskiness of various dams, but to stimulate further discussion of the topic.

*Event Tree.*—Fig. 14 presents a simple event tree, in which each branch point represents some characteristic of the dam that has a major effect upon safety. The words used to denote these so-called events must be interpreted liberally.

To begin with, the word *dam* is shorthand for the combined embankment plus foundation plus abutments. The phrase *dam cracks* would include any happening that might make the flow of water enough to begin internal erosion within the embankment, foundation or abutments; e.g., cracking of the core, existence of open joint within an abutment, incomplete foundation cutoff, or a channel along a conduit or retaining wall. The *filter* includes all provisions for preventing the growth of internal erosion. Similarly, the word *drain* encompasses all design details intended to relieve pore pressures within the downstream portions, i.e., both internal and foundation drains. Provisions for detecting the onset of piping or high-pore pressures include not only any physical measur-

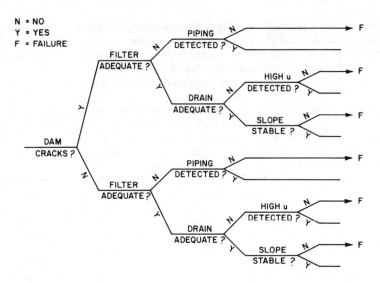

**FIG. 14.—Event Tree for Evaluation of Geotechnical Aspects of Dam Safety**

ing systems, but also procedures for making, interpreting, and acting upon observations concerning the physical state of the dam. The "no" branch at *high u detected* (Fig. 14) means that pore pressures high enough to cause failure are not observed or acted upon in time to prevent a stability failure; the "yes" branch means that steps are taken to reduce these pore pressures before failure can occur. The phrase *slope stable* refers to the possibility of a slope failure when pore pressures are essentially equal to those assumed for design.

With such broad interpretations, a simple event tree may be used to identify the important features that must be evaluated in order to characterize overall safety. It provides a systematic approach to accounting for multiple lines of defense against failure—a concept that Casagrande emphasized repeatedly in his practice and teaching. Such a liberal treatment of the nodes of an event tree runs counter to a fundamental principle of operations research; however, the aim here is to give some structure to what must be a subjective process.

The event tree of Fig. 14 is intended to cover those aspects of safety that fall within the province of the geotechnical engineer concerned with the dam itself. The possibility of overtopping or spillway failure during floods is not included, nor is any possible concern about landslides into the reservoir. The effects of earthquakes are not explicitly mentioned in this examination, but could be included when each of the branch point probabilities is enumerated. These probabilities, whose evaluation will now be considered, are the probabilities during the entire life of the dam.

*Branch Point Probabilities.*—The probabilities of the various "events" in the tree of Fig. 14 may be evaluated partly through analysis, but mostly by subjective judgment based upon careful review of pertinent considerations.

To indicate how this might be accomplished, Fig. 15 lists a number of factors that influence the likelihood of cracking of the core. By bringing

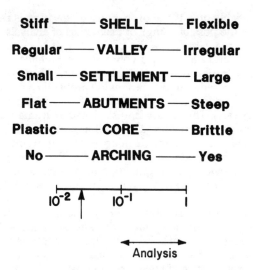

FIG. 15.—Factors Affecting Cracking of Core

in the foundation and abutments, other items might be added to this list, including the thoroughness of the geological and subsurface investigations, the care with which grouting is done, etc. Also indicated in Fig. 15 is the range of probabilities that I believe are pertinent to the "event" of cracking. I doubt that it is ever meaningful to speak of a probability of cracking less than $10^{-2}$; a risk at least this large is inherent when even the best engineering is done. Considering the various factors, an experienced engineer should be able to rate any dam along this scale of probabilities. If a number of factors are unfavorable and the engineering is mediocre, the probability of cracking might approach unity. A combination of favorable natural conditions plus beneficial design features, such as arching the dam upstream and providing a flexible core material, would lead to a rating at the low end of the scale. Analysis, such as procedures for computing strains within the core (28), may be of some help in assigning the probability. However, because of an inadequate data base, the probability computed by any such analysis is

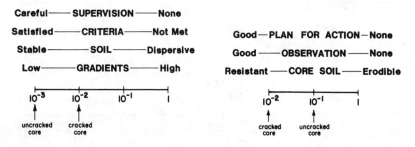

FIG. 16.—Factors Affecting Performance of Filter

FIG. 17.—Factors Affecting Detection of Piping

accurate and reliable only if the probability of cracking is relatively high. Fig. 15 indicates the possible range of usefulness of analysis for this event. An arrow shows my estimate of the probability of cracking, if engineers follow what it often called "good conservative practice."

As further illustration of the evaluation of branch point probabilities, Figs. 16 and 17 provide similar ranges of probability as well as lists of pertinent factors governing performance of the filter and detection of piping, respectively. These lists emphasize that human and organizational factors must be considered as well as design details. The rate at which a possible failure might develop is an important consideration. Thus, use of a highly erodable core may mean that any observation system will fail to provide warning in time to prevent a piping failure (although it might permit warning soon enough to minimize loss of life downstream).

*Hypothetical Examples.*—To illustrate how the overall evaluation might proceed, Fig. 18 shows branch point probabilities corresponding to "good, conservative design practice." By following through each branch to its end, the probability for a specific sequence of events is found. Summing these various probabilities gives the overall probability-of-failure. In this case, "failure" means a potentially catastrophic release of water to downstream of the dam.

Several comments are in order about the individual branch point probabilities. Note that the probabilities that the filtration and drainage systems are adequate, and other values as well, are different depending upon whether or not the dam "cracks"; any such cracking is a serious event that must sorely tax the design details and human chains called upon to minimize the effects of any such crack. Note also that a probability of essentially zero has been assigned to a serious slope instability

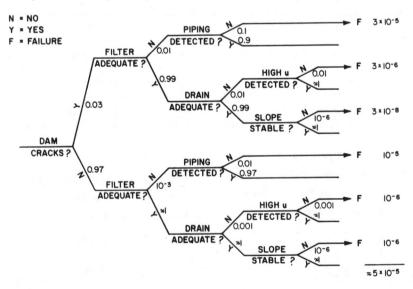

**FIG. 18.—Estimated Lifetime Probabilities Applicable for Good Conservative Practice**

in the case where the core and cutoffs, plus drainage systems, have actually functioned to keep pore pressures to the low values anticipated during design. This is one situation where design based upon analysis can virtually guarantee success. The paths through the event tree ending in "slope stable" represent a dam performing as the designer intended.

For this hypothetical example, the overall lifetime probability-of-failure is $10^{-4}$. If one were to assume a life of 100 yr, the average annual probability-of-failure would be $10^{-6}$ (although the probability doubtless is not constant over the lifetime, being highest during the initial filling of the reservoir). This is smaller than the historical rate of failure for dams ($10^{-4}$) mentioned at the outset of this section, partly because this analysis does not consider overtopping and partly because a dam designed by good, conservative practice should be better than the average. Obviously the probabilities used in this example represent considerable judgment on my part, and should be the topic of further thought and discussion. (It should also be remarked that using a probability of $3 \times 10^{-3}$, for example, does not suggest an ability to be precise about the number, but rather reflects the difficulty in choosing between $10^{-2}$ and $10^{-3}$.)

As a counter-example, Fig. 19 evaluates the hypothetical case of a dam with no filter and an erodable core, constructed against abutments containing open joints that are difficult to treat adequately by grouting. These are circumstances that should cause engineers to assign a rather high overall risk of failure to the dam.

*Closing Comments.*—The rating system discussed in this section is intended to: (1) Provide a logical framework for rating the relative safety of a dam; (2) allow for use of judgment and experience, plus results from analysis when appropriate; and (3) yield quantitative ratings for use in

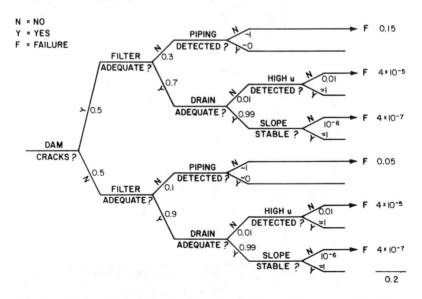

**FIG. 19.—Lifetime Probabilities for Hypothetical Case of Dam on Jointed Foundation and Erodible Core**

cost-benefit analyses. The last objective is vital, and unless there is this purpose, a rating system need be neither quantitative nor cast in terms of probabilities.

**General Comments Concerning Risk Evaluation.**—A numerical risk evaluation such as that illustrated by the preceding examples is not an end point in itself. Rather, it must be coupled to other information so as to decide whether a project is safe enough or appropriate for its purpose.

One way to do this is to incorporate the consequences of failure into a cost/benefit analysis. All of the costs and expected benefits are uncertain to some degree, and inclusion of another very uncertain cost does not in principle alter the nature of such an analysis. Refs. 34, 46, and 51 illustrate how this approach may be used in connection with the cost/benefit analysis of dams. Of course, as mentioned earlier, it is very difficult to evaluate a cost associated with possible loss of life. Engineers should not be solely responsible for such studies, since matters affecting society as a whole are involved; but they should participate to reflect a proper appreciation of the risks that are involved.

Another approach is to contrast the evaluated risk with an allowable risk. At least insofar as geotechnical projects are concerned, unfortunately, there are no standards of allowable risk. Engineers are often asked to suggest allowable risks, although, once again, they should not be requested to assume that responsibility alone. One answer is to compile observed risks from natural and man-made events. One such compila-

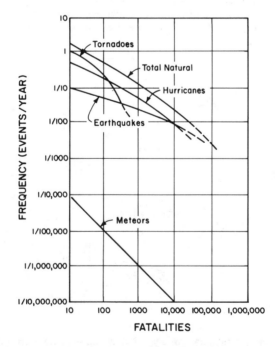

FIG. 20.—Fatalities Due to Natural Disasters (39)

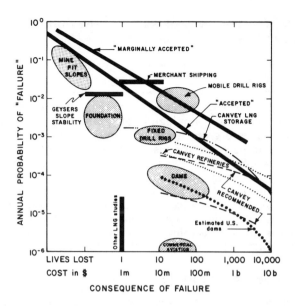

FIG. 21.—Risks for Selected Engineering Projects (from Baecher, Private Communication)

tion, involving the annual rate of single events causing various numbers of fatalities, is shown in Fig. 20. Another set of information, involving primarily evaluated risk to structures and other civil engineering projects, appears in Fig. 21. These various results give some indication of accepted risks and, thus, of allowable risk.

The prospect of assigning a numerical value to risk presents a rather frightening dilemma to engineers and their clients. On the one hand, such quantification is essential for cost/benefit studies and other decision-making activities. On the other hand, there is a danger that criteria for allowable risk might become fixed and inflexible, thus demanding a precise evaluation of risk beyond what can realistically be achieved. The latter danger is real, and when it happens, important discussions about the objectives and priorities of society are reduced to unprofessional squabbling over the details of an analysis. The many arguments following an initial attempt to evaluate the risk of nuclear plant failures—the so-called WASH-1400 report (39)—emphasize the troubling nature of this dilemma.

Rasmussen, who was the prime mover in the study leading to the WASH-1400 report, has provided a thoughtful analysis of the use of quantitative safety criteria (38). For such criteria to be successful, i.e., to be accepted by those concerned, he suggests that four requirements must be met:

1. The criteria must be defined on a logical and understandable basis.
2. There must be reasonable, acceptable methods for demonstrating that the criteria are met.
3. The criteria must reduce risk over that inherent in current practice.

4. The fixes required by imposing the criteria must not lead to severe economic penalties.

It seems doubtful at best that these requirements can all be satisfied in the case of an earth dam.

Whatever the difficulties, however, the need for quantitative risk assessment will not disappear. Much is now being written on this matter, and it is the subject of many high-level studies (31). I am reminded of an explanation, once given me by a friend, of the three laws of thermodynamics—using poker as a metaphor: (1) The first law says you can't win; (2) the second law says you must lose; and (3) the third law says you can't get out of the game. Applying this to risk analysis in geotechnical engineering: (1) Not enough is known about soil or rock and its behavior in situ to do an accurate evaluation of risk; (2) we will be criticized no matter how we do the analysis; but (3) we must proceed to make such studies. Learning how to do so in a meaningful and responsible manner is still a major challenge to the profession.

## FINAL COMMENTS AND CONCLUSIONS

The main points from the foregoing discussion may be summarized as follows:

1. There are many recent papers discussing the application of probability theory to soil and rock mechanics, often involving transference of material developed in other fields of engineering.
2. Very few papers discuss the use of this theory in connection with specific engineering projects.
3. While the overall theory is complex, there are simple but powerful tools that may be used to understand the relation between uncertainty and safety.
4. Numerical results from a reliability analysis can be no better than the underlying deterministic model and the quality of the statistical parameters from the data base.
5. Reliability theory may be used to improve consistency in the safety of different slopes. It is important to separate the effects of spatial variability and systematic errors.
6. When there is no standard for a safety factor, but the problem is well understood and there is an adequate data base, reliability theory may be used to guide selection of a safety factor consistent with the degree of safety in other problems.
7. If economics is the only consideration, optimization techniques are useful for guiding the choice of a safety factor.
8. Careful and detailed formulation of a systems reliability problem aids greatly in understanding the problem, even if computed numerical results are of doubtful accuracy.
9. Where risk must be very low, it is not possible to quantify the actual risk accurately by analysis alone. However, the framework of risk evaluation can help guide subjective evaluations.

These conclusions are oriented primarily toward problems in which the major loading is the pull of gravity upon the soil or rock and supported structure. Where extreme environment loadings are included in the design requirement, the overall probability of failure involves both the probability of the extreme event and the risk of failure should that event occur. This overall risk may be very small and yet be evaluated with some confidence.

**Other Applications.**—As mentioned previously, there are other types of applications of probability theory in geotechnical engineering. These potential applications are every bit as important as those considered in this paper, but space limitations preclude a detailed discussion.

One such use is in connection with the exploration and sampling of soil and rock masses. A number of papers have dealt with the statistical interpretation of penetration resistance and the probabilistic modeling of soil profiles (18,45,47). In a different class of applications, probability theory may be employed effectively to choose the spacing and pattern of bore holes so as to minimize the chance of missing a suspected weak zone or buried cavern (5,16). This approach has found particular application in the design of slopes for open pit mines (8). Statistical analysis of joint surveys has progressed far, has been studied intensely by a large number of researchers, and has been used in practice in the design of pit slopes (37). Other uses include the search for and evaluation of burrow areas during the planning and design phases of an earth dam, and the planning and execution of an inspection program to ensure that a compacted embankment meets the intent of the specifications. Many of the references to such work are very mathematical, and there is a need to make the procedures and results understandable to the engineer who lacks an advanced understanding of probability. In all such applications it is necessary, of course, to couple any probabilistic considerations closely to the insights and understandings provided by geology or soil mechanics, or both.

The phrase *Bayesian updating* refers to a method of updating the probability of a given event as additional information is obtained. To illustrate this approach, suppose that an engineer has evaluated the probability of failure for a slope. Included within this analysis will be probabilities for the possible pore pressures at points within the slope. Now actual pore pressures are observed at several points, and the engineer wishes to use this information to improve the estimate of the probability-of-failure. Bayesian updating provides the steps necessary for this purpose (21). Veneziano and Faccoli (53) have used Bayesian methods to update estimates (i.e., maps) of soil properties. These same techniques may also be used to study the value of obtaining more information, and are potentially of great use in the design of observational systems. Indeed, Bayesian updating may be viewed as a formalization of the observational approach advocated by Terzaghi, Peck (35), and others.

**Use of Probability in Planning and Design.**—The probabilistic approaches discussed in this paper might be used throughout the history of a project—from the time of initial conception and detailed planning on through design, construction, and eventual operation. However, certain of the approaches are most useful at specific stages in a project's history. This is indicated by the matrix in Fig. 22.

| METHODOLOGY \ STAGE OF PROJECT | CONCEPTION | PLANNING | DESIGN | CONSTRUCTION | OPERATION |
|---|---|---|---|---|---|
| EXPLORATION | | ⊠ | ⊠ | | |
| RELIABILITY | | ⊠ | | | |
| OPTIMIZATION | | ⊠ | | | |
| RISK EVALUATION | ⊠ | | | | ⊠ |
| UPDATING | | | | ⊠ | ⊠ |

**FIG. 22.—Use of Probability-Based Methodologies During Various Phases of Project**

This figure emphasizes that techniques such as reliability, optimization, and risk evaluation are likely to be of greatest usefulness during the conception and planning stages. (If the safety of an existing structure is being studied, e.g., during the recent and ongoing national program to inspect old dams, the methods of risk evaluation may again be appropriate.) It is during these stages that decisions are made about whether the benefits of a project outweigh the costs, including costs of possible failure, and, thus, about whether or not to proceed with a project. It is also at these times that the criteria for design are established. Probabilistic thinking is very appropriate—indeed essential—in these stages. Once a decision is made to proceed and the design criteria are set, engineers will then almost always—for various practical and legal reasons—proceed to design the facility in a deterministic fashion and, especially for a critical facility such as an earth dam, will be properly conservative in the interpretation of the design criteria.

Engineers often complain that they play too small a role in the planning and criteria-setting stages of major projects. In fact, there is indeed a great need during these stages for the best possible information bearing on the trade-offs between construction costs and risks of failure. This is exactly the type of information that engineers should be able to provide. However, engineers have been conditioned to provide safe designs against specific criteria, and relatively few are either prepared or inclined to think in terms of a risk of failure. This, however, is exactly what they must do if they are to participate in the vital early stages of a project.

In some ways this challenge is particularly suited to geotechnical engineers, who are accustomed to relying heavily upon experience and judgment, rather than depending totally upon analysis. During the planning stages of a project, there is seldom the possibility of extensive laboratory testing and detailed analytical studies. Experience and judgment are, thus, of the greatest importance. However, engineers must be able to quantify their knowledge in a form that permits effective communication with others involved in decision making who lack their technical background. Fragility curves and performance probability matrices are among the devices that are proving useful in this connection. Moreover, engineers must be prepared to discuss uncertainty in their evaluations of uncertainty. These matters are not simple, and much research can be done to provide better means for communicating wisdom and knowledge about risks.

**Randomness versus Determinism.**—One final thought: Use of probability theory has been criticized as a cover-up for lack of knowledge or inadequate measurement (e.g., Ref. 36). There is validity to this observation, but not necessarily to the criticism.

Take, for example, the matter of variations in shear strength from point-to-point within the ground. With some ideal device for measuring strength that is absolutely accurate and can be inserted into the ground without disturbing it, plus unlimited time and funds, it is theoretically possible to determine this variation and include it in a deterministic description of the soil profile. To the extent that we fail to reach this ideal, and introduce randomness to describe those remaining variations that we do not measure, we are replacing actual determinism by fictitious randomness. Those who use probability theory should indeed keep this fact in mind. However, the use of probability theory to account for such remaining uncertainties can also be viewed as good hardheaded engineering. Achieving a proper balance between the benefit of having a more accurate deterministic model and the cost of obtaining it is the type of juggling act that engineers are expected to perform.

In similar fashion, once we have put an earth dam in place, we have predetermined whether or not piping will develop upon first filling of the reservoir. It is also more or less true that nature knows when and where the next earthquake will strike, and how large it will be. Possibly the date of the next 1,000-year flood is also fully predetermined at this moment, although, at this scale, there is possibly some real randomness in the affairs of the solar system. Certainly engineers should not use probabilistic approaches as a substitute for measurements and interpretations that they can and should make, with good practice. On the other hand, probability theory—when properly used—can be a powerful tool for achieving a good balance between the consequences of uncertain knowledge and the cost of reducing that uncertainty.

**Status Report.**—In the introduction, I indicated that this paper would be a status report upon progress toward the utilization of probabilistic concepts and methods in geotechnical engineering, and I posed the following two questions:

First, *How can probability theory be used in geotechnical engineering practice?* I hope that the several examples presented herein provide initial answers to this question.

Second, *Is it possible today to evaluate risk?* Here there are several answers:

1. If a relatively large probability-of-failure (0.05 or more) under design loading conditions is tolerable, then this risk can be evaluated with accuracy sufficient for decision-making purposes. This situation applies primarily when only economic losses, and not life safety, are of concern.

2. If a very small probability-of-failure (say <0.001) under design loading conditions is required, the actual risk cannot today be evaluated accurately by analysis. However, conducting a formal evaluation of probability-of-failure can help greatly in understanding the risk and what might best be done to reduce it.

If the design loading condition includes extreme environmental loads with a small likelihood of occurrence, then the overall probability-of-failure from such loadings may be very small indeed.

There are numerous areas of research that might be pursued profitably to increase the usefulness of probability theory. I would not dare to attempt even a partial listing. However, the greatest need at this time is for published case studies of projects wherein probabilistic analysis has influenced engineering decisions directly or indirectly. Only in this way will the profession learn how to make effective use of this potentially powerful tool.

In the introduction I suggested that Casagrande, in his Terzaghi lecture, aimed to call attention to the nature of risks in geotechnical engineering and to urge that the existence of such risks be recognized and dealt with openly. Society and the profession have come a long way in these directions since 1969. My own objectives have been to: (1) Reduce some of the mystery surrounding probability theory; (2) call attention to the work that has been done to bring probabilistic concepts and methods to bear upon geotechnical engineering; (3) give examples of first and sometimes tentative steps toward the use of this theory in connection with practical problems; and (4) encourage the profession to seize all possible opportunities to employ available methodology as an aid to actual engineering decisions—and to publish case studies for the benefit of the profession. It is in the spirit of these aims that I hope this paper will be received.

### Acknowledgments

This paper results from interactions with many colleagues, beginning with Allin Cornell, who first introduced me to probabilistic thinking. I have benefited greatly from discussions and collaborations with Gregory Baecher, Erik Vanmarcke, and Daniele Veneziano. I thank these three, plus Charles Ladd, for careful readings of my drafts of this paper; they strove mightily to help me avoid errors arising from my limited grasp of probability theory. I also owe great gratitude to T. William Lambe and Allen Marr, who encouraged and supported the use of probabilistic thinking and analyses over the course of a number of consulting projects. Finally, my thanks to Peter Wroth and Peter Vaughn, who first suggested that I present some of these ideas in a lecture to the Geotechnical Section of the Institution of Civil Engineers, way back in 1976.

### Appendix I.—Probabilities and Expected Costs for Great Salt Lake Crossing

The total cost, $TC$, for the crossing is given by the equation

$$TC = C_1 + C_2(F_d) + \Sigma \Delta C(F_d, \sigma_F, p_o) \quad \quad (12)$$

**TABLE 4.—Values of $\Delta C$ Per Segment**

| $F$ (1) | Before actual failure (2) | After actual failure (3) |
|---|---|---|
| >1.0 | — | — |
| 0.9–1.0 | 0.16 | 0.8 |
| 0.8–0.9 | 0.4 | 0.8 |
| 0.7–0.8 | 0.72 | 1.6 |
| >0.7 | 1.2 | 1.6 |

in which $F_d$ = the safety factor used for design, which is assumed equal to the mean value of the actual safety factor; $\sigma_F$ = the standard deviation for the actual safety factor, $F$ (the random variable); $p_o$ = the probability that an incipient failure is detected in time to take corrective action; and $\Delta C$ = the cost of corrective action (either before or after an actual failure) for one segment. These corrective costs are summed over all segments.

For the example, $C_1 = 7$ and $C_2 = 36$. (All costs are in millions of 1950 dollars.) For $\Delta C$, step-wise relations are used for simplicity to represent the actual continuous relationship between the cost of corrective action and the actual safety factor. Values of $\Delta C$ per segment are given in Table 4. Table 4 identifies seven *cost states*, $\Delta C_i$, into which the cost of a segment may fall, depending upon the values of $F$ and $p_o$ (which is taken as 5/6).

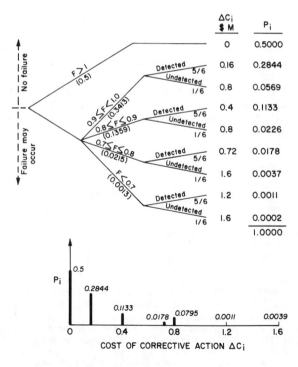

**FIG. 23.—Various Costs for $F_d = 1$ and $\sigma_F = 0.1$**

**TABLE 5.—Probability of Exceeding $51,000,000 Total Cost, for $\sigma_F = 0.15$**

| $\sigma_{TC}$, in millions of dollars (1) | $F_d$ (2) | $E[TC]$, in millions of dollars (3) | $\{51 - E[TC]\}/(\sigma_{TC})$ (4) | $P[TC < 51]$ (5) | $P[TC > 51]$ (6) |
|---|---|---|---|---|---|
| 1.844 | 1.00 | 49.88 | 0.607 | 0.73 | 0.27 |
| 1.531 | 1.05 | 49.39 | 1.052 | 0.85 | 0.15 |
| 1.234 | 1.10 | 49.39 | 1.216 | 0.89 | 0.11 |
| 0.935 | 1.15 | 50.08 | 0.984 | 0.84 | 0.16 |
| 0.708 | 1.20 | 51.13 | −0.184 | 0.43 | 0.57 |

For any $F_d$, a simple event tree may be used as an aid in evaluating the probabilities, $p_i$, for the various costs; this is shown in Fig. 23 for the case of $F_d = 1$ and $\sigma_F = 0.1$. The standard normal variable in this case is $(F - F_d) \div \sigma_F$. For $F_d = 1$, this variable is 0 if $F = 1$; while for $F = 0.9$, it is −1. Looking in a table of standard normal probability, the corresponding values are 0.5 and 0.1587. Thus the probability, $p_i$, that $0.9 \leq F \leq 1.0$ is $(0.5 - 0.1587) = 0.3413$. From these results, the probability mass function for corrective costs may be constructed as shown. The expected cost of corrective action for all 30 independent sections is then $E[C_c] = \Sigma \, (\Delta C_i \cdot p_i) = \$5,240,000$. The variance of this cost, Var $[C_c] = 30\{\Sigma \, (\Delta C_i^2 \cdot p_i) - E^2 [C_c]\} = \$1,990,000$, and the standard deviation, $\sigma_{TC}$, = the square root of the variance or $1,410,000.

The expected total cost, $E[TC] = \$7,000,000 + \$36,000,000 + \$24,000,000 = \$48,240,000$. For the limiting total cost of $51,000,000, the standardized normal variable is $(51 - 48.24) \div 1.41 = 1.95$. Using a table for cumulative normal probability, the probability that the total cost will be less than $51,000,000 is found to be 0.97. For $\sigma_F = 0.15$, the calculations for the probability of exceeding $51,000,000 are shown in Table 5. These calculations assume that $\sigma_F$ is known; whereas in reality this parameter is itself uncertain. With additional calculations, this uncertainty can also be taken into account.

## APPENDIX II.—REFERENCES

1. Ang, A. H-S., and Cornell, C. A., "Reliability Basis of Structural Design and Safety," *Journal of the Structural Division*, ASCE, Vol. 100, No. ST9, 1974.
2. Applications of Statistics and Probability to Soil and Structural Engineering, International Conferences: Hong Kong, 1971; Aachen, West Germany, 1975; Sydney, Australia, 1979.
3. ASCE National Convention, Boston, 1979, Reliability Analysis and Geotechnical Engineering (Preprint 3600).
4. Athanasiou-Grivas, D., ed., *Probability Theory and Reliability Analysis in Geotechnical Engineering*, Rensselaer Polytechnic Institute, 1977.
5. Baecher, G. B.,, "Analyzing Exploration Strategies," *Site Characterization and Exploration*, C. H. Dowding, ed., ASCE, 1979.
6. Baecher, G. B., "Simplified Geotechnical Data Analysis," presented at the Aug.–Sept., 1982, NATO Advanced Study Institute on Reliability Theory and Its Application in Structures and Soil Mechanics, held at Bornholm, Denmark.
7. Baecher, G. B., "First Order Geotechnical Reliability," *International Journal of Numerical and Analytical Methods in Geomechanics*, in press.

8. Baecher, G. B., and Einstein, H. H., "Slope Reliability Models in Pit Optimization," *APCOM*, Tucson, 1979, pp. 501–512.
9. Baecher, G. B., Pate, M. E., and de Neufville, R., "Risk of Dam Failure in Benefit-Cost Analysis," *Water Resources Research*, Vol. 16, No. 3, 1980, pp. 449–456.
10. Benjamin, J. R., and Cornell, C. A., *Probability, Statistics and Decision for Civil Engineers*, McGraw Hill Book Co., Inc., New York, N.Y., 1970.
11. Casagrande, A., "Role of the 'Calculated Risk' in Earthwork and Foundation Engineering," *Journal of the Soil Mechanics Division*, ASCE, Vol. 91, No. SM4, July, 1965, p. 1.
12. Cornell, C. A., "First Order Uncertainty Analysis of Soil Deformation and Stability," *Proceedings of the 1st International Conference on Applications of Statistics and Reliability to Soil and Structural Engineering*, Hong Kong, 1971.
13. deMello, V. F. B., "Reflections on Design Decisions on Practical Significance to Embankment Dams," 17th Rankine Lecture, *Geotechnique*, Vol. 27, No. 3, 1977, pp. 279–355.
14. Ditlevsen, O., "Basic Reliability Concepts," presented at the Aug.–Sept., 1982, NATO Advanced Study Institute on Reliability Theory and Its Application in Structures and Soil Mechanics, held at Bornholm, Denmark.
15. Duncan, J. M., and Houston, W. N., "Estimating Failure Probabilities for Levees in the California Delta," presented at the May, 1981, ASCE National Conference, held at New York, N.Y.
16. Einstein, H. H., and Baecher, G. B., "Probabilistic and Statistical Methods in Engineering Geology," presented at the 1981, 30th Geomechanics Colloquium, held at Salzburg, Austria; and *Supplementum 12 Rock Mechanics*.
17. Ellingwood, B., Galambos, T. V., MacGregor, J. G., and Cornell, C. A., "A Probability-Based Load Criterion for Structural Design," *Civil Engineering*, July, 1981.
18. Fardis, M. N., and Veneziano, D., "Estimation of SPT-N and Relative Density," *Journal of the Geotechnical Engineering Division*, ASCE, Vol. 107, No. GT10, 1981, pp. 1345–1349.
19. Fardis, M. N., and Veneziano, D., "Statistical Analysis of Sand Liquefaction," *Journal of the Geotechnical Engineering Division*, ASCE, Vol. 107, No. GT10, 1981, pp. 1361–1377.
20. Freudenthal, A. M., "Safety and Probability of Structural Failure," *Transactions*, ASCE, Vol. 121, 1956, p. 1337.
21. Hachich, W. C., "Seepage-Related Probability of Embankment Dams," thesis presented to the Massachusetts Institute of Technology, at Cambridge, Mass., in 1981, in partial fulfillment of the requirements for the degree of Doctor of Philosophy.
22. Haldar, A., and Tang, W. H., "Probabilistic Evaluation of Liquefaction Potential," *Journal of the Geotechnical Engineering Division*, ASCE, Vol. 105, No. GT2, 1979, pp. 145–163.
23. Harr, M. E., *Mechanics of Particulate Media: A Probability Approach*, McGraw Hill, New York, N.Y., 1977.
24. Hoeg, K., and Muraka, R. P., "Probabilistic Analysis and Design of a Retaining Wall," *Journal of the Geotechnical Engineering Division*, ASCE, Vol. 100, No. GT3, 1974, pp. 349–366.
25. Kay, J. N., "Safety Factor Evaluation for Single Piles in Sand," *Journal of the Geotechnical Engineering Division*, ASCE, Vol. 102, No. GT10, 1976, pp. 1093–1110.
26. Keeney, R. L., and Raiffa, H., *Decision Analysis with Multiple Objectives*, John Wiley and Sons, Inc., New York, N.Y., 1976.
27. Lambe, T. W., Consulting files.
28. Leonards, G. A., "Flexibility and Cracking of Earth Dams," *Journal of the Soil Mechanics and Foundations Division*, ASCE, Vol. 89, No. SM2, 1963, p. 47.
29. Leonards, G. A., "Investigation of Failures," *Journal of the Geotechnical Engineering Division*, ASCE, Vol. 108, No. GT2, 1982, pp. 185–246.

30. Lumb, P., "Application of Statistics in Soil Mechanics," *Soil Mechanics: New Horizons,* I. K. Lee, ed., Newnes-Butterworth, 1974, pp. 44–112.
31. Matsuo, M., and Kuroda, K., "Probabilistic Approach to Design of Embankments," *Soil and Foundations,* Vol. 14, No. 2, 1974, p. 1.
32. Meyerhof, G. G., "Concepts of Safety in Foundation Engineering Ashore and Offshore," *Proceedings of the 1st International Conference on the Behavior of Offshore Structures,* Trondheim, Norway, Vol. 1, 1976, pp. 501–515.
33. Newmark, N. M., "Fifth Rankine Lecture: Effects of Earthquakes on Dams and Foundations," *Geotechnique,* Vol. XV, 1965, pp. 139–160.
34. Pate, M.-E., "Risk-Benefit Analysis for Construction of New Dams: Sensitivity Study and Real Case Applications," *Report R81-26,* Massachusetts Institute of Technology, Cambridge, Mass., 1981.
35. Peck, R. B., "Advantages and Limitations of the Observational Method in Applied Soil Mechanics," Ninth Rankine Lecture, *Geotechnique,* Vol. 19, No. 2, 1969, pp. 171–187.
36. Peck, R. B., "Where Has All the Judgement Gone?," *Fifth Bjerrum Lecture,* Norwegian Geotechnical Institute, 1980.
37. Priest, S. D., and Hudson, J., "Estimation of Discontinuity Spacing and Trace Length Using Scanline Surveys," *International Journal of Rock Mechanics and Mining Science,* Vol. 18, 1981, pp. 183–199.
38. Rasmussen, N. C., "Setting Safety Criteria," presented at the 1979 MIT Symposium on Dam Safety, held at Cambridge, Mass.
39. *Reactor Safety Study, WASH-1400,* U.S. Nuclear Regulatory Commission.
40. *Risk and Decision Making; Perspectives and Research,* National Research Council, National Academy Press, 1982.
41. Sangrey, D. A., and D'Andrea, R. A., "Safety Factors for Probabilistic Slope Design," *Journal of the Geotechnical Engineering Division,* ASCE, Vol. 108, No. GT9, 1982, pp. 1101–1118.
42. Seed, H. B., and Idriss, I. M., "Simplified Procedure for Evaluating Soil Liquefaction Potential," *Journal of the Soil Mechanics and Foundations Division,* ASCE, Vol. 97, No. SM9, 1971, pp. 1248–1273.
43. Seed, H. B., and Idriss, I. M., "Evaluation of Liquefaction Potential of Sand Deposits Based on Observations of Performance in Previous Earthquakes," ASCE, Preprint 81-544, 1981.
44. Smith, E. S., discussion of "Role of *Calculated Risk* in Earthquake and Foundation Engineering," *Journal of the Soil Mechanics and Foundation Engineering Division,* ASCE, Vol. 92, No. SM2, 1966, p. 185.
45. Tang, W. H., "Probabilistic Evaluation of Penetration Resistances," *Journal of the Geotechnical Engineering Division,* ASCE, Vol. 105, No. GT10, 1979, pp. 1173–1191.
46. Vanmarcke, E. H., "Decision Analysis in Dam Safety Monitoring," *Proceedings of the Conference on the Safety of Small Dams,* Engineering Foundation, ASCE, 1974.
47. Vanmarcke, E. H., "Probabilistic Modeling of Soil Profiles," *Journal of the Geotechnical Engineering Division,* ASCE, Vol. 103, No. GT11, 1977, pp. 1227–1246.
48. Vanmarcke, E. H., "Reliability of Earth Slopes," *Journal of the Geotechnical Engineering Division,* ASCE, Vol. 103, No. GT11, 1977, pp. 1247–1266.
49. Vanmarcke, E. H., "Probabilistic Stability Analysis of Earth Slopes," *Engineering Geology,* Vol. 16, 1980, pp. 29–50.
50. Vanmarcke, E. H., *Random Fields: Analysis and Synthesis,* M.I.T. Press, Cambridge, Mass., 1983.
51. Vanmarcke, E. H., and Bohmenblust, H., "Risk-Based Decision Analysis in Dam Safety," *Report R82-11,* Massachusetts Institute of Technology, Cambridge, Mass.
52. Veneziano, D., "Statistical Estimation and Data Collection: A Review of Procedures for Civil Engineers," *Proceedings of the 3rd International Conference on Applications of Statistics and Probability to Soil and Structural Engineering,* Sydney, Australia, 1979.

53. Veneziano, D., and Faccioli, E., "Bayesian Design of Optimal Experiments for the Estimation of Soil Properties," *Proceedings of the 2nd International Conference on Applications of Statistics and Probability to Soil and Structural Engineering*, Aachen, West Germany, 1975.
54. Whitman, R. V., Biggs, J. M., Brennen, J. E., III, Cornell, C. A., de Neufville, R. L., and Vanmarcke, E. H., "Seismic Design Decision Analysis," *Journal of the Geotechnical Engineering Division*, ASCE, Vol. 101, No. ST5, 1975, pp. 1067–1084.
55. Wu, T. H., "Uncertainty, Safety, and Decision in Civil Engineering," *Journal of the Geotechnical Engineering Division*, ASCE, Vol. 100, No. GT3, 1974, pp. 329–348.
56. Wu, T. H., and Kraft, L. M., "The Probability of Foundation Safety," *Journal of the Soil Mechanics and Foundations Engineering Division*, ASCE, Vol. 93, No. SM5, 1967, pp. 213–230.
57. Wu, T. H., Thayer, W. B., and Lin, S. S., "Stability of Embankment on Clay," *Journal of the Geotechnical Engineering Division*, ASCE, Vol. 101, No. GT9, 1975, pp. 913–948.
58. Yegian, M. K., and Whitman, R. V., "Risk Analysis for Ground Failure by Liquefaction," *Journal of the Geotechnical Engineering Division*, ASCE, Vol. 104, No. GT7, 1978, pp. 921–938.
59. Yucemen, M. S., and Tang, W. H., "Long Term Stability of Soil Slopes: A Reliability Approach," *Proceedings of the 2nd International Conference on Applications of Statistics and Probability to Soil and Structural Engineering*, Aachen, West Germany, 1975.

# EVALUATING CALCULATED RISK IN GEOTECHNICAL ENGINEERING[a]

Discussion by Herbert Klapperich,[2] Ulrich Sturm,[3] and Stavros A. Savidis[4]

The author presented a comprehensive paper that will help disseminate the ideas of modern reliability theory and applied probability, and their use in the decision-making process. It is especially valuable since the use of these approaches lags behind in soil and rock mechanics, compared to efforts in structural engineering. This discussion is related to "Applications of Reliability Theory," and is in the context of "Stability of Slopes."

One important point concerns the reliability index β (see Table 2) and its selection in connection with geotechnical problems. Contrasted to the use of a factor of safety (FS)—an empirical deterministic safety definition-reliability theory allows for a much more realistic decision concerning stability, as the author outlined in his paper. Additionally, uncertainties in the selected mechanical model, with its different parameters, must also be taken into consideration.

In Germany, a new safety concept is being debated in the context of the "Eurocode." Following the proposal in Ref. 60, we find a value of β = 4.7 related to the structural safety concept. With this value of β and the same variation coefficients as in Table 2, much larger mean safety factors would, of course, be required. The observation in the paper concerning the case of β = 2, $V'_c = 0.3$, which implies the need for a mean safety factor larger than typical safety factors in geotechnical practice, should be underscored. In the writers' opinion, the best way to find a realistic β-value in geotechnical problems is to conduct a large number of test calculations with parameter variation for different standard problems, such as slope stability, foundation failure, earth pressure, etc. In addition to the coefficients of variation, other factors, e.g., those of autocorrelation, have significant influence on the β-value (61). To solve the questions raised by the author, experimental investigations with large-scale in-situ testing of the behavior of soil as a whole, as well as the influence of different parameters, can be helpful. An important parameter is the friction angle, determined either by laboratory testing or by back calculations of in-situ failure tests.

In an overview of design practice, the triad—the mathematical model, real parameters (geometry, material data, load), and safety definition—affect each other. A good approximation for a mechanical model is the kinematic element method, which delivers equations of limit state directly, quite useful for the reliability method (see Ref. 62). While "classical safety definitions" refer only to deterministic values, reliability theory enables one to consider the probability of failure.

For the foundation failure problem, Fig. 24 shows β, determined with

---

[a] February, 1984, by Robert V. Whitman (Paper 18569).
[2] Technical University Berlin, Federal Republic of Germany.
[3] Research Asst., Technical Univ. Berlin, Federal Republic of Germany.
[4] Prof., Demokritos Univ., Thrazien, Greece.

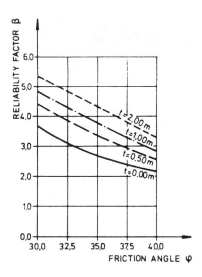

**FIG. 24.—Reliability Factor $\beta$**

the kinematic element method (KEM) as a function of friction angle $\phi$ and foundation embedment $t$ using the equation of limit state from DIN 4017 (German standard)

$$P_f = a' \cdot b' (c \cdot N_c \cdot v'_c \cdot \chi_c + \gamma_1 \cdot t \cdot N_t \cdot \chi_t \cdot v'_t + \gamma_2 \cdot b' \cdot N_b \cdot \chi_b \cdot v'_b) \dots \dots (13)$$

using the safety definition $\eta_P = P_f/P$.

### Appendix.—References

60. "Grundlagen zur Festlegung von Sicherheitsanforderungen für bauliche Anlage," DIN Deutsches Institut für Normung, Beuth Verlag, Berlin—Köln, Germany, 1981.
61. Pientinger, B., "Auswirkung der räumlichen Streuung von Bodenkennwerten," *Beiträge zur Anwendung der Stochastik und Zuverlässigkeitstheorie in der Bodenmechanik*, Heft 2, 1983.
62. Sturm, U., Savidis, S., Berner, K., and Klapperich, H., "Probabilistic Analyses Applied to Foundation Failure and Slope Stability," *ICASP 4*, Florenz, 1983.

### Closure by Robert V. Whitman,[5] F. ASCE

Klapperich, Sturm and Savidis referred to the use of reliability approaches in the development of codes and standards in Europe. Several European countries are indeed more advanced in this regard than the U.S., at least in the geotechnical area. However, these developments are regarded with mixed emotion by European engineers, as illustrated in

---

[5]Prof., Dept. of Civ. Engrg., Massachusetts Inst. of Tech., Cambridge, Mass. 02139.

the Bjerrum Lecture by Mortensen (63).

The writer agrees that the safety definition, the choice of a model for analysis and the evaluation of parameters characterizing the soil are closely intertwined. However, the most important step is assessment of the model error and the errors arising from sampling disturbance, and the way in which the strength of soil is defined and measured. Often these uncertainties are not evaluated realistically, leading to unreasonably high values for the reliability index.

### APPENDIX.—REFERENCE

63. Mortensen, K., "Is Limit State Design a Judgement Killer?" (6th Laurits Bjerrum Memorial Lecture), *Publication 148*, Norwegian Geotechnical Inst., Oslo, Norway.

# THE EIGHTEENTH TERZAGHI LECTURE

Presented at the American Society of Civil Engineers
1982 Annual Convention

October 28, 1982

J. BARRY COOKE

# INTRODUCTION OF THE EIGHTEENTH TERZAGHI LECTURE

By Ernest T. Selig

The eighteenth Terzaghi lecturer is James Barry Cooke. Following graduation from the University of California, Mr. Cooke joined the engineering department of Pacific Gas and Electric Company. There, for 22 years, he was involved with the engineering of 18 new hydroelectric developments and the operating problems of 70 hydroelectric developments, 110 dams, and 130 miles of tunnels. For the past 22 years, Mr. Cooke has been a consultant on hydroelectric projects throughout the world—Europe, Africa, South America, Southeast Asia, Australia, the United States, and Canada.

Currently among his consulting activities are the Guri project in Venezuela, the Dartmouth Dam in Australia, the Khao Laem Project in Thailand, the Revelstoke Dam on the Columbia River in British Columbia, and the James Bay Project in Quebec. Mr. Cooke still sets a grueling pace, having traveled to New Orleans for this Terzaghi lecture from Australia via Canada and having left for Chile after the lecture.

A registered professional engineer, Mr. Cooke has also served as a member and officer of ASCE committees, the U.S. Committee on Large Dams, and the International Committee on Large Dams.

Mr. Cooke was elected to the National Academy of Engineering in 1979. Among his other professional honors and awards are:

1. ASCE Rickey Medal for advancement of the art of hydroelectric engineering.
2. ASCE James Laurie Prize for meritorious contribution to engineering science.
3. ASCE Thomas A. Middlebrooks Award for meritorious contribution to soil mechanics.
4. Citations by Engineering News Record for service to the construction industry.
5. Golden Beaver Award from the construction industry for outstanding achievement in heavy engineering construction.

Mr. Cooke has published a number of papers on rockfill dams. This is the subject of his latest contribution, the Terzaghi lecture titled "Progress in Rockfill Dams." The written paper makes available to the profession some of the experience he has gained over the years. The Geotechnical Engineering Division appreciates his willingness to make this valuable contribution.

# PROGRESS IN ROCKFILL DAMS

## By J. Barry Cooke,[1] F. ASCE

**ABSTRACT:** Karl Terzaghi, a major contributor to progress in embankment dams, made significant contributions to the understanding of properties of rockfill and to the improvement of design of the concrete-face rockfill dam. His contributions are noted and followed by a history of parallel development of earth core and concrete-face rockfill dams and the impact of compacted rockfill on the concrete face rockfill dam. Emphasis in this paper is directed towards the concrete face type of rockfill dam, since the earth core type is already well-covered in the literature. Recent progress in the design and construction of the concrete-face rockfill dam is outlined and documented. Design trends and performance data are presented.

## INTRODUCTION

Karl Terzaghi is best known in the field of soil mechanics and, of course, as one of our finest engineers. Additionally, he made significant contributions to progress in rockfill and rockfill dams. He accelerated thinking important to the development of concrete-face rockfill dams in two discussions in the ASCE Symposium on Rockfill Dam, published in the only separate and supplemental Transaction Volume in ASCE history, Vol. 104, 1960 (1).

Among the significant points presented in his ASCE discussions on rockfill and concrete-face rockfill dams were: (1) The characteristic segregation and lack of compaction of dumped rockfill; (2) the loss of strength of various rocks upon saturation; (3) the benefits of compacted rockfill, specifically to permit higher concrete face rockfill dams; (4) the potential usability of weaker rock with compaction; (5) the value of performance observations; and (6) the development of cutoffs by grouting from a dowelled toe slab as opposed to "the brutal practice of blasting a cutoff trench" (1). All of his published 1960 comments were noted and put into practice by the profession.

On the subject of earth and earth core rockfill dams, Terzaghi's contributions on the related aspects of soil mechanics, filters, and foundations are many and well-known (28). One important thought of his was stated in a letter to me on subjects including Kenney Dam (designed 1950) expressing a preference for the sloping core design, as follows:

> "Vertical cores are likely to crack, but the seepage losses through cracks can usually be reduced to tolerable values by a thick layer of fine but cohesionless sand covering the upstream face, combined with an adequate graded filter on the downstream face of the core, because the sand will be washed into the cracks—and retained by the downstream filter—at my request this is a feature in Kenney Dam design."

---

[1]Consulting Engr., J. Barry Cooke, Inc., San Rafael, Calif.

Note.—Discussion open until March 1, 1985. To extend the closing date one month, a written request must be filed with the ASCE Manager of Technical and Professional Publications. The manuscript for this paper was submitted for review and possible publication on April 19, 1984. This paper is part of the *Journal of Geotechnical Engineering*, Vol. 110, No. 10, October, 1984. ©ASCE, ISSN 0733-9410/84/0010-1383/$01.00. Paper No. 19206.

This feature is now incorporated in most earth core rockfill dam designs, particularly in earthquake zones, and is sometimes referred to as a "crack filler" zone.

I was not a Harvard student of Terzaghi, but a student of his nevertheless. I worked with him on the Aswan Dam on the Nile, Cowichan Dam in Canada, founded on quicksand, and on other projects. We corresponded and met in connection with his book on tunnels, which were, unfortunately, never completed. Because of the enormous respect which he deservedly earned, I am honored to be a Terzaghi Lecturer.

My subject is "Progress in Rockfill Dams." With design advancements and the performance records of recent decades, the concrete-face rockfill dam has experienced great recent progress and is featured in this lecture.

In the ASCE Symposium on rockfill dams in 1939 (2), the definition of a rockfill dam was, "A dam consisting of loose (dumped) rockfill with slopes on both faces closely approximating natural slopes, with an impervious facing on the upstream side between which and the rockfill there should be placed a cushion of dry rubble." The earth core rockfill type of dam emerged later, in 1940. Great progress has since occurred on both types of rockfill dams.

## Evolution of Modern Rockfill Dam

Rockfill dam and construction practices have changed markedly over the past century, and during the past couple of decades, large rockfill dams—particularly the concrete-face rockfill dams—have evolved from an experimental stage to the design of first choice for many dam building agencies throughout the world. Indeed, during the last decade, most of the new dams over 100-ft (30-m) high have been rockfills. However, in contrast to the development of earth dams, the present state-of-the-art of modern rockfill dam design and construction technology has evolved more as a result of field observation of construction practices and performance evaluation than by theory and laboratory testing. This applies more to the concrete-face rockfill dam than to the earth core rockfill dam.

The evolution of contemporary rockfill dam technology may be illustrated by considering its progress as falling into three distinct eras (Table 1). The nineteenth and early twentieth century may be considered as the era of the dumped rockfill, followed by two decades of aggressive experimentation with new designs and construction techniques, beginning in the early 1940's. The last 20 yr or so constitute a third contemporary period in which fairly standardized earth core and concrete-face rockfill dam designs have been applied with refinements and with increasing confidence to even larger and higher dams and progressively more difficult sites.

## Early Period (1850–1940): Dumped Rockfill Dams

Although the use of rock in dams dates back to antiquity, the modern rockfill dam is generally considered to be a product of the California gold rush. The miners of the California Sierras, because of knowledge of blasting and availability of rock, used rockfill dams to store water for dry season sluicing of placer ore deposits. These dams were usually tim-

**TABLE 1.—Summary History of Rockfill Dams**

| Periods in evolution of dam technology (1) | Concrete Face | | | Earth Core | | |
|---|---|---|---|---|---|---|
| | Year (2) | Height, in feet (3) | Number or names (4) | Year (5) | Height, in feet (6) | Number or names (7) |
| Early period (1850–1940) | | | | | | |
| Dumped rockfill | 1850 | 75 | Many timber face | 1850 to 1940 | | No earth core |
| | 1925 | 100 | 8 dams | | | |
| | 1925 | 275 | Dix River | | | |
| | 1930 | 330 | Salt Springs | | | |
| | 1930–40 | 200 | Many | 1940 | | Begin earth core |
| Transition period (1940–1965) | | | | | | |
| Dumped | 1950–55 | 200 | A number | 1940 | | Begin Kenney, Watauga |
| | | 200–300 | PG&E dams | 1950 | 250–400 | |
| | 1955 | 370 | Paradela | 1955 | 250–420 | 30 dams |
| | 1965 | 490 | Exchequer | 1960 | 510 | Goschenen |
| Compacted | 1955–65 | 200 | Several | 1955–65 | 300–400 | Ambuklao, Brownlee, Lewis Smith |
| Modern period (1965–1984) | | | | | | |
| Compacted rockfill | 1965–70 | 200–300 | Many | 1965–84 | 300–600 | Many major dams |
| | 1971 | 320 | Cethana | | 800 | Oroville,[a] Mica,[a] |
| | 1974 | 460 | Anchicaya | | 850 | Chivor |
| | 1980 | 530 | Areia | | 1,000 | Chicoasen Nurek[a] |

[a]Earth core gravel shell dams.

ber faced, dumped rockfills up to 25-m high, with very steep [0.5:1 to 0.75:1 (H:V)] slopes and with a "skin" of hand placed rocks to maintain the face slope during dumping (2,5). [See Fig. 1(a).]

During the period 1920–1940, many rockfills exceeding 100 ft (30 m) were built. Notable examples of early high dams are the 275-ft (84-m) Dix River Dam in Kentucky and the 330-ft (101-m) Salts Springs Dam in California. Rockfill dams were of the impervious face type until earth core designs began to be developed about 1940.

The dumped rockfill concrete-face dams performed safely, but leakage became an increasing problem as the dams became higher.

## Transition Period (1940–1965): Era of Engineered Rockfill Dam

**Concrete Face Rockfill Dams.**—The extensive use of dumped rockfill in concrete-face rockfill dam construction up to the early 1960's exposed

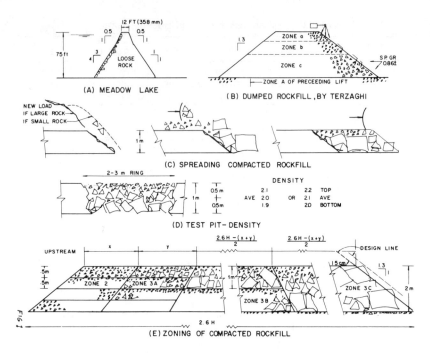

**FIG. 1.—Sketches—Dumped and Compacted Rockfill**

certain limitations and potential problems associated with that method of dam construction for high dams [higher than 300 ft (91 m)].

Availability of suitable rockfill materials had long been accepted as one limitation. Dumped rockfill was widely recognized as requiring rock with high unconfined compressive strength, since stable performance of the fill required high strength at the contacts of rock points and edges. Well-engineered specifications traditionally required use of hard, sound, durable rock of quality that would ordinarily pass ASTM specifications for concrete aggregate. Sluicing by hydraulic monitors, using from one to as many as three volumes of water to one volume of rockfill, at pressures typically of 100 psi (690 kPa), was usually specified to assure clean rock-to-rock contact.

Rockfill compressibility was yet another limitation on the use of dumped rockfill. Typically, dumped rockfill was placed and sluiced in comparatively thick, 60–200 ft (18–60 m) lifts. Even with rock of the best quality, rockfill settlement under reservoir loading was frequently large enough to result in joint movements and cracking of the concrete facing, with subsequent leakage of stored water. The rockfill still retained its high strength, and the permeability of the large, interlocked rocks at or near the base of the dam was demonstrably several orders of magnitude greater than the recorded rates of leakage (Ref. 7, p. 87), as well as greater than the maximum credible leakage, so dam safety was not impaired. However, leakage did in some cases create economic loss (5). Since major rockfill dams constructed during this period were concrete faced, it was important for the proper performance of dams that rockfill settlement somehow be limited in the future to amounts that could be tolerated by the concrete facing.

The Pacific Gas and Electric Company's (PG&E) extensive construction of concrete-face rockfill dams prior to and during this period has afforded the profession a unique opportunity of studying the long-term performance of dumped rockfill dams. Six PG&E rockfill dams, ranging from 75–140 ft (23–43 m) in height, with slopes in the range of 0.5H:1V–1.3H:1V, have now given between 50–105 yr of service (1,2,5). The historical stability of these PG&E dams, with dumped rockfill slopes as steep as 0.5H–0.75H:1V, constitutes a prototype field test demonstrating the existence of greater than 45° friction angles for dumped rockfill at low confining pressures (18).

The excellent performance of these dams probably provided the impetus and the necessary confidence for the construction of higher concrete-face rockfill dams ranging from 150–360 ft (46–110 m) in height during the period 1930–1950. The 150-ft (46-m) Lower Bear River No. 2 Dam, completed in 1952 with 1H:1V upstream slope, has neither experienced leakage nor required maintenance. The 245-ft (75-m) No. 1 Dam, while requiring little maintenance, does leak 4 cfs (110 L/s), which is, nonetheless, less than the required fish release. However, other dumped rockfill dams exceeding 200 ft (61 m) in height developed face damage and leakage problems. Salt Springs, at 328 ft (100 m) and Paradela, at 360 ft (110 m), exhibited serious enough leakage and maintenance problems to lead to a temporary suspension of construction of the concrete-face rockfill dam for more than 300 ft (92 m) in height during the 1960–1965 period.

The last high dam, New Exchequer (1958), during this period combined compacted, 4 and 10 ft (1.2 and 3 m) layer, and dumped rockfill, 60 ft (18 m) lift. However, face problems and leakage occurred due to insufficient use of compacted rockfill and the use of joint design practice that was traditional at that time but is now obsolete.

A transition developed to sluiced rockfill in thinner layers of 10–12 ft (3–3.6 m) subsequently compacted by construction truck and dozer traffic. This modified form of rockfill was specified at Goschenen, Little Grass Valley, Miboro, and other later dams. Goschenen (1) rockfill was derived from talus of competent rock, while Miboro (1) was constructed of granite of varying quality, both dams having rockfill high in percentage of fines (less than #100 mesh). Both dams have performed well.

**Earth Core Rockfill Dams.**—While a detailed discussion of the design of the core, filters, and foundation treatment of earth core rockfill dams is considered outside the scope of this paper, the performance of this type of rockfill dam will be considered in relation to the concrete-face dam. Its introduction during this period was successful in coping with the leakage problem in concrete-face dumped rockfill dams, and a new and successful type of dam was developed.

History shows a logical change, beginning in 1940, from concrete faces to earth cores in dumped rockfill dams, mainly because of leakage experience with the concrete-face dumped rockfill, coupled with the increasing need for higher dams, in which the leakage problem was clearly becoming aggravated. It was beginning to be evident to designers of high dams that the deformation characteristics of dumped rock and concrete had simply not proven compatible. On the contrary, the flexibility of earth cores with filters rendered them capable of accommodating the large settlement of dumped rockfill. In the 1950's, acceleration of knowledge and application of soil mechanics principles, especially with respect to core foundation treatment and filter design, made earth core rockfill dams safer and increasingly popular. The introduction of earth cores provided a new lease on life for dumped rockfill, since the earth core was more deformable and thus compatible with dumped rockfill. Indeed, with the successful completion of the earth core Goschenen Dam in 1960, dumped rockfill in lifts of 10 ft (3 m) had exceeded 500 ft (152 m), a height that was unprecedented at that time.

In the early, inclined core type of dams developed by J. W. Growdon (1), filters and earth core were placed on the inclined upstream slope of dumped rockfill embankments, the rock having been dumped downslope from the full height of the dam. Other early dams had central cores, with dumped rockfill placed and sluiced in 30 ft (9 m) layers. Examples of both types include Kenney and Cherry Valley Dams (1).

The 325-ft (99-m) high Kenney Dam (1), completed in 1952, is one of the notable, sloping core, dumped rockfill dams of the Growdon type. With W. G. Huber as engineer, and K. Terzaghi, J. W. Growdon, I. C. Steele, and C. Dunn on the board, it was a progressive and practical type of dam design for that period. Abutment-core contact treatment consisted of a 3-ft (1-m) thick concrete grout cap and a grout curtain. Four filter zones—three downstream and one upstream—were used, the latter being a sand "crack-stopper" layer suggested by Terzaghi. Riprap was automatically created on the surface of the quarry-run, dumped rockfill. The lift thickness was 30 ft (9 m). The settlement history was satisfactory, with 2.5 ft (0.8 m) total crest settlement in 30 yr (1%) and 0.02 ft/hr (0.6 cm/yr) in the last 15 yr. Crest settlement plots as a smooth curve along the axis, as well as along benches, a favorable characteristic of rockfill, whether dumped or compacted. Unfortunately, from a construction cost standpoint, Kenney Dam was constructed just before the introduction of compacted rockfill, and about 30% of the quarried rock that had to be discarded as being too small for dumped rockfill, would have been acceptable today as excellent material for compacted rockfill.

The 315-ft (96-m) high Cherry Valley Dam (1), completed in 1956, has also experienced no cracks in its central core zone and has given excellent performance. In its 26-yr life (up to 1982), it has settled a total of 1.2 ft (0.4 m), or 0.3% of maximum height, and in the last 15 yr, has settled at a rate of 0.015 ft/yr (0.5 cm/yr). The bulk of the settlement occurred in the thick central decomposed granite core, while for the Kenney Dam, with a sloping core, the greater crest settlement represents the settlement of the dumped rockfill supporting the core.

**Transition from Dumped to Compacted Rockfill.**—The transition to compacted rockfill for both earth core and concrete-face dams occurred during the period 1955–1965. The need for higher dams of both earth core and concrete-face types, the unavailability of high quality rock at many embankment dam sites, and the development and field proving out of heavy, smooth drum, vibratory rollers were among the principal contributory factors to the transition to compacted rockfill.

Since 1960, earth core rockfill dams using compacted rockfill have progressed rapidly in height, in large steps between 500 and 1,000 ft (150 and 300 m). Development of heavy equipment lowered the cost of handling and placing rockfill. The need for siting dams at poorer sites no doubt accelerated development and adoption of various types of embankment dams, but more importantly, the introduction of the smooth drum, vibratory roller also allowed use of lower quality rock in rockfill.

Widespread acceptance of high concrete-face rockfill dams, however, did not come readily. It took time to get over the high leakage experience with high dumped rockfill concrete-face dams. It also took time to gain experience with compacted rockfill dams, and to develop effective and economical toe slab and concrete-face designs.

Terzaghi himself accelerated thinking that was important to the development of the concrete-face rockfill dam in two discussions in the ASCE Symposium on Rockfill Dams, published in the only separate and supplemental Transaction Volume in ASCE history, Volume 104, 1960 (1). Terzaghi (1) described high, dumped rockfill lifts as being segregated into three zones; an upper zone of sluiced small rock, a central zone of medium size rock with some smaller rock in voids, and a lower zone of large rock, this lower zone being the most compressible [Fig. 1(b)]. He did not like the high compressibility characteristic of dumped rockfill and noted the loss of strength of various rock types upon saturation. As a remedy, he suggested compacted rockfill, which he believed would improve the performance of concrete-face dams, permit their adoption for higher dams, and permit use of weaker rock. This discussion was related specifically to the concrete-face rockfill dams.

At about the time of this 1958 ASCE Symposium on rockfills, fundamental changes in rockfill specifications became widespread. The general specifications of 60–200 ft (18–61 m) lifts, had changed to 10 ft (3 m) on some dams. At Ambuklao in 1955, most of the dumped rock was changed to 2 ft (0.6 m) layer rockfill due to the low strength and small size of some of the available rock (6). The same change was made at Brownlee (1958) (1) due to coarse, filter size basalt. Both began as dumped rockfill, earth core dams, and were completed with both dumped and compacted rockfill. Growdon's classic sloping core, dumped rockfill dam type (Nantahala in 1940 and others), evolved into a zoned type dam, of compacted and dumped rockfill, a typical example being at Lewis Smith, in 1960. Although two major dumped rockfill dams—the thinner-lift Goschenen Dam and the high lift downstream shell of Exchequer Dam— were completed as dumped rockfills in the early 1960's, the transition from dumped to compacted rockfill was virtually complete by 1965.

## Modern Period (1965–1982): Widespread Use of Compacted Rockfill Dams of Earth Core and Concrete Face Type

**General.**—In the ASCE Rockfill Dam Symposium of 1960 (1) and in the Eighth Congress on Large Dams in Edinburgh in 1964 (3,4), a rockfill dam was still loosely defined as "a dam that relies on rock, either dumped in lifts or compacted in layers, as a major structural element." The definition includes earth core and concrete-face rockfill dams, as well as other impervious membrane dams. Though dumped rockfill is still effectively used in outer zones of earth core rockfill dams, in the downstream toes of embankment dams, in cofferdams, and in underwater placement, compaction has become so universal in rockfill dams that a more contemporary definition of a rockfill dam would be "a dam that relies on compacted rockfill as the main structural element," compacted rockfill being considered as rockfill placed in layers up to 6.6 ft (2 m) and compacted by smooth drum vibratory roller.

The historical development of compacted rockfill dams, prior to 1964, has been well-documented (3,4,6), and only a few pertinent aspects will be discussed. The transition from dumped to compacted rockfill was rapid and inevitable. There was an increasing need for improved and higher concrete-face rockfill dams, but the traditional requirement for high strength and large size rock for dumped rockfill dams had limited this type of rockfill dam to sites where such rock was available. The increasing need to utilize dam sites with foundation conditions that were neither suitable for the construction of concrete dams, nor adaptable for earth dams, and the need to use locally available weak rock at many of these sites all contributed to the development of compacted rockfill. The advent of the vibratory roller facilitated the successful and economic use of weak rock, compacted in layers. With this equipment, high strength and low compressibility of rockfill composed of weak rocks are achieved by compacting the rockfill in relatively thin layers, often supplemented by application of water, to achieve high density. By contrast, the strength and low compressibility for rockfill composed of strong rock are achieved by the rock strength, and the wedging effect caused by vibratory compaction. The wedging factor is of particular importance to the concrete-face rockfill dam, as will be examined in the following paragraphs.

One further advantage of compacted rockfill over dumped rockfill is also worth mentioning here. An advantageous feature of rockfill dams is their ability to withstand passage of flood water through and over the uncompleted dam. For probable flow over the rockfill, reinforced rockfill is necessary (Ref. 7, p. 87). There has been experience with two major washouts of dumped rockfill during construction and one well-known case of overtopping failure after completion. At Hell Hole Dam, which is 200 ft (61 m) high, dumped rockfill lifts failed when the flow through the rockfill exceeded 15,000 cfs (430 $m^3/s$) (Ref. 7, p. 87). The sloping core for the dam had not yet been placed, and an early storm prematurely caused unplanned storage. The permeability in the voids of the large segregated rockfill at the base of the rockfill was great. The dam was later completed to its full 410 ft (125 m) height. A similar event for compacted rather than dumped rockfill would have had less leakage and would probably not have failed.

**Characteristics of Compacted Rockfill.**—Figs. 1($c$–$d$) show the placement and resulting character of compacted hard rockfill, shown for a 3 ft (1 m) layer. The resulting rockfill is not homogeneous, but segregated, which results in the development of desirable properties in respect to density, strength, and permeability.

Fig. 1($c$) shows dumping of the layer being placed. The dozer, with raised blade, pushes large rocks ahead where the full thickness of lift can accept them. Smaller rocks fall under the blade, and when spread, provide a relatively smooth surface for compaction and construction traffic. The vibratory roller is effective in compacting the smaller rocks and fines

A. KANGAROO CREEK  B. TALBINGO
C. DARTMOUTH  D. CHIVOR

**FIG. 2.—Examples of Compacted Rockfill Layers (Note: Chivor was Renamed Esmeralda)**

in the upper zone and in tight wedging of the larger rocks in the lower zone. The wedging action caused by the vibratory roller is particularly effective in developing horizontal rock-to-rock contact. The smooth fill surface is desirable for density, strength, and low vertical permeability. It should never be scarified. The horizontally-oriented and well-compacted surface of surficial rocks is desirable.

A layer of rockfill, particularly of the type composed of high strength rock particles, typically appears as shown in Fig. 1(d). The difference in density of top half and bottom half is about 10–12 lb/cu ft (160–192 kg/m$^3$). Note that the upper half of each layer is of much smaller size material than the lower half. Horizontal permeability of these layers is much greater than vertical permeability.

Fig. 1(e) shows the conventional zoning of a concrete-face rockfill dam. It indicates increasing permeability principally in the thicker bottom areas of each of four zones, and for the full width of the dam, as well as the desirable perched water table effected by the fine and semi-pervious surface of each layer. The perched water tables, assuming a source of water, avoid a high phreatic line with its consequent high pore pressures.

The four photos in Fig. 2 show end views of rockfill placed in one meter layers. Three show rock sizes ranging up to full layer thickness. It is evident that a single rock of full layer thickness will not permit maximum compaction of immediately adjacent rockfill; however, it also happens that the single rock of full layer thickness is incompressible and will "attract" vertical load when the layer is loaded, so undue settlement will not occur. Thus, it is not necessary to specify a maximum rock size of less than layer thickness. Further, field inspection of rock size is not practicable.

Fig. 2(a) shows use of poor quality Schist rock at Kangaroo Creek (8) where water was liberally used. The surface became muddy during construction, but large rocks remained as rocks within the layer. Fig. 2(b) shows the typical embedment of a large rock and a fine rockfill surface in the 3.3 ft (1 m) layer of rhyolite at Talbingo Dam. Fig. 2(c) shows the many large rocks at Dartmouth Dam and the typical smooth rock surface. Fig. 2(d) shows various sizes of rock in a 3.3 ft (1 m) layer at Chivor Dam, including a 3.3 ft (1 m) rock and view of nearby impervious surface holding water. These photos show the nature of compacted rockfill as sketched in Fig. 1. All these dams have given excellent performance.

**Use of Water in Compacted Rockfill.**—Use of water during fill placement (on the order of 20% volume of rock, far less than for dumped rockfill) is beneficial for any quality of rock, but is especially desirable for types of rocks that lose strength upon saturation, as pointed out by Terzaghi (1). However, it is sometimes not feasible to add much water due to the turbidity of excess water draining from the fill, especially where downstream water uses or fish resources place limits on turbid runoff. In such cases, the weak rock may be placed in thinner layers, about 2 ft (0.6 m), or even less, instead of 3.3 ft (1 m), and compaction coverage increased to 6 instead of the standard 4 compaction passes, to achieve satisfactory density. Strength and low compressibility of rockfill composed of weak rocks are attained by high density, whereas for hard rock density is lower, and its strength is from tight interlocking of the competent rocks. Rockfill of some weak rocks requires provisions for drainage (8,9).

**Rock Quality in Compacted Rockfill.**—There is no rigid specification for rock quality or gradation in compacted rockfill. Rock that fails to meet concrete aggregate specification tests is still acceptable rock for use as rockfill when properly handled. Soft rock, such as siltstone (9), some sandstones (9), schists (8), argillite, and other potentially weak rocks can be used, as demonstrated by successful past experience.

Indeed, if blasted rockfill is strong enough to support construction trucks and the 10-ton vibratory roller when wetted, it may be considered to be suitable for use in compacted rockfill. In the case of compacted rockfills made from weak rock types, zones of pervious rockfill often have to be incorporated in the dam design to provide for internal drainage if it becomes necessary following breakdown of the weak rock during compaction. However, the added cost of the internal drainage provisions is more than offset by savings made in handling weak rock compared with harder rock.

It is, of course, necessary to know the rock source before writing a specification for any given job. If the rock is hard, a satisfactory general specification is "quarry-run rock—the maximum size shall be that which can be incorporated in the layer and provides a relatively smooth surface for compaction, not more than 50% shall pass a 1 in. (2.5 cm) sieve, and not more than 6% shall be clay-sized fines."

The modulus of compressibility of rockfill, as determined within the prototype dam by water level settlement gages, or cross arms, is a useful measure of rockfill quality. Experience on many projects indicates moduli ranging from 3,000–20,000 psi (21–138 MPa) depending on the rock, the rock grading, layer thickness, compaction, and other factors. Rockfill modulus is a useful parameter for both earth core and concrete-face types of rockfill dams; it gives a basis for comparing performance of existing dams and establishing design and placement procedures for new dams. Of significance to higher dams is that the modulus of compressibility of compacted rockfill increases with pressure (10,20). However, it should be noted that the frictional shear strength decreases progressively and moderately with increasing pressure.

Moduli, as determined by water level settlement devices, of as high as 22,500 psi (150 MPa) has been attained by Murchison Dam for well-graded rhyolite, use of water and 8 passes of the 10-ton vibratory roller on a 3.3 ft (1 m) layer (27).

**Use of Gravel as Compacted Rockfill.**—Natural gravel falls outside of the usual definition of rockfill but can be considered to be similarly useful as a dam shell material. It is more economically handled. It can be used in combination with rockfill. Its modulus of compressibility is 5–10 times higher than that of compacted rockfill, and about 40 times that of dumped rockfill. Because face slab movements of rockfill dams increase as the square of the dam height, gravel shell concrete face dams can safely be built to much greater height than the present highest dam, Areia, which is 525 ft (160 m) high.

Typical vertical compression moduli of gravel, based on field measurements during construction, are 53,000 psi (365 MPa) for Oroville, and 80,000–100,000 psi (551–689 MPa) for W.A.C. Bennett Dam. Oroville was compacted by a 3.5-ton, static weight, smooth drum, vibratory roller and Bennett by a 6-ton vibratory roller. At the time of their construction those were the available sizes.

Layer thicknesses specified for compacted gravel vary between 1–3 ft (0.3–0.9 m) depending on height of dam, size of gravel, and percentage minus no. 200 mesh material. Usually no water is added. In fact, for dirty gravel (7–12% minus no. 200 mesh) excessive moisture causes a springy fill, though acceptable density can still be achieved. With dirty gravel, permeability assumes importance in the design. Chimney drains, abutment drains, filters, and intermediate drainage layers, may all have to be considered.

**Modern Earth Core Compacted Rockfill Dams.**—As stated elsewhere, the advent of compacted rockfill has encouraged the design and construction of higher rockfill dams of both the concrete-face and earth core types. The two highest earth core rockfill dams at this time (1982) are Esmeralda in Colombia at 777 ft (235 m), completed in 1975, and Chicoasen in Mexico at 856 ft (261 m), completed in 1980. These are heights above the low point in the foundation, both dams having approximately 180 ft (55 m) of riverbed excavation. The experience with Chivor and Chicoasen has been excellent. It is noted that Nurek, at 1,000 ft (305 m), Oroville and Mica at 800 ft (244 m), and Bennett at 600 ft (183 m), are all gravel shell dams. The experience with these high earth core dams has been excellent and the rockfill and gravel experience can be transferred to concrete-face dams.

Esmeralda Dam (13,14), Chivor Project, Colombia (Fig. 3) is a moderately sloping, earth core rockfill completed in 1975. Its performance

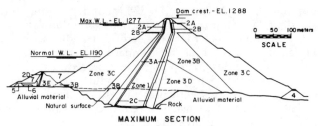

**FIG. 3.—Esmeralda Dam—Section and Zoning**

has also been excellent. Crest settlement on first filling was 3.3 ft (1 m), but no cracks were detected in the core, and abutment gallery leakage was only 3 cfs (90 L/s). The rockfill was necessarily of a variety of types, including limestone, quartzite, phyllite, slate, argillite, and some gravel. The design of the 840 ft (256 m) Guavio Dam, also in Colombia, currently (as of 1982) under construction is similar to that of the Esmeralda Dam.

Chicoasen Dam (14,15,16) has also given excellent performance. The use of an outer zone of dumped rockfill at Chicoasen is a desirable, although infrequently used feature.

**Modern Concrete Face Rockfill Dams.**—Analysis of the excellent field performance of high, earth core, compacted rockfill dams provides not only useful data on the behavior of rockfill under high pressures, but has also restored the confidence of dam designers in concrete-face rockfill dams, since most of the desirable features of the rockfill in the high earth core type was directly utilized in the development of higher concrete-face rockfill dams.

Table 2, which gives design data on large dams constructed or planned in the past 45 yr, shows a definite change to the use of compacted rockfill in the past 15 yr. The design trends for dams in this period, is also evident. These trends include thinner face slabs, less reinforcing, improved joint design, higher dams, and more frequent use.

Prototype measurements demonstrating the low compressibility of compacted rockfill have provided the data base for the design of such high, concrete-face rockfill dams as Cethana (19,20), Anchicaya (10), and Areia (21). Each of these dams has, in turn, contributed to the state-of-the-art of rockfill dam design through improved performance measurements. Murchison Dam (27) has subsequently provided valuable performance data, and its performance is excellent.

**Exceptionally Cold Weather Sites.**—With the emphasis placed, in recent years, on worldwide natural resource development, it seems inevitable that more rockfill dams will be constructed in the future in areas of the world that experience extremes of climates. In this connection, it is pertinent to make a few remarks about the performance of concrete-face rockfill dams already constructed in these areas.

Four concrete-face dams have demonstrated the feasibility of rockfill dams in cold climates and under severe freeze-thaw conditions. These dams were Outardes 2 in Northern Canada, Cabin Creek at elevation 12,000 ft (3,660 m) in the Rockies, Courtright at elevation 8,200 ft (2,500 m) in the California Sierras, and Golillas at elevation 10,000 ft (3,000 m) in the Andes, where temperatures reach minus 20° C. The coldest temperatures are at Outardes 2, where the winter range is minus 25–35° C, and the summer day range is 20–25° C. Outardes 2 was constructed, in part, in cold weather, and in 3 yr, the 180-ft (55-m) high dam has settled only 0.015 ft (0.5 cm). No cracks have occurred, and there is no leakage. The concrete faces at Courtright and at Cabin Creek show no temperature cracks or freeze-thaw damage after their respective service lives of 26 and 17 yr. No ice-plucking or ice thrust damage has been reported. Courtright has had face damage and leakage problems as a 300 ft (98 m) dumped rockfill dam.

From this data base on the performance of compacted rockfill dams in areas subject to extremes of weather, it may be concluded that rockfill dam design criteria do not need any major modifications to accommodate these extreme weather conditions. The measures to prevent freeze-thaw damage, air entrainment and pozzolan in the concrete, have proved to be successful.

**Seismic Resistance of Compacted Rockfill.**—Compacted rockfill dams have an inherently high resistance to seismic loading (22,25). According to Seed (25), "Two rockfill dams have withstood moderately strong shaking with no significant damage, and if the rockfill is kept dry by means of a concrete facing they should be able to withstand extremely strong shaking with only small deformations." The two dams involved were the 420-ft (128-m) high Miboro earth core rockfill, placed in 10–13 ft (3–4 m) layers, and the 275-ft (84-m) high Cogoti concrete face rockfill. More recently, in 1979, Infiernillo Dam, 470-ft (145-m) high, developed crest accelerations on the order of $0.35\ g$ with 5 in. (13 cm) maximum crest settlement, and with only nominal longitudinal and transverse cracking at the crest. All movement was in the top 26 ft (8 m) of the dam (17). Rockfill compaction of Infiernillo was by 4 passes of a D8 on a 3 ft (0.9 m) layer, a procedure that was customary just before the days of the vibratory roller.

For the 330 ft (110 m) Fortuna Dam, a dynamic seismic analysis was performed for 1.3H:1V upstream and 1.4H:1V downstream slope, for $0.4\ g$ (ground acceleration), using a number of recorded major earthquake accelerograms. The results indicated a loosening of the crest area rockfill that could result in several feet of settlement, but no movement of the face slab below water level was indicated.

At Cethana, Anchicaya, Areia (21), and Murchison (27), it has been

TABLE 2.—Partial List of CFRD Dams

| Name (1) | Height, in meters (2) | Location (3) | Year completed (4) | Slopes US (5) | Slopes DS (6) |
|---|---|---|---|---|---|
| Morena | 54 | California | 1895 | 0.5–0.9 | 1.3 |
| Strawberry | 50 | California | 1916 | 1.2 | 1.3 |
| Dix River | 84 | Kentucky | 1925 | 1.0–1.2 | 1.4 |
| Salt Springs | 100 | California | 1931 | 1.1–1.4 | 1.4 |
| Cogswell | 85 | California | 1934 | 1.35 | 1.6 |
| Malpaso | 78 | Peru | 1936 | 0.5 | 1.33 |
| Cogoti | 75 | Chile | 1939 | 1.6 | 1.8 |
| Lower Bear Number 1 | 71 | California | 1952 | 1.3 | 1.4 |
| Lower Bear Number 2 | 50 | California | 1952 | 1.0 | 1.4 |
| Paradela | 112 | Portugal | 1955 | 1.3 | 1.3 |
| La Joie | 87 | Canada | 1955 | 1.1 | 1.5 |
| Pinzanes | 67 | Mexico | 1956 | 1.2 | 1.3 |
| Courtright | 98 | California | 1958 | 1.0–1.3 | 1.3 |
| Wishon | 82 | California | 1958 | 1.0–1.3 | 1.4 |
| San Idelfonso | 62 | Mexico | 1959 | 1.4 | 1.4 |
| New Exchequer | 150 | California | 1966 | 1.4 | 1.4 |
| Cabin Creek | 76 | Colorado | 1967 | 1.3 | 1.3 |
| Fades | 70 | France | 1967 | 1.3 | 1.3 |
| Rama | 110 | Yugoslavia | 1967 | 1.3 | 1.3 |
| Kangaroo Creek | 59 | Australia | 1968 | 1.3 | 1.4 |
| Pindari | 45 | Australia | 1969 | 1.3 | 1.3 |
| Pindari Raised | 75 | Australia | P | 1.3 | 1.3 |
| Cethana | 110 | Australia | 1971 | 1.3 | 1.3 |
| Alto Anchicaya | 140 | Colombia | 1974 | 1.4 | 1.4 |
| Le Rouchain | 60 | France | 1976 | 1.4 | 1.4 |
| Little Para | 54 | Australia | 1977 | 1.3 | 1.4 |
| Golillas (Chuza) | 130 | Colombia | 1978 | 1.6 | 1.6 |
| Outardes 2 | 55 | Canada | 1978 | 1.4 | 1.4 |
| Winneke (Sugarloaf) | 85 | Australia | 1979 | 1.5 | 2.2 |
| Bailey, R.D. | 95 | W.Va., USA | 1979 | 2.0 | 2.0 |
| Areia | 160 | Brazil | 1980 | 1.4 | 1.4 |
| Neveri-Turimiquire | 115 | Venezuela | 1981 | 1.4 | 1.5 |
| Mackintosh | 75 | Australia | 1981 | 1.3 | 1.3 |

## over 50 Meters High with Design Data

| Fill-rock type (7) | Face area, in square meters (8) | Slab equation $m + CH$ (9) | Reinforced each way, as a percentage (10) | Toe slab width, in meters (11) | Zone 1, in meters (horizontal) (12) |
|---|---|---|---|---|---|
| DR-Granite | | 0.23 + 0.003H | 0.5 | Trench | |
| DR-Granite | | 0.23 + 0.003H | 0.5 | Trench | |
| DR-Limestone | | | 0.5 | Trench | |
| DR-Granite | 10,900 | 0.3 + 0.0067H | 0.5 | Trench | |
| DR | | | | | |
| PR & DR | | | | | |
| DR-Gravel | | | | | |
| DR-Granite | 5,800 | 0.3 + 0.0067H | 0.5 | Trench | |
| | 2,800 | 0.3 + 0.0067H | 0.5 | Trench | |
| DR-Granite | 55,000 | 0.3 + 0.00735H | 0.5 | Trench | |
| DR | | Shotcrete | | Trench | |
| DR | | | | | |
| DR-Granite | 6,700 | 0.3 + 0.0067H | 0.5 | Trench | |
| DR-Granite | 20,000 | 0.3 + 0.0067H | 0.5 | Trench | |
| CR-DR | | | | | |
| Main DR-Meta Andesite Supporting Zone-CR | | 0.3 + 0.0067H | 0.5 | | |
| CR | | | 0.5 | | |
| CR-Granite | 16,500 | 0.35 + 0.0042H | 0.5 | 4.0 | |
| CR | | | | | |
| CR-Schist | 8,000 | 0.3 + 0.005H | 0.5 | 3.7 | 3.6 |
| CR-Rhyolite | 16,400 | 0.48 + 0.002H | 0.81 | 3.0 minimum (2.6 × 0.085H) | 4.5 |
| CR-Rhyolite | | | | | |
| CR-Quartzite | 23,700 | 0.3 + 0.002H | 0.5$^a$ | 3–5.36 | 3 + 3 |
| CR-Hornfeld | 22,300 | 0.3 + 0.003H | 1 + 0.5 | 7.0 | Varies-10 (toe) |
| CR-Granite | 16,000 | 0.35 + 0.0042H | 0.5 | 4.0 | |
| CR-Shaley Dolomite | 10,200 | 0.3 + 0.0029H | 0.5 | 4.0 | 4.0 |
| CG-Gravel | 14,300 | 0.3 + 0.0037H | 0.4 | 3.0 | 4.0 |
| CR-Gneiss | 8,375 | 0.3 | 0.45 | 3.05 | |
| CR-Sandstone | 82,500 | 0.3 + 0.002H | 0.5 | 0.1H minimum 6 m | 5.0 |
| CR-Sandstone | 65,000 | 0.3 + 0.003H | 0.5 | 3.05 + 0.019H | 4.3 |
| CR-Basalt | 139,000 | 0.3 + 0.0034H | 0.4 | 4, 5.5, 7.5 | 5.0–7.10 |
| CR-Limestone | 53,000 | 0.3 + 0.002H | 0.5 | 3.5–7.50 | 5.0 |
| CR-Greywacke | 27,100 | 0.25 | 0.5$^a$ | 3.0–3.86 | 3 + 3 |

TABLE 2.—

| (1) | (2) | (3) | (4) | (5) | (6) |
|---|---|---|---|---|---|
| Mangrove Creek | 80 | Australia | 1981 | 1.5 | 1.6 |
| Mangrove Creek Raised | 105 | Australia | P | 1.5 | 1.6 |
| Shiroro | 130 | Nigeria | 1982 | 1.3 | 1.3 |
| Yacambu | 150 | Venezuela | 1982 | 1.5 | 1.5 |
| Murchison | 89 | Australia | 1982 | 1.3 | 1.3 |
| Awonga | 47 | Australia | 1982 | 1.3 | 1.3 |
| Awonga, Raised | 63 | Australia | P | | |
| Fortuna | 65 | Panama | 1982 | 1.3 | 1.4 |
| Fortuna, Raised | 105 | Panama | 1983 | 1.3 | 1.4 |
| Glennies Creek | 67 | Australia | 1983 | 1.3 | 1.3 |
| Salvajina | 145 | Colombia | 1983 | 1.5 | 1.5 |
| Bastayan | 75 | Australia | 1983 | 1.3 | 1.3 |
| Boondooma, Stage I | 63 | Australia | 1983 | 1.3 | 1.3 |
| Boondooma, Stage II | 73 | Australia | P | | |
| Khao Laem | 105 | Thailand | 1984 | 1.4 | 1.4 |
| Bejar | 71 | Spain | 1984 | 1.3 | 1.3 |
| Terror Lake | 58 | Alaska | 1985 | 1.5 | 1.4 |
| Alsasua | 50 | Spain | 1985 | 1.3 | 1.4 |
| Kotmale | 97 | Sri Lanka | 1985 | 1.4 | 1.45 |
| Batang A1 (Main Dam) | 85 | Sarawak | 1985 | 1.4 | 1.4 |
| Batang A1 (Lima Saddle) | 60 | Sarawak | 1985 | 1.4 | 1.4 |
| Batang A1 (Bekatan Saddle) | 70 | Sarawak | 1985 | 1.4 | 1.4 |
| Lower Pieman | 122 | Australia | 1986 | 1.3 | 1.3–1.5 |
| Iruru | 50 | Peru | 1986 | | |
| Ita | 123 | Brazil | UD | 1.3 | 1.3 |
| Cirita | 85 | Indonesia | UD | 1.3 | 1.4 |
| Segredo | 145 | Brazil | UD | 1.3 | 1.3 |
| Acena | 65 | Spain | UD | 1.3 | 1.3 |
| Kaliwa | 100 | Philippines | UD | | |
| Machadinho | 124 | Brazil | UD | 1.3 | 1.3 |
| Split-Rock | 67 | Australia | UD | 1.3 | 1.3 |
| La Miel | 180 | Colombia | UD | 1.5 | 1.5 |
| Dinkey Creek | 115 | California | P | | |

*Continued*

| (7) | (8) | (9) | (10) | (11) | (12) |
|---|---|---|---|---|---|
| CR-Siltstone and Sandstone | 29,100 | 0.375 + 0.003H | 0.35 | 3, 4, 5 | 4.0 |
| CR-Siltstone and Sandstone | 33,400 | 0.3 + 0.003H | 0.35 | 3, 4, 5 | 4.0 |
| CR | 150,000 | 0.3 + 0.003H | 0.4 | 6.0 | 7.0 |
| CG-Gravel | 13,000 | | 0.4 | | |
| CR-Rhyolite | 16,200 | 0.3 | 0.5[a] | 3–4.6 | 1 + 5 |
| CR-Meta Sediments (Siltstone-Sandstone) | 30,000 | 0.3 + 0.002H | 0.55[b] | Existing concrete dam | 3.0 |
| CR-Andesite | 22,000 | 0.411 + 0.003H | 0.5 | 4.0 | 5 + 0.02H |
| CR-Andesite | | 0.411 + 0.003H | 0.5 | 4.0 | 5 + 0.02H |
| CR-Welded Tuff | 24,500 | 0.3 | 0.43 | 3–4 | 4.0 |
| CR-Dredger Tailings | 50,000 | 0.3 + 0.0031H | 0.4 | 4.0–8.0 | 5.0 |
| CR-Rhyolite | 18,600 | 0.25 | 0.5[a] | 3–3.8 | 3 + 3 |
| CR-Rhyolite | 25,000 | 0.3 | 0.4 | 3.5–5.5 | 3.5 |
| CR-Limestone Karstic | 140,000 | 0.3 + 0.003H | 0.5 | 4.6 (Gallery) | 3.5 |
| CR-Granite | 19,140 | 0.35 + 0.003H | 0.4 | 3-H/15 | 4 |
| CR-Greywacke Argillite | | 0.3 + 0.003H | 0.4 | | |
| CR-Limestone | 13,850 | 0.3 + 0.003H | 0.4 | 4.5 (Gallery) | 3 |
| CR-Charnockite | 60,000 | 0.3 + 0.002H | 0.5[a] | 3–8 | 3 + 3 |
| CR-Dolerite | 65,000 | 0.3 | 0.5 | 4.6 | 3.5 |
| CR-Dolerite | 15,000 | 0.3 | 0.5 | 4.6 | 3.5 |
| CR-Dolerite | 42,500 | 0.3 | 0.5 | 4.6 | 3.5 |
| CR-Dolerite | 35,000 | 0.3 + 0.001H | 0.5[a] | 3–9.0 | 3 + 3 |
| Basalt | 110,000 | 0.3 + 0.00334H | 0.4 | | 3.0–6.0 |
| CR-Breccia, Andesite | | 0.3 + 0.003H | 0.4 | 4, 5, and 7 | 5–9 |
| CR-Basalt | | | | | |
| CR-Gneiss | 21,800 | 0.3 + 0.003H | 0.4 | 3-H/15 | 3 |
| Basalt | 100,000 | 0.3 + 0.00334H | 0.4 | | 3.0–6.0 |
| CR-Greywacke Gravel Breccia, Siltstone | | 0.3 | 0.35 | | 4.0 |
| CR-Diorite | | 0.3 + 0.003H | | | |

**TABLE 2.—**

| (1) | (2) | (3) | (4) | (5) | (6) |
|---|---|---|---|---|---|
| Musa | 150 | New Guinea | P | 1.5 | 1.5 |
| Poza de los Ramos | 98 | Spain | P | 1.3 | 1.5 |
| Poza de los Ramos (Second Stage) | 136 | Spain | P | 1.3 | 1.5 |
| Uruguay | 73 | Argentina | P | 1.33 | 1.3 |
| Piray-Guazu II | 73 | Argentina | P | 1.3 | 1.3 |
| Cuesta Blanca | 80 | Argentina | P | 1.4 | 1.5 |

[a]Percentage reinforced on design area + 10 cm.
[b]Percentage reinforced on design area + 12 cm.
Note: P = proposed; UD = under design; US = upstream slope; DS = down-
= compacted gravel; and PR = placed rock.

determined that the modulus of rockfill compressibility, calculated from face movement upon reservoir filling, is about 3 times that measured vertically within the rockfill by water level settlement gages, i.e., the face movement was one-third of that calculated using the vertical modulus. Thus, the rockfill appears to be substantially stiffer in the horizontal direction than in the vertical direction. This suggests a higher threshold of shear deformation and lower strains than are used in dynamic seismic calculations. Another factor not taken into account in many current analyses is the different properties in the upper and lower half of each rockfill layer. Both of these factors improve seismic resistance of compacted rockfill.

Concrete-face rockfill dam zoning will safely store high flood waters before the concrete face is placed and, consequently, if the face were damaged. All of the high strength and zoned rockfill is downstream from the reservoir water. A well-designed concrete-face rockfill dam can be considered to offer a high resistance to severe seismic loading. Damage is acceptable for a maximum credible earthquake. It is breaching that must be avoided.

**Pumped Storage Cyclic Operation.**—The loading cycle of the upper reservoir of pumped storage projects has provided some interesting data on cyclic resistance of rockfills. For the upper reservoir of pumped storage projects, there was an initial question of the effect of repeated large daily cycles of reservoir level on settlement of concrete-face rockfill. The 100–140 ft (31–43 m) high Taum Sauk Dam constructed of dumped rhyolite rockfill, completed in 1963, with daily operation range of 80 ft (24 m) for the 140 ft (43 m) section, settled 1.0, 0.31, and 0.13 ft (0.3, 0.1, and 0.04 m) in the first, second, and third 5-yr periods. While this is more than the normal annual cyclic settlement performance for dumped rockfill, it is nevertheless within tolerable limits, since it has not adversely affected performance. Cabin Creek, a compacted rockfill completed in 1967, having a 250 ft (76 m) height, has a daily operating cycle of 90 ft (27 m). Settlement after 10 yr was 0.4 ft (0.12 m), 0.02% of its height, while leakage has never exceeded 0.9 cfs (28 L/s). These data suggest that daily cyclic loading does not affect the performance of concrete-face rockfill dams.

*Continued*

| (7) | (8) | (9) | (10) | (11) | (12) |
|---|---|---|---|---|---|
| CG-Gravel | | | | | |
| CR | | | | | |
| | | | | | |
| CR | | | | | |
| CR-Basalt | 38,400 | 0.3 + 0.005H | 0.5 | 3–5 | Variable 3.48–5.5 |
| CR-Basalt | 38,000 | 0.3 + 0.005H | 0.5 | 3–4.0 | 3.60 |
| CR-Gneiss | 50,000 | 0.3 + 0.003H | 0.5 | 3.5–4.5 | 3.50 |

stream slope; DR = dumped rockfill; CR = compacted rockfill; H = height; CG

## CURRENT CONCRETE-FACED ROCKFILL DAM PRACTICE

**Design Practice.**—The foregoing description makes it evident that the design of a concrete-face rockfill dam is essentially empirical and based on experience and judgment (22,23). Table 2 lists dams over 165 ft (50 m) in height in chronological order, with some physical and design data. The sudden change from dumped to compacted rock in about 1962 has already been noted, and the adoption of compacted rock at that time is also associated with abandonment of toe trenches in favor of toe slabs, increasing use of lower strength rock types, larger face areas, thinner

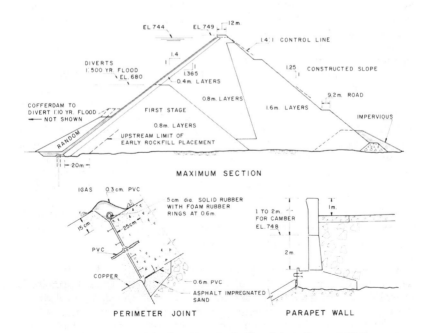

**FIG. 4.—Areia Dam Section—Perimeter Joint—Parapet Wall**

A. ROCKFILL PLACEMENT - LOOKING DOWNSTREAM

B. ROCKFILL PLACEMENT - LOOKING UPSTREAM

C. PLACEMENT OF FACE ZONE CRUSHED ROCK

D. VIBRATING FACE CRUSHED ROCK

FIG. 5.—Areia Dam Photos of Rockfill Construction

face slabs, and lower percent reinforcing of facing. The table does not show the elimination of horizontal joints and compressible vertical joints that was a feature introduced at Cethana, and has remained a design practice since then.

The 525-ft (160-m) high Areia Dam (Fig. 4) shows current design practice for a high concrete-face rockfill dam.

The photos in Figs. 5–6 show the following features of modern rockfill dam construction: (1) The high rockfill on the right abutment was placed before river diversion, and, in the foreground, rockfill is rising to the 1:500 yr flood level while toe slab placing and grouting continue; (2) the 1:500 yr level has been reached, and rockfill is being placed downstream; (3) placement of the zone under the concrete face; (4) vibration of the face; (5) sand is sprayed over emulsified asphalt to provide a firm and nonerodible surface, and toe slab, face form, and reinforcement proceed; (6) starter slabs are screeded by hand methods, in preparation for full-slab-width, mechanized screed; (7) first stage concrete slab is completed, and earth material placement covering slab is underway; and (8) completed dam, with aerated spillway in operation. These and other aspects of the construction of compacted rockfill dams will be examined later.

Fig. 7 shows longitudinal profiles of selected high rockfill dams. The dam shape does not appear to be critical, since the water load principally causes only perimeter joint opening, and most of the slab just floats on the rockfill. For steep abutments, however, special attention needs to be paid to perimeter joint design, to accommodate greater offset movements, as, for example, at Chuza and Yacambu.

E. TOE SLAB AND COMPACTED ROCKFILL FACE

F. STARTER SLABS AND EARLY FACE CONCRETE

G. COMPLETED FIRST STAGE CONCRETE

H. AERIAL VIEW OF PROJECT

**FIG. 6.—Areia Dam Photos of Concrete-Face Construction**

**Toe Slab Foundations.**—The most desirable foundation for a toe slab is hard, nonerodible rock that is groutable. However, faults, weather-stained rock with clay filled seams, and even rock containing zones susceptible to possible erosion and piping are acceptable with proper engineering. Among corrective methods that may be adopted are: excavation of erodible material and use of concrete backfill; widening the toe slab upstream or downstream locally; adding rows to the grout curtain; and use of a diaphragm wall or concrete backfilled backhoe trench or downstream filters, or both. Placement of upstream core material on the face and toe slab is often used to provide a supplementary impervious

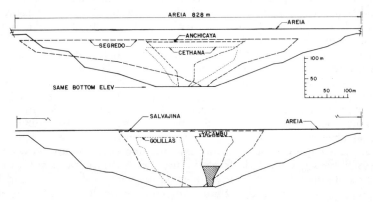

**FIG. 7.—Longitudinal Profiles of Some Compacted Concrete-Face Rockfill Dams**

zone or crack or joint sealing material. Where earth is used upstream, it is placed on a foundation prepared in the same way as for an earth core. Most concrete-face dam sites justify some local toe slab foundation treatment.

Design and construction excavation practice for the toe slab consists of excavating a trench to the nominal trench depth that exploration has indicated will provide a reasonably competent foundation. Local imperfections are then taken care of by one or more of the preceding special measures. Resloping or additional excavation is seldom necessary, except perhaps locally.

The toe slab is dowelled to well-cleaned rock prior to grouting and is placed in 20–26 ft (6–8 m) lengths. The width of toe slab is determined by judgment, and varies with quality of rock and height of dam (see Table 2). Reinforcing is positioned in one layer near the top, and design slab thickness is at least that of face slab thickness. The toe slab construction joints do not have to coincide with face slab joints, but rather with the natural or excavated rock topography. The face slab pulls away from the toe slab under movement of the rockfill when the reservoir is filled, and all the perimeter joints open. In this way, the face becomes structurally independent of the toe slab. The width of toe slab is much less than the usual core/foundation contact width for earth core dams, since the toe slab foundation is usually either nonerodible or is fully protected against piping by the aforementioned special measures, and the concrete is nonerodible.

**Section and Stability.**—Since no rockfill dam has ever failed because of inadequate stability, it is hardly possible to recommend a verified realistic method of stability analysis, though wedge and circle analyses have been used. With all the rockfill being downstream from the water loading, the ratio of vertical weight to horizontal force is greater than 6:1 for 1.3H:1V slopes. The major proportion of water load goes into the foundation upstream from the dam axis. Neither sliding nor uplift is a problem. Design of slopes has therefore been based on a traditional 1.3H:1V precedent that evolved for hard rock rockfill and hard rock foundations, with occasional variations. The historical 1.3H:1V slopes were altered to 1.4H:1V by judgment for the 525 ft (160 m) Areia Dam (Fig. 4) out of respect for its unprecedented height, and by the knowledge that rockfill shear strength under the higher internal pressure is lower. For Areia, the downstream rock slopes are actually 1.25H:1V between haul roads on the face, 1.4:1 being an average control slope. For weaker rockfill or less competent foundations, flatter slopes have been used, and have been based on stability analysis. For example, at Cabin Creek, fill downstream from the axis is compacted earth over a rockfill drainage zone, and the downstream slope is 1.75H:1V.

Design variables in compacted rockfill, for different rocks and dam heights, include varying layer thickness, use of water, and number of passes of vibratory roller. Zoning on the main rockfill is established so as to meet requirements of compressibility and permeability. Low compressibility is generally desirable in any concrete-face dam, but special abutment zones of particularly low compressibility are desirable for the higher dams. Fortunately, zoning for compressibility is automatically compatible with requirements for permeability and economy. Normal rockfill zoning provides safety during construction, in the event of flooding before the concrete face is placed. Any credible leakage after the face is placed has been shown to be negligible compared to the ability of rockfill to accept leakage safety (Ref. 7, p. 87).

**Upstream Face Rockfill.**—With the advent of compacted rock, the older practice of crane-placed rockfill under the concrete face was abandoned. In the early compacted dams, it was believed that about a 13 ft (4 m) horizontal width zone of 2–10 in. (5–25 cm) particle size rockfill would be a good base for the face slab when the face, compacted by a vibratory roller, was drawn up the slope. The theory was that nothing could erode in the event of concrete face leakage. However, such a material had no binder, and tended to loosen when forms and reinforcing were being placed. It was also more pervious than desirable in event of a flood before concrete face placement. It was used until, at Cethana, unprocessed tunnel muck and some quarry-run, fine rockfill from thin-bedded quartzite was used beneath the face. This material provided a dense, semi-pervious, and nonerodible base, and became the precedent for future dams. Present practice is to use a finer, well-graded material with maximum size in the 3–1-1/2 in. range (10.2–3.8 cm), well-graded down to sand size, with 5–15% passing the no. 100 mesh. It was attempted on some projects to obtain such material by using select, quarry-run, jaw-crusher-run material, but it was often too coarse. Present specifications have a grading envelope that may require more processing than just crusher-run material.

The face-supporting zone is usually 12 ft (4 m) in horizontal width at the top and remains at that width for dams under 330 ft (100 m) in height. For Areia, it was widened to 33 ft (10 m) at the base. The thicker width provides more of this relatively incompressible, more semi-pervious material for the case where the partially completed dam may act as a cofferdam. Placement is in 14–20 in. (0.4–0.5 m) layers with four passes of 10 static ton, smooth drum, vibratory roller. On the face, the roller is first pulled up the slope without vibration for several passes, and then the face is given 4–6 passes with vibration. At the abutments this premium zone may be flared for high dams or steep abutments.

**Face Slab.**—The face slab has no horizontal joints, and vertical joints are cold joints with waterstops for most of the face. A simple horizontal construction joint with reinforcing going through is used when it is necessary to interrupt placement. During construction, moderate compression develops in more than 90% of the face due to rockfill settlement. With no normal water load on the face, the rockfill cannot transfer significant load to the face and, consequently, to the perimeter joint. When water load is applied, downstream movement of the embankment pulls the concrete face away from the perimeter joint, resulting in some tension areas in the face (20). Near the top, the perimeter joint opening occurs due to the lateral component of rockfill movement toward center of dam. The slab floats on the rockfill, with moderate compressive stress in more than 90% of the area and with acceptable tension for a short distance in a direction approximately normal to the perimeter. These are just favorable natural occurrences, and permit use of nearly any abutment shape (Fig. 7). For a nearly vertical abutment (Chuza, Yacambu) the perimeter joint must be designed for more vertical offset since the water load is transmitted through a greater column of rockfill. This is partly taken care of by thickening, at the abutments, the high modulus zone under the face.

The trend in face design has been toward thinner slab thickness and lower percentage reinforcing, as shown in Table 2. Both economical trends are justified by the uniformity and low compressibility of compacted rock. The face slab thickness formula of $t = 1 + 0.003H$ in feet (0.3 + 0.003H in meters) is economical and is desirable for maximum flexibility. Adequate life for high gradients has been established. The reinforcing no longer needs to function to allow for distribution of load onto crane-placed and dumped rock, but rather serves only as temperature steel. The first departure from the traditional dumped rockfill practice of 0.5% reinforcing steel each way was a reduction to 0.4% at Areia. A percentage of 0.35 has since been used for several dams. On some new designs, 0.3% is planned in slabs on lower dams, where stresses are conservatively estimated to be in compression, although 0.4% is still used near the perimeter. There has been no incident of face cracking in compacted rockfill dams, except for some minor horizontal shrinkage cracks, which are of no consequence.

The concrete facings of two early dams—McKay in 1925 and Buck's Creek in 1928—and those of many other dams exceeding 165 ft (50 m) in height were designed with no joints and continuous reinforcing (24). They have not shown any appreciable distress and continuous horizontal reinforcing is acceptable. Continuous vertical pours and reinforcing, with cold joints as required, is now standard and successful practice. Continuous reinforcing through vertical face panel joints—as recently used at Turimiquire and Yacambu (Venezuela), Khao Laem (Thailand), and Batang Ai (Indonesia)—is still considered optional. The concrete face is essentially floating on the rockfill monolithically with low compressive stresses.

Dense and impervious concrete, having 28-day strengths typically between 3,000 psi and 3,500 psi (20 MPa and 24 MPa), functions satisfactorily under high hydraulic gradients. A higher strength concrete (higher cement content) is generally not considered desirable, since it is subject to greater shrinkage cracking and since soundness and impermeability are more important than strength.

**Face Slab Placement in Stages.**—For high dams, placement of the face slab in two or more stages of elevation may be sometimes desirable and economically justifiable. This procedure has been used on some lower dams, but its successful use during the construction of the Areia Dam has demonstrated the feasibility of this placement sequence in high dams.

The upstream, trapezoidal, 318-ft (97-m) high rockfill of Areia was quickly placed to give 1:500 yr flood protection without the concrete face (see Fig. 4) (21).

The remaining height and volume of rockfill followed. All the rockfill face was placed to design line. When the total concrete face was completed, check surveys were made up the face, determining the slope location at 3.3 ft (1 m) vertical intervals to compare design line to existing line. The lower 130 ft (40 m) was exactly on design line. Downstream movement had been 0–2 ft (0.6 m) between 130–260 ft (40–80 m) in height, then decreasing to zero at crest height of 525 ft (160 m). All readings were on a very smooth curve. The top of the 260 ft (80 m) of the earliest placed rockfill and concrete had moved downstream normal to the face for a distance of 2 ft (0.6 m) during the period of remaining rockfill placement. During reservoir filling, that point at 260 ft (80 m) in height, the center of the 525-ft (160-m) high dam of 2,720 ft (828 m) crest length, moved downstream 2.6 ft (0.72 m), a negligible amount in the center of a 1,500,000 sq ft (139,000 sq m) surface area with 2,720 ft (828 m) crest length.

There is detailed evidence to support the feasibility of concrete placement in stages convenient to the contractor. The starter, or trapezoidal slabs on the perimeter joint, may also be placed on a schedule convenient to the contractor. An added advantage of the acceptability of this construction technique occurs when it becomes necessary to raise concrete-face rockfill dams.

**Construction Scheduling.**—For dam projects in general and for hydro projects in particular, construction at minimum cost and within the shortest possible schedule is highly desirable, and, in this connection, compacted rockfill dams with upstream concrete membranes have certain important advantages over other types of dams.

Except in cold climates where low temperatures and heavy snow may limit the construction year, adverse weather, especially rainfall, does not affect a full year construction schedule for concrete-face rockfill dams, unlike earth core rockfills. Also, the removal of cutoff excavation and grouting from the critical path affords both the designer and the contractor a certain degree of flexibility that assures a secure cutoff and has schedule and economic advantages.

Apart from the production of good simple design and clear specifications, other features of a concrete-face rockfill dam construction that contribute to a fast and economic schedule include: well-planned river handling; avoidance of multiple handling of rockfill by arranging direct placement of material from required excavation in the dam; and adequate contractor capability, which can be limited to the provision of a few types of major equipment and the possession of conventional construction know-how for excavation and for rockfill and concrete placement.

**River Handling.**—A major consideration in river handling during construction is the amount of stored water that could be released by failure of the cofferdam or partially completed dam, and the consequences of overtopping failure to third parties and the project. In the normal case of seasonal flooding, it is usual to design for two diversion flood risks: (1) About 3–10 yr return period, when overtopping would not release enough storage to create a significant flood; and (2) about a 200–1,000 yr return period, when a large flow can be temporarily stored, and overtopping failure would be equivalent to a dam failure.

Fig. 4 shows the approach adopted during the construction of the Areia Dam (21) where a 10-yr return period risk was taken for the initial closure cofferdam, and a 500-yr return period risk adopted for a stage of the partially completed dam. Overtopping of the low upstream cofferdam, which would only flood the toe slab work, is generally acceptable. The main "cofferdam," which is formed subsequently by rapid construction of the upstream portion of the main dam, is zoned to store water without the concrete face and with acceptable leakage. A tem-

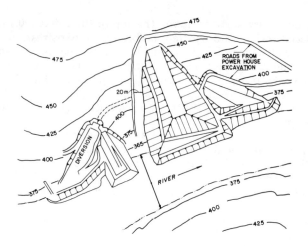

FIG. 8.—Concrete-Face Rockfill Prior to Diversion

porarily flooded toe slab can always be pumped dry after the flood without causing any delays in the construction schedule.

Fig. 8 shows a typical diversion procedure that could be used on major and wide rivers, and on riverbeds of exposed bedrock or acceptable dense and strong gravels. In this case, the temporary upstream rockfill toe was 66 ft (20 m) downstream from the toe slab, permitting work to proceed on the main rockfill and on the toe slab at the same time. Rockfill placed below water level is dumped sound rockfill on rock. This construction procedure permitted the contractor to begin diversion tunnel and powerhouse excavation simultaneously, using excavated rockfill directly in the dam, and leaving a restricted channel until diversion is made. Thus, the riverbed toe slab is placed downstream from a low, high risk cofferdam. The powerhouse was on the critical path, and no stockpiling of rockfill was necessary.

River handling at sites that are subject to flash flooding or year-round flooding poses special problems. In Australia, where such conditions are common, the development of the "reinforced rockfill" concept has provided a solution to this problem. The use of this technique at the Googong Dam enabled the partially completed dam to be overtopped by 10 ft (3 m) in 1976 without significant damage to the dam.

**Rockfill Placement.**—The rockfill zone immediately under the face should provide a semi-pervious zone that, when placed, has minimum segregation and a nonerodible and firm face to resist rainfall erosion before concrete placement, and should provide a firm working surface. Such a material should be dense and have a high modulus of compressibility. It should be desirably semi-pervious, at about $10^{-3}$ to $10^{-4}$ cm/s. Particle size should grade from 1-1/2 or 3 in. down to 5–15% passing the no. 100 sieve. For some rock types it may be necessary in specifications to require a very fine crusher (VFC—Barber Greene or equivalent) or to obtain adequate fines by blending fines with a coarser crusher-run material.

In placing such a zone at the edge of a steep slope, 1.3–1.4H:1V, it is not practical to windrow by motor patrol or dozer to the face. A practical procedure is to windrow to 2 ft (0.6 m) from edge, and then, from occasional dumped loads, to pick unsegregated material up with wide bucket Gradall, followed by setting the bucket edge on the face plane, rotating it to dump the material, pulling the bucket up in plane of face, and, finally, tamping with the bucket. Segregation is avoided, and a minimum amount rolls down the face.

The face rock is subject to rainfall erosion and to disturbance by workers in constructing the face. Several methods have been used after face vibration. At Areia, a spray coat of emulsified asphalt was covered by a shotcrete-applied dry sand, and additional face vibration was applied. At several dams, 2–3 in. (5–7.5 cm) concrete is roughly screeded in 10–13 ft (3–4 m) applications. Shotcrete, 1–2 in. (2.5–5 cm), has also been used. The contractor is responsible for maintenance of the rock face and can select a method of protection.

Rockfill placement procedures are well-known. The zoning is important, in that low compressibility should be provided where needed, and economy and drainage by thicker downstream zone lifts can be achieved where low compressibility is not needed.

Special rockfill placement at the downstream face (or on both faces of an earth core rockfill dam) can be accomplished economically to produce a good appearance or riprap. An example is Chivor, as constructed by Impregilo, and shown in Fig. 9. Fig. 9(a) shows large rocks from selected dumped loads being dozed to face. In Fig. 9(b), a second layer of large rock has been dozed up on a ramp of smaller quarry-run rockfill to the 6.6 ft (2 m) level of the next layer. The dozing puts the high point or

A. DOZING FACE ROCK

B. FACE ROCK DOZED TO 2 M LAYER

C. COMPLETED DOWNSTREAM FACE

D. COMPLETED DAM

**FIG. 9.—Esmeralda Dam—Chivor Project—Photos of Rockfill Face**

edge of the rock at the design line, with no protrusion greater than 6 in. (15 cm). Figs. 9(a–b) show the surface achieved.

Local oversteepening of slopes saves rock and permits design and construction flexibility. Slopes steeper than 1.3H:1V may be used adjacent to haul roads and will not be subject to any ravelling. For the 525-ft (160-m) high Areia, 1.25H:1V slopes were used with road centered on a 1.4H:1V control slope. For example, a 1H:1V slope was used on the outer fill slope of the Bilbao-Biarritz Autopista (26). The instrumented limestone fill showed no spreading of the 160-ft (50-m) high 1H:1V slope. To obtain steeper than 1.3H:1V slopes, large rocks must be dozed to the face.

## FACE SLAB CONSTRUCTION

**Face Slab.**—A single slip form with 4 ft (1.3 m) contacting face can travel at 6.6–10 ft (1–3 m) per hour on a 1.3H:1V slope, placing 2.5 in. (6 cm) slump concrete. Delivery of concrete can be by bucket, pumpcrete or chute. Chutes have been used at Areia, Chuza, and Khao Laem and are most often used on such high dams where crane placement is not possible. A finishing platform is carried on the screed to allow wood float finishing as necessary. Where a horizontal, cold, construction joint becomes necessary, care is required to avoid honeycomb concrete. The excess concrete beyond design line has ranged between 3 and 5 in. (7.5 and 12.5 cm).

**Joints.**—The installation of waterstops needed to make joints watertight and the placement of dense concrete at the joints needs to be done with utmost care, with close supervision if good watertight performance of these joints is to be ensured.

**Parapet Wall.**—The provision of parapet walls on the crest of rockfill dams has certain important advantages. Not only does the provision of the parapet wall replace a layer of rockfill that is not really required for stability, but the use of the high parapet walls, typically 10–13 ft (3–4 m) in height, provides the convenience of a wide working crest for the concrete face slip form placement. The vertical wall also provides best wave control.

**Instrumentation and Surveillance.**—The installation and monitoring of appropriate instrumentation and the proper interpretation of readings obtained from such instrumentation have to become accepted as an integral part of the design of dams. Information yielded by such an instrumentation program often provides verification that the performance of the dam is within acceptable limits and facilitates comparison between actual and predicted behavior of the dam. For concrete-face dams, the main value of an instrumentation program is to contribute to the data base for the design of future dams. Instrumentation often consists of the installation of joint meters on the perimeter joints, settlement gages, face and rockfill crest movement monuments, and leakage weirs. Water level type settlement gages are used internally to determine the modulus of compressibility of the rockfill. Inclinometers are used on dams higher than about 394 ft (120 m), where movements normal to the face are measureable.

The principal means of surveillance for rockfill dams is leakage measurement. If the downstream toe is submerged, the downstream cofferdam, or a low impervious zone in the downstream toe, can be used to collect seepage and support a weir, as was done at Areia (21). Excessive face leakage, should it occur, can be located by sonic equipment, using a hydrophone and sound level meter. Such leakage may be reduced to an acceptably low level by placing dirty fine sand under water, by caulking of joints, or other means. The use of an approximate filter grading under the face, means that placement of a dirty sand on a leakage location can seal the leak.

## Future Higher Concrete Face Rockfill Dams

Future design of concrete-face compacted rockfill dams in excess of the present 525 ft (160 m) in height will have to address the important question of face movements under water load. For dams between 525 and 1,000 ft (160 and 300 m) in height, the current site and material requirements are a nonerodible rock foundation, with special treatment as required, and a reasonably high compressive modulus rockfill. Since face movements are directly proportional to the square of the height and inversely proportional to the modulus of compressibility, the face movements of higher dams can be predicted. The modulus of existing high dams and their movements are known, and the desired modulus for a future dam can be attained by specifying the compaction to achieve what is required (27).

Esmeralda Dam, which is 777 ft (235 m) in height, and Chicoasen Dam, which is 856 ft (251 m) in height, are both of the earth core type, are well-instrumented, and have, by their excellent performance to date, established the general feasibility of constructing rockfill dams to this height. This experience with high earth core type rockfill dams, when combined with the experience already gained on the face movements of existing lower concrete-face dams, assist in providing a basis for the design of higher concrete-face rockfill dams.

A design concept for a proposed 800 ft (244 m) concrete-face compacted rockfill dam is shown in Fig. 10. This design employs a combination upstream earth core and concrete-face dam in the lower 400 ft (122 m) and a simple concrete-face rockfill in the upper 400 ft (122 m). Perimeter joint movement would still have to be predicted and designed for. At Areia, which is 525-ft (160-m) high, maximum perimeter joint offsets and openings were under 2 in. (5 cm) for a low modulus, basalt rockfill (21). At and near the crest, several vertical joints opened. The vertical joints that were expected to open had been constructed with multiple waterstops. The horizontal tension zone in the rockfill, where such waterstops had been placed, can be predicted with reasonable accuracy.

The upstream core feature in the lower 400 ft (122 m) is in the narrow lower part of the canyon (see Fig. 10). It does not represent a major cost feature and should not affect schedule. At Anchicaya (10) and Areia (21), some earth material was used as a concrete face crack healer, assuming cracks might occur or joints might leak. The concept proposed here uses a thicker earth zone, with wider foundation treatment, and a processed filter under the concrete face, thus providing both earth core and con-

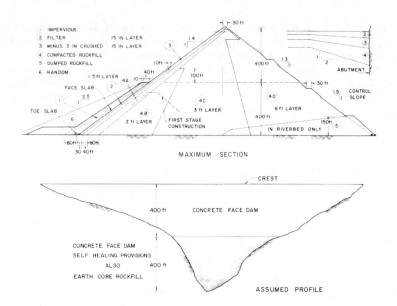

**FIG. 10.—Section of Conceptual 800-ft (245-m) Concrete-Face Rockfill Dam**

crete-face features in the lower 400 ft (122 m).

The upstream core for the 800-ft (244-m) high combination embankment can reasonably be expected to sustain a higher gradient than that for conventional earth core rockfills. Though many of today's central core dams have thicknesses yielding gradients of 3 or 4, the Growdon-Alcoa dams have gradients of 9, and Kenney and Brownlee, 7. This experience suggests that a core thickness yielding a gradient of about 10 would be reasonable where a concrete face and filter back up exists. Foundation treatment would be the same as for the earth core rockfill dam, even though backed up by nonerodible dowelled toe slab.

The layer thickness and compaction of rockfill zones for the 800-ft (244-m) high dam would be carefully planned to provide maximum modulus of compressibility where high modulus is needed. Economy could be achieved by using thicker layers where high modulus is not needed and greater permeability is fundamentally desirable. Desired relative permeabilities are favorably matched with compressibility requirements. Dumped rockfill proposed for zone 5 provides necessary filtering at top and high drainage at the bottom. In that zone, compression arising from the lower modulus of dumped rockfill would not affect concrete face movement.

Conditions that would especially favor the construction of high concrete face rockfill dams include requirements for maximum seismic resistance, high precipitation or cold weather conditions, scarcity of or unfavorable properties of earth material, need for minimum risk in river handling, shortest schedule, and low cost.

## CONCLUSION

I have presented a chronicle of modern rockfill dam design, culminating in a description of the practice of the concrete-face rockfill dam design as it exists today. It has been a story of evolution that has brought the concrete face rockfill dam to a significant place in contemporary dam engineering. Certainly, it should now be considered in all dam and layout alternative studies.

What I have described here is a process that is more evolutionary than revolutionary, in which engineering progress came about principally as a consequence of cautious trial and error. The concrete-face rockfill dam was not "invented," but rather developed by contributions of many engineers, beginning in the early days with J. Growdon, L. H. Harza, I. C. Steele, and K. Terzaghi. More recent progress in developments of the modern concrete face rockfill dam can be similarly credited to many organizations and engineers: The Hydroelectric Commission of Tasmania, SMEC and other Australia organizations, Ingetec of Colombia, COPEL and ELETROSUL of Brazil, Kaiser Engineers, and Electric Generating Authority of Thailand (EGAT). Among the engineers who must be included in credits for current progress in concrete-faced rockfill dams are J. L. Sherard, T. M. Leps, J. W. Libby, Victor F. B. de Mello, N. L. Pinto, C. S. Ospina, B. Materon, I. L. Pinkerton, J. K. Wilkins, M. D. Fitzpatrick, and J. Hacelas.

Editing for publication by Richard Meehan and ASCE GE personnel and staff is gratefully acknowledged.

## Appendix.—References

1. Cooke, J. B., "Symposium on Rockfill Dams," *Transactions,* ASCE, Vol. 104, 1960.
2. Galloway, J. D., "Symposium—The Design of Rockfill Dams," *Transactions,* ASCE, Vol. 104, 1939, pp. 1–92.
3. "High Rockfill Dams—Q31," *Eighth ICOLD Congress Papers,* Vol. III, pp. 579–894, and *Discussions,* Vol. V, 1964, pp. 438–553.
4. Cooke, J. B., "Design, Methods of Construction, and Performance of High Rockfill Dams," *Eighth ICOLD Congress,* Vol. IV, 1964, pp. 595–621.
5. Davis, C. V., "Rockfill Dam," Chapter 6, *Handbook of Applied Hydraulics,* 2nd Ed., I. C. Steele, ed., McGraw Hill Book Co., Inc., New York, N.Y., 1952.
6. Fucik, E., Montford, and Edbrooke, F. E., "Ambuklao Rockfill Dam, Design, and Construction" *Transactions,* ASCE, Vol. 125, Part I, 1960, pp. 1207–1227.
7. Hirschfeld, R. C., and Poulos, S. J., *Embankment Dam Engineering,* John Wiley and Sons, Inc., New York, N.Y., 1972.
8. Good, R. J., "Kangaroo Creek Dam—Use of a Weak Schist as Rockfill for a Concrete Face Rockfill Dam," *Twelfth ICOLD Congress,* 1976, Vol. 1, Q44-R33, pp. 645–665.
9. Mackensie, P. R., and McDonald, L. A., "Use of Soft Rock in Mangrove Creek Dam," *20th ANCOLD General Meeting,* 1980.
10. Regalado, G., et al., "Alto Anchicaya Concrete Face Rockfill Dam—Behavior of the Concrete Face," *Fourteenth ICOLD Congress,* 1982, Vol. IV, Q55-R30, pp. 517–535.
11. Seed, H. B., "Foundation and Abutment Treatment for High Embankment Dams on Rock," ASCE, Vol. 98, No. SM10, Paper 9269, Oct., 1972.
12. Davis, C. V., and Sorensen, K. E., "Concrete Face Rockfill Dams," Chapter 19, I. C. Steel and J. B. Cooke, eds., *Handbook of Applied Hydraulics,* 3rd Ed., McGraw Hill Book Co., Inc., New York, N.Y.
13. Sierra, J., et al., "Chivor Dam—Design—Rockfill—Performance," *Proceedings of the Fifth Panamerican Conference on Soil Mechanics and Foundation Engineering,* Buenos Aires, Argentina, Vol. II, 1975, pp. 277–325.

14. Hacelas, J. E., and Ramirez, C. A., "Interaction Phenomena Observed in the Core and Downstream Shell of Chivor Dam," *Ninth International Conference on Soil Mechanics and Foundation Engineering*, Specialty Session 8, June, 1977.
15. Moreno, E., and Alberro, J., "Behavior of the Chicoasen Dam: Construction and First Filling," *Fourteenth ICOLD Congress*, 1982, Vol. 1, Q52-R9, pp. 155–182.
16. Alberro, J., and Moreno, E., "Interaction Phenomena in the Chicoasen Dam: Construction and First Filling," *Fourteenth ICOLD Congress*, 1982, Vol. 1, Q52-R10, pp. 183–202.
17. Marsal, R. J., "Monitoring of Embankment Behavior," *Fourteenth ICOLD Congress*, 1982, Vol. 1, Q52-R84, pp. 1451–1466.
18. Leps, T. M., "Review of Shearing Strength of Rockfill," *Proceedings*, ASCE, Vol. 96, No. SM4, July, 1970.
19. Wilkins, J. K., et al., "The Design of Cethana Concrete Face Rockfill Dam," *Twelfth ICOLD Congress*, Vol. III, 1973, pp. 25–42.
20. Fitzpatrick, M. D., Liggins, T. B., and Barnett, R. H. W., "Ten Years Surveillance of Cathana Dam," *Fourteenth ICOLD Congress*, Vol. 1, Q52-R51, 1982, pp. 847–865.
21. Pinto, N. L. de S., Materon, B., and Lago Marquez, P., "Design and Performance of Foz do Areia Concrete Membrane as Related to Basalt Properties," *Fourteenth ICOLD Congress*, Vol. IV, Q55-R51, 1982, pp. 873–905.
22. Wilson, S. D., and Marsal, R. J., "Current Trends in Design and Construction of Embankment Dams," ASCE, prepared for ICOLD, C. S. Ospina, Chairman, and ASCE, W. F. Swiger, Chairman, 1978.
23. Peck, R. B., "Where has all the Judgement Gone," Fifth Laurits Bjerrum Memorial Lecture," National Research Council of Canada, 1980.
24. Sherard, J. L., "Earth and Earth-Rock Dams," John Wiley and Sons, Inc., New York, N.Y., 1963.
25. Seed, H. B., et al., "Performance of Earth Dams During Earthquakes," *Journal of the Geotechnical Engineering Division*, ASCE, No. GT7, Paper 13870, July, 1978, p. 992.
26. Miranda, M. A. R., et al., "Comportamento de un Gran Pedraplen Construido con un Talud de 45 Grados," *Revista de Obras Publicas*, June, 1977, pp. 479–490.
27. Cole, B. A., "Concrete Faced Rockfill Dams in the Pieman River Power Development," *Paper C1385*, Institution of Engineers, Australia, submitted Mar. 24, 1982.
28. Terzaghi, Karl, "From Theory to Practice in Soil Mechanics," *Selections from the writings of Karl Terzaghi*, John Wiley and Sons, New York, N.Y., 1960.

# Progress in Rockfill Dams[a]

## Discussion by Ranji Casinader[2]

Cooke's choice of subject is in keeping with his enthusiasm for rockfill dams, in particular, concrete faced rockfill dams, and the encouragement and assistance he has imparted to rockfill dam engineers in many parts of the world. The lecture stands as the most up-to-date statement on the state of the art.

There are two points mentioned in his lecture concerning concrete faced rockfill dams that I would like to develop from my own experience.

The first concerns the toe slab foundation which, as the lecturer points out, should most desirably be "hard, non-erodible rock that is groutable." However, in the future, concrete faced rockfill dams could be proposed, and to some extent already have been adopted at sites where these ideal conditions are not satisfied. Successful details can be developed at these sites by the adoption of special measures, including those mentioned by Cooke.

Winneke Dam in Australia, constructed between 1975 and 1980, is one such site. The depth of weathering of the siltstone foundation rock is such that the toe-slab could not economically be founded on "groutable rock." Although the highly weathered rock substance at toe slab foundation level was itself non-erodible, joints and particularly bedding plane joints contained erodible and often dispersive material. Cut-off details adopted at this site are shown in Fig. 11 and may be summarized as follows:

- The conventional toe slab, width one-tenth of the head of water at any point, was extended downstream by a concrete slab of 150 mm minimum thickness, so that the total width of toe slab and "foundation concrete" was one-half of the head of water at any point.
- The foundation downstream of the "foundation concrete" was protected by inverted filters for a further distance equivalent to one-half of the head of water.
- A two-row grout curtain was adopted, with air and water flushing prior to grouting with neat cement.

It will be noted that the total cut-off width adopted was similar to that usually adopted for an earth core dam.

The second point I should like to elaborate on is the incidence of cracking of the concrete face. As the lecturer points out, compaction of rockfill has generally eliminated cracking of the face due to rockfill settlement, and only minor cracking due to shrinkage occurs.

Recent close examination of some Australian dam faces, with a 20 times magnification eyepiece, has revealed a fairly regular pattern of shrinkage cracks of a minor nature. Crack spacing varies from 1–7 m, and the majority of cracks are less than 0.2 mm in width. The writer believes that similarly detailed examination of other dams will show similar results.

---
[a]October, 1984, by J. Barry Cooke (Paper 19206).
[2]Consultant in Water Resources, Engrg., Melbourne, Australia.

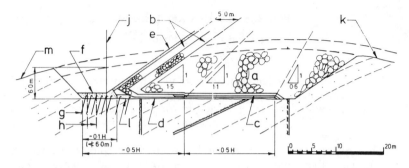

FIG. 11.—Cut-Off Details: (a) Rockfill; (b) Transition Zones; (c) Filters; (d) Foundation Concrete; (e) Concrete Facing; (f) Plinth; (g) Anchor Bars; (h) Grout Curtains; (j) Plinth Reference Line; (k) Foundation Stripped to Highly Weathered Rock; (l) Buttress; (m) Original Ground Surface; (H) Hydraulic Head at Foundation Level

It is difficult to assess what proportion of the leakage through the dams might be due to these cracks, as generally face and foundation leakage are measured together, but total leakage through concrete-faced dams is rarely significant.

Future developments of concrete face design and construction should concentrate on minimizing the phenomenon of shrinkage cracking, which has a bearing not only on leakage but also on the durability of the face. Design and construction aspects which should be given further consideration are:

- Use of minimum cement content in the concrete and/or cement replacement by pozzolans, but without sacrificing durability.
- Adoption of a low water-cement ratio but without sacrificing workability—use of plasticizers and super-plasticizers may be considered.
- Use of low shrinkage aggregates, with testing of aggregates to meet specified shrinkage limits.
- Review of the positioning of reinforcing steel in the face slab, e.g., should it be placed a constant distance from the water face rather than centrally in the slab as is currently usual.
- Correct and sufficient curing of concrete after placement.

### Discussion by W. L. Chadwick,[3] F. ASCE

The author has produced a valuable record of the historical development of rockfill dams and of the evolution of both earth core and concrete-face rockfill designs as dictated by experience during use. That history shows remarkable improvements in design and performance. The author's citing of the use, by miners in the Sierra Nevada Mountains, of timber-faced rockfills, calls attention anew to the innovations of 130 yrs ago. Rockfills of that type still give excellent service on the Southern

[3]Consulting Engr., Claremont, CA.

California Edison Company hydro system in the Sierra Nevada. Man caused fires have occurred but even then caused no risk to the safety of the dam. These early rockfills stand at 3/4 H to 1 V downstream, and 1-1/4 H to 1 V on the upstream or timber-face side.

The author's citing of the Salt Springs Dam of the Pacific Gas & Electric Company, built in 1931, illustrates the adaptability of the rockfill design to construction, in virtually a bare rock environment, where local building materials, other than quarried rock, are unusually scarce, and mountain and distant transportation were expensive. The concrete face rockfill supplied this need ideally with a membrane which could be built of crushed native rock and a relatively small volume of hauled-in cement. Not only was earth material scarce, but the earth core rockfill had not been developed. Since that time, as the author mentions, many developments in both types of rockfill dams have been made as a result of experience-driven judgment. The author has played a large part in that evolution.

The author's noting of the experience with concrete-faced rockfills in very cold climates is impressive. The writer's own early experience with the concrete damage, caused by up to 200 freeze-thaw cycles/yr in the upper elevations of the Sierra Nevada Mountains, makes this experience seem truly remarkable: Evidence of progress, in this case use of air entrainment, pozzolan and lower water-cement ratio.

After the difficulties experienced with leakage at New Exchequer in 1966 it seemed to the writer at that time that this type of dam had a heavy liability in repairs. However, it has been interesting to see the developments in rockfill compaction and concrete face design which essentially eliminates the problem of leakage.

Both the earth core and concrete faced rockfills are here to stay.

### Discussion by Claude A. Fetzer[4]

The R. D. Bailey Dam, which was completed by the Huntington District, US Army Corps of Engineers in 1979, illustrates two points in the author's paper: (1) Reinforcement of the downstream slope is needed for overtopping; and (2) compacted rockfill dams can withstand through seepage for non-overtopping impoundments.

A typical section of the 315-ft (96-m) high dam is shown in Fig. 12. The central rolled random rock section was composed mostly of shale and other soft rock, and it had a much lower permeability than the compacted hard sandstone. The 15-ft (5-m)-thick bottom drain was continued up the abutments, thereby separating the rolled random rock from the bedrock in the abutments. A temporary cofferdam constructed of 60-ft (18-m)-high sheet-pile cells was provided upstream of the dam to divert the river through the outlet tunnel in the left abutment while the permanent cofferdam was constructed.

During the initial stage of constructing the permanent cofferdam, the

---

[4]Consulting Geotechnical Engr., Cincinnati, OH; formerly, Chf., Geotechnical Br. Ohio River Div., Corps of Engrs.

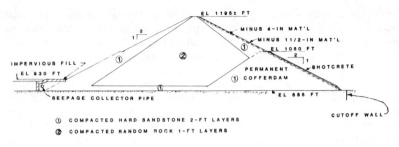

FIG. 12.—R. D. Bailey Dam—Maximum Section, April 1, 1977

height of fill of the 160-ft (49-m) section adjacent to the left abutment was restricted to El. 952.5 ft (290 m) while the remainder of the cofferdam was being constructed to El. 1,035 ft (315 m). The idea was to restrict the potential storage upstream until the gap at the left abutment could be filled in very quickly when the risk of major storms was at the minimum. In mid-March 1975, a storm occurred that overtopped the upstream temporary cofferdam and subsequently eroded out the low section of the permanent cofferdam, which was not reinforced. No damage occurred downstream. At the time of the flood, the remainder (the right side) of the permanent cofferdam had been constructed to El. 976 ft (297 m), and hydrologic studies indicated that the water would have overtopped the cofferdam at that level. It was decided to raise the top of the cofferdam to El. 1,050 ft (320 m), thereby raising the statistical frequency of overtopping from once in 15 yrs to once in 100 yrs on an annual basis.

The permanent cofferdam and the remainder of the embankment to El. 1,195 ft (364 m) were completed in 1975, 1976, and early 1977 without major flood incidents except for overtoppings of the upstream temporary cofferdam. By April 1, 1977, the 3-in. (7-cm)-thick shotcrete had been placed on the upstream face of the permanent cofferdam up to El. 1,050 ft (320 m). The shotcrete had been placed over a 1-ft (30-cm)-thick compacted layer of minus 1-1/2-in. (4-cm) crushed sandstone, which was underlain by a 14-ft (4-m)-wide compacted zone of minus 4-in. (10-cm) crushed sandstone. Both materials extended up the slope above the shotcrete.

A major storm started at 2100 hrs on April 3, 1977. The accumulated rainfall at the dam site in 27 hrs was 4.96 in. (12.6 cm). The rising pool overtopped the temporary upstream cofferdam at 1530 hrs on April 4, 1977, and the water level against the dam rose 140 ft in 20 hrs. The stored water crested at El. 1,055.9 ft (322 m) at 1730 hrs on April 5, 1977. Flow from the downstream seepage collection system gradually increased with the rising reservoir, but it was soon overtaxed. Seepage broke out on the El. 930-ft (283 m) downstream berm at 2330 hrs on April 4, 1977, when the reservoir water surface reached El. 1,003 ft (306 m). As the reservoir continued to rise, several boils broke through the El. 930-ft (283-m) impervious berm, sheet flow developed at the toe of the rolled rock fill, and large concentrated flows developed at the abutments. At 1600 hrs on April 5, 1977, with the reservoir at El. 1,055 ft (322 m), seepage broke out on the downstream slope in an area about

El. 970 ft (296 m), and in another area about El. 945 ft (288 m) at distances of 45 ft (13 m) and 150 ft (46 m), respectively, from the left abutment. It was estimated that the flow from the seepage collection system was 50 cfs (1.4 cms) and that the uncontrolled flow from the toe of the dam was 200 cfs (6.2 cms).

The stored water had several access points into the rockfill embankment. After the flood, a continuous crack was found in the shotcrete near El. 960 ft (292 m). A separation in the shotcrete and large holes were found at the break in the slope, at approximately El. 910 (277 m). Several holes were found in the shotcrete at the top of bedrock exposing the sandstone fill. Two sinkholes were found in the shotcrete near the left abutment. The minus 4-in. (10-cm) material was exposed along the abutment contacts above El. 1,050 ft (320 m). The central random rock zone acted as seepage barrier across the valley section, but the seepage could have been passed below and along the sides of the central random rock zone through the more pervious rock drain.

Diagonal tension cracks were found in the minus 1-1/2-in. (4-cm) material near the crest at the abutments, but no other visible damage was found. Partial saturation of the rockfill embankment probably accelerated the rate of settlement, but the measured settlements were not excessive.

The rapidity with which major storms develop in many mountainous and other areas indicates that cofferdams should be designed not to fail from overtopping at all levels. The cost of reinforcement of the full downstream slope of rockfill cofferdams is usually insignificant compared to the potential damage to the downstream area from a sudden washout and to the cost of the cofferdam repairs and associated delays.

### Discussion by M. D. Fitzpatrick[5]

Barry Cooke's excellent paper clearly and concisely traces progress in rockfill dams. The writer would like to acknowledge the author's significant contribution to the development of rockfill dams during the last 30 yrs and particularly in the field of concrete-face rockfill dams. No one has done more than he towards the rapid dissemination to dam engineers throughout the world of ideas and advancements resulting in improved performance.

Although the writer agrees with the author's contention that design of concrete-face dams is essentially empirical and based on experience and judgment, it is considered that the time has been reached when further progress should be assisted by an analytical approach. It is suggested that the experience accumulated to date is not adequate for extrapolation to much higher dams such as proposed conceptually in the paper. For example, there is insufficient data on face strain to correlate maximum values with any major parameter such as dam height or rockfill modulus.

If, in the absence of any relationship established by correlation, maximum compressive strain is assumed to be directly proportional to dam

---

[5]Chf. Civ. Engr., The Hydro-Electric Commission, Tasmania, Australia.

height squared, then, based on Cethana, the maximum compressive strain in the author's conceptual dam would have a magnitude of $2,000 \times 10^{-6}$ (ratio of height squared is 5 and maximum compressive strain in Cethana (29) at 10 yrs was $400 \times 10^{-6}$).

As this value lies in the failure range for concrete, between $1,000 \times 10^{-6}$ for rapid loading and $3,000 \times 10^{-6}$ for gradual loading, it must be regarded as unacceptable. The inference from this is then that a greater rockfill modulus would be required than was obtained at Cethana. A similar result is obtained working from the maximum compressive strain of $665 \times 10^{-6}$ in Foz do Areia face (30) and making an allowance for increase with time due to creep settlement.

Although insufficient for correlation studies, available data on face strains from the few dams where it has been measured can be used to test the validity of finite element models; of course this also requires appropriate definitive relationships for the "stress-strain" behavior of rockfill. In the writer's view, a research effort to establish a model calibrated with measured behavior will be essential to make further significant progress on the question of face movements under water load, particularly in conjunction with very high dams and the raising of existing dams.

In their analysis of Alto Anchicaya Dam using the finite element method, Sigvaldason, et al. (31) went a long way towards this goal. Their work with two-dimensional models demonstrated the following:

- Substantial increase in rockfill stiffness under water load
- Limited load transfer to the concrete slab from rockfill subject to extensional strain
- Close agreement with measured normal deflection along the abutments of Alto Anchicaya

These results indicate the validity of their approach. However, considerable research is required to make their approach more general. They point out that their definition of rockfill "stress-strain" behavior in terms of volumetric and deviatoric components needs to be made more general and used in the analysis of three-dimensional finite element models based on isoparametric formulation.

In summary, the writer is emphasizing the point that progress will be more soundly based if design is guided not only by experience but also by analytical methods which have been tested against prototype observations. It is recommended that research along these lines be strenuously followed.

## Appendix.—References

29. Fitzpatrick, M. D., Liggins, T. B., and Barnett, R. H. W., "Ten Years Surveillance of Cethana Dam," *14th ICOLD Congress*, 1982, Q. 52, R51.
30. Pinto, N. L., Materon, B., and Lagos Marques, P., "Design and Performance of Foz do Areia Concrete Membrane as Related to Basalt Properties," *14th ICOLD Congress*, 1982, Q.55, R.51.
31. Sigvaldason, O. T., Benson, R. P., Mitchell, G. H., and Eichenbaum, H., "Analysis of the Alto Anchicaya Dam Using the Finite Element Method," prepared for presentation at the International Symposium, Criteria and Assumptions for Numerical Analysis of Dams, Swansea, UK, Sept., 1975.

## Discussion by E. M. Fucik,[6] F. ASCE

The author has presented a very complete and useful history of the developments in the design of rockfill dams. The increase in the maximum height of these dams, from the 275 ft of Dix River in 1925 to 1,000 ft at present is a striking example of the courage and ingenuity of their designers.

The writer had the pleasure of working for many years with L. F. Harza, who was the designer of the Dix River Dam, located in Kentucky. This dam was by far the highest rockfill dam which had been undertaken up to that time, and was also an early example of the use of a concrete facing for watertightness. The dam showed a considerable amount of leakage after it was constructed and early attempts to permanently stop this leakage were unsuccessful. Harza attributed this leakage to cracks in the concrete facing, but recently (within the past five yrs), the leakage has been reduced to a negligible amount by placing a reverse filter (larger stone first, then finer material) against the toe of the dam, covering a fault that ran down the center of the river channel. So it appears that the concrete facing on this early example was, and is, a success.

Harza was also involved in extending the limits of height in rockfill dam design when he designed the Ambuklao Dam in the early 1950s. This dam is a center earth core rockfill dam located on the Agno River in the Philippines. It is 435 ft high, and at the time it was built, was the highest rockfill dam in the world.

The development of compacted rockfills which reduced settlement and permitted extensive use of concrete faced rockfill dams has greatly broadened the feasibility of this type of dam and the writer is confident its use will continue to increase.

The author is to be congratulated for his comprehensive and interesting treatment of this very important segment of dam design.

## Discussion by Jorge E. Hacelas[7] and Carlos A. Ramírez[8]

The Salvajina concrete face gravel/rockfill dam, 485 ft (148 m) in height, recently completed in Colombia, may amplify some of the significant points raised by the author in his outstanding lecture. Design data on Salvajina is included in Table 2. Crest length to height ratio is 2.2.

**Zoning and Instrumentation of Salvajina Dam.**—Fig. 13 shows the maximum section of the Salvajina dam as well as the embankment instrumentation. Table 3 shows the characteristics of the materials.

**Construction Settlements.**—The maximum construction settlement in the gravel fill (Zone 2) was 1.3 ft (40 cm) and occurred in the middle

---

[6]Chmn. Emeritus, Harza Engineering Co., Chicago, IL.

[7]Prin. Assoc. Engr., Resident Engr., Salvajina Hydroelectric Project, INGETEC S.A., Bogotá, Colombia.

[8]Civ. Engr., Chf., Technical Office, Salvajina Hydroelectric Proj., INGETEC S.A., Bogotá, Colombia.

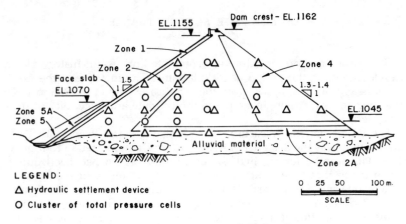

FIG. 13.—Section through Salvajina Dam

and 1/3 elevation of the dam. This represents 1.4% of the underlying layers. In contrast, the maximum construction settlement in the rockfill (Zone 4) was 2.4 ft (73 cm).

**Stresses.**—The value of the octahedral stress ($\sigma_{oc}$) has been adopted as a measure of the stress level within the dam fill. The magnitude and orientation of the principal stresses acting on the maximum section of the dam is also indicated in Fig. 14. The relationship between these stresses within the gravel fill was of the order of 10, while for the rockfill it was 2, which shows the striking difference in response of gravels and rockfill under similar gravity load conditions.

**Modulus of Compressibility.**—The load vertical strain plots for the gravel fill yielded the gravity load moduli that are shown in Table 4.

TABLE 3.—Zoning of Dam

| Zone (1) | Description (2) | Maximum size (in.) (3) | Thickness of layers (ft) (4) | Compaction (5) |
|---|---|---|---|---|
| 1 | Processed gravels ($k = 10^{-3}$ cm/s from lab tests) | 4–6 | 1.5 | Four passes of vibratory roller (10 ton). Eight passes in sloping direction (5 ton), vibration only in the upward direction. |
| 2 | Unprocessed gravels | 12 | 2.0 | Four passes of vibratory roller (10 ton). |
| 2A | Processed gravels (uniform material) | 12 | 2.0 | One pass of vibratory roller (10 ton). |
| 4 | Soft rockfill ($k = 3 \times 10^{-2}$ cm/s from field tests) 40–80% passing 1 in. sieve. | 24 | 3.0 | Four to six passes of vibratory roller (10 ton). |

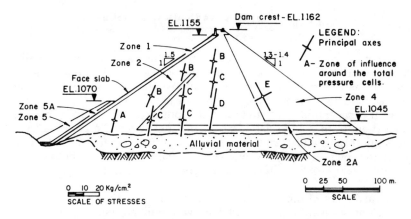

**FIG. 14.—Principal Stresses—Salvajina Dam**

These records indicate that for the same stress level, the gravity load modulus of the gravel fill turned out to be 7 times higher than that of the compacted rockfill. This, in turn, shows that gravel is remarkably incompressible; and the writers concur with the author's conclusion that "gravel shell concrete face dams can safely be built to much greater height than the present highest dam, Areia, which is 525 ft (160 m) high." This also demonstrates that soft rockfill can safely and economically be used in combination with gravels in those zones of the downstream shell that carry little water load and where low compressibility is not needed. The relationship between the gravity load moduli and $\sigma_{oc}$ as observed in Salvajina Dam is shown in Table 4.

For 62% of reservoir filling, the ratio between the water load modulus and the gravity modulus is 3, for a stress level of 35–30 psi (0.24–0.28 MPa), at El. 1,060, corroborating thus the author's statement. The writers attribute this behavior to the fact that upon changing the direction of loading, the "confinement" stress in the embankment, parallel to the concrete face, is larger (0.57 $\sigma_1$) than the "confinement" horizontal stress corresponding to the vertical gravity load (0.17 $\sigma_1$).

**TABLE 4.—Relationship between Gravity Load Moduli and $\sigma_{oc}$ as Observed in Salvajina Dam**

| Zone of influence around cluster of pressure cells[a] (1) | $\sigma_{oc}$ (lb/sq in.)[b] (2) | Gravity load moduli (lb/sq in.) (3) |
|---|---|---|
| A | 21–36 | 7,100–14,200 |
| B | 36–50 | 14,200–21,350 |
| C | 50–71 | 21,350–28,500 |
| D | 92–107 | 42,600–56,800 |
| E | 92–107 | 7,100–8,500 |

[a] See Fig. 14.
[b] $\sigma_{oc}$ adopted as a measure of the stress level at several locations within the dam fill.

## Discussion by A. Clive Houlsby,[9] F. ASCE

The interesting review presented by the author was of course not able to go into detail on every aspect of concrete faced rockfill dams. Some grouting aspects of these dams are therefore enlarged upon by the writer as a constructive augmentation of the subject.

As the paper indicates, foundation grouting of this type of dam is carried out from the toe slab. This should ideally remove grouting from the critical path and permit less hurried grouting with more attention given to problem areas. However, at some dams things have not worked out this way. Problems have arisen from the following:

1. Movement of the toe slab during grouting.
2. Grout leaks into the embankment where fill has been placed ahead of the grouting.
3. Fill materials rolling down on to the grouting crews and plant where the fill has proceeded substantially ahead of the grouting.

The desirability of tailoring the construction program of a dam to minimize these problems depends on how important the grouting is. On many sites, the rock is so competent that the grouting is only nominal and little weight need be afforded it. There are, however, sites where the opposite applies, particularly in poor rock where grouting to a high standard is needed to handle several head drop in seepage, over a short distance. In these circumstances, the writer has found it prudent for completion of the grouting before any rockfill is placed alongside. Some comment on this in relation to the points previously listed is as follows.

**Point 1.**—Some toe slabs or their foundations move during grouting even though dowelled. Modest grouting pressure, if able to penetrate to any extent in the concrete rock interface, can develop sufficient force to cause movement. Maybe only part of a slab lifts, or a crack develops in it or sometimes weaker sections of foundation locally go downwards, having been jacked against an unmoved slab which is securely held on both sides of the depressed zone. Sometimes there are combinations of these. Whatever way it is, the first manifestation is grout leaking out from the edge of the slab upstream or downstream, or from cracks freshly broken in it. The prevalence of this is very approximately assessed by the writer from experience with grouting of a reasonable number of these dams, as at a few percent of slabs. When it happens there is both the need to seal the joint made by the movement, and to carry out the grouting below it to achieve the desired standards without producing repetition of the problem. The latter can be difficult and is the reason for a preference by the writer to do the grouting before placement of the toe slab if at all possible, even if a final shallow tightening up operation is needed directly under the slab when it is later placed.

**Point 2.**—The possibility of grout leaking into the embankment has been a consistent source of concern in those dams where grouting has not been completed before fill placement. The concern comes from the

---

[9]Sr. Surveillance Engr. and Engrg. Specializing in Grouting, Water Resources Commission, NSW Australia.

uncertainty, i.e., if a hole is taking grout fast, is it going into cracks in the rock or just escaping into the rockfill? Such uncertainty has led to more holes than would otherwise be necessary on some dams and has interfered with judicious decisions about thickening during the grout application—if the fast take is due to a leak it needs different thickening treatment than a genuine leak in rock. Best grouting is obtained at this type of dam when, as in core type dams, the foundation is clear for 10 ft (3 m) upstream and downstream of the grouting so that any surface leaks can be seen and identified as such.

**Point 3.**—Several dams have been constructed with the fill above the toe slab to a considerable height before completion of the grouting. This means that rocks are constantly rolling down the face from newly placed layers together with material from face preparation activities. At Glennies Creek Dam (listed in Table 2 of the paper), the drills, grouting crews and gear had to operate in a continuing deluge of such materials. At Khao Laem Dam, a bamboo fence was built to try to catch much of this material, but many stones skipped and hopped down the face in the course of their descent and jumped the bamboo, including one whose target was the writer's head!

**Discussion by A. Marulanda,[10] M. ASCE and C. S. Ospina,[11] F. ASCE**

The paper provides an excellent history of the development of the rockfill dam, with special emphasis on concrete-face dams, by one of the most experienced engineers in the field. J. Barry has been involved with rockfill dams for the past 45 yrs.

Several high rockfill dams have been built or are under construction in Colombia and the writers have participated in the design and construction of most of them. Cooke has been a consultant to all the Colombian rockfill dams.

The design of rockfill dams is an empirical art and each site yields further lessons and pushes the art a little further.

There are some general comments on the design of concrete-face rockfill dams that deserve further emphasis:

1. A concrete-face rockfill dam is suitable for unyielding rock foundations. In this sense, it competes with concrete dams and rockfill impervious core dams. Yielding foundations would create problems with no easy solutions.

2. It is a very stable type of dam, well-suited for areas subjected to earthquake shocks.

3. The percolation path is very short and, therefore, very high gradients exist. This requires very thorough treatment of the foundation and abutments. This is critical compared to a center core rockfill.

Very careful investigations of the abutments are needed to locate pos-

[10]Partner and Head of Geotechnical Div., INGETEC S.A. (Consulting Engrs.) Bogotá, Colombia.
[11]Partner and General Mgr., INGETEC S.A. Bogotá, Colombia.

sible water paths. This is even more important in the tropics where the pattern of rock weathering is erratic and sometimes very deep. Due to these high gradients, in cases of less than excellent foundations, the writers consider it essential to provide several lines of defense for protection of the abutments, such as concrete cutoffs, extensive grouting and filter zones between the rock abutment and the main rockfill. Very few concrete face rockfill dams have been built with defective foundations.

4. The connection of the concrete face with the foundation and the abutment is a weak link. The shape of the perimeter is very important. Sharp breaks require special treatment. Very steep abutments create problems. Seals do not work with equal efficiency in three dimensions. The movement of the dam and, thus, of the concrete face, as the dam settles, may result in the shearing of the seals between the face and the plinth.

Once a concrete face rockfill dam is built and tested for leakage, it is a very safe dam. Leakage does not affect the stability and, if the value of water is not a factor, as in the case of a flood control dam, it may not be worthwhile to reduce the leakage.

The critical danger for all rockfill dams is overtopping. This may be solved by reinforcing the downstream face. Cooke may comment on this in his closure.

### Discussion by Bayardo Materon,[12] M. ASCE

The description presented in the paper should be of great interest to all engineers engaged in the design and construction practice of future rockfill dams. The author has made a very useful contribution in summarizing the modern trends of concrete face rockfill dams.

The author mentions that a fast and economic schedule could be obtained in the construction of a concrete face dam when "multiple handling of rockfill by arranging direct placement of material from required excavation in the dam" is avoided.

The writer has had the opportunity to participate in the construction of three major modern concrete face dams, and in all three, the construction scheduling has been met due to the simplicity of planning and the practical and economical way of developing ramps within the dam. At the high Foz do Areia Dam (21), where a volume of 18,300,000 yd$^3$ (14,000,000 m$^3$) had to be placed, the civil contractor developed a series of ramps, in both abutments and within the dam [Fig. 5(a)], which speed up the placing of rockfill directly from the simultaneous excavation of the Powerhouse, Power Intake and Spillway. Average production of almost 654,000 yd$^3$ (500,000 m$^3$)/month was obtained during a period of two yr, reaching a peak of 876,000 yd$^3$ (670,000 m$^3$) month and completing the dam in a period of 30 months when the original schedule considered 34 months.

Another interesting point by the author is the adopted use of horizontal reinforcing steel running through the vertical joints of the face

[12]Consulting Engr., São Paulo, Brazil.

slab, optional to the standard practice of having in these joints only water stops. The writer has observed that openings of the joints, closer to abutments, are related to the height of dam, shape of valley, width of canyon and quality of rockfill (27).

In Alto Anchicayá Dam [Table 2 (10)], it was observed that most openings occurred at the perimetral joints. However, the second joint of the left abutment and the second and third of the right abutment opened slightly at crest level. The valley of this dam is very narrow—920 ft (280 m) with a width-height ratio of 2 and a very well compacted high modulus rockfill—as determined during the construction. Rockfill moduli of compressibility were ranging from 14,000—24,000 psi (100–170 MPa).

In Foz do Areia Dam [Table 2 (21)], it was observed that at higher levels, the perimetral joint did not open. However, the first eight tension joints, close to both abutments, presented different magnitudes of opening ranging between 0.08 in. (2 mm) to 1-1/4 in. (32 mm). The valley is very wide, 2,720 ft (828 m) with a width-height ratio of 5.2 and low modulus basalt rockfill, as noted by the author. Rockfill moduli of compressibility were ranging between 4,300 and 8,000 psi (30–55 MPa).

The writer sees advantages to monolithically building the face slab when the field conditions and the rockfill produce controllable movements at the perimetral joint. However, the writer wonders if in valleys like Foz do Areia, with compressible uniform basalt, the effect of passing the horizontal reinforcement through the vertical joints probably will result in transferring to the perimetric joint a large amount of undesirable movement. One possible alternative in those sites where large potential movements are expected is to restrict the use of continous reinforcing to joints located in the compression area in the face, to allow the joints close to abutments to move freely.


### Discussion by A. H. Merritt,[13] M. ASCE


The author stated that the most desirable foundation for the toe slab of a concrete face rockfill dam is hard non-erodible rock that is groutable. It was also emphasized that with proper engineering, weathered and seamy rock, fault zones, and materials susceptible to possible erosion and piping are acceptable.

The writer offers the following observations and guidelines concerning foundation treatment from two recently constructed concrete face rockfill dams. These observations support the author's view that because few foundation rocks are of uniformly good quality, some type of remedial treatment should be anticipated in most cases.

Hard, non-erodible, and groutable rocks are commonly found along glaciated valleys or those with steep sides where erosion progresses more or less equally with weathering. In these areas, the softer weathering products do not develop to any appreciable depth and the toe slab usually can be constructed in a cut of minimum dimensions. Extensive ex-

---

[13]Consulting Engr. Geologist, Andrew H. Merritt Inc., 7726 SW 36 Ave., Gainesville, FL 32608.

cavation is to be avoided where possible as the resulting high cuts can lead to slope stability problems. The depth of the cut is chosen to give a width of toe slab varying between 0.04H–0.06H as indicated in Table 2 for foundations in good rocks.

One should be aware, however, that the steeper abutments in hard rocks commonly contain well-developed stress relief joints oriented parallel to flow. These geologic features can lead to significant seepage unless recognized early and specifically provided for in the design and execution of the grout curtain beneath the toe slab. Seepage approaching 1 $m^3$/s was recently measured exiting from one dam which was determined to be by passing the toe slab through stress relief joints. Fortunately, these joints are of nominal width and only contain a thin amount of erodible material. Hence the flows do not continue to increase with time and they are generally groutable. Because the rockfill is quite pervious such flows do not pose safety problems for the dam.

In regions of more subtle relief with deeper weathering, soft erodible and non-groutable materials may be consistently found to depths upwards of 20 m and deeper depending upon local conditions. During the design period, exploratory borings are drilled around the perimeter of the proposed dam and the depth of the toe slab and position of the reference line is established. A misrepresentation of the depth of weathering could eventually require a re-excavation of the abutment slopes to set the reference line deeper. Such changes could lead to extra costs and possible schedule delays, thus appropriate conservation is justified in this aspect of the design.

There are few uniform guidelines regarding the application of the results of borings to the selection of appropriate toe slab foundation. The general procedure is to attempt to determine the amount of overburden or thickness of residual soil and the degree of rock weathering. Because the contacts of soil, weathered rock and hard rock are gradational in nature for typical weathering profiles, considerable judgment is required in assessing suitable foundation elevation. On one recently completed concrete face rockfill dam, experience showed that the toe slab should be located at a depth corresponding to rock with uniform core recoveries greater than 75% and RQD values greater than 50–60%.

These criteria can be accepted as representing the upper horizon of rock which lies below the saprolite and "transition zone." Below these weathered materials, the water tests generally indicate a normal rock joint permeability, i.e., groutable conditions. Weathering of the rock below this depth should be confined largely to alteration along some joints or foliation planes or layers of initially low strength rock and should not have notably altered the minerals between joints in hard rock.

According to this description, some potentially erodible joints can be anticipated at foundation grade, and because typical weathering profiles are noted for their variability, adverse geologic conditions are the rule and special remedial measures can be anticipated.

Various defensive design measures can be adopted. In order to increase the seepage path, the width of the toe slab may be increased over that used for good rock (approx 0.05H). Widths varying between 0.2H–0.4H have been used for local clay seams or pockets of saprolite. When the rock is intensely jointed and some joints are weathered, an intermediate width of toe slab of 0.08H–0.15H may be advisable. Deformable cutoffs connected to the toe slab have been constructed in hand excavated trenches in fault gouge or wide saprolite zones. The area of low pressure consolidation grouting is commonly increased to attempt to control seepage by filling all near surface pervious fractures caused by stress relief or excavation procedures. This is in preference to increasing the scope of the grout curtain because of the difficulty of achieving a reliable cutoff in fine grained materials with normal grouting methods. The length of the seepage path can be extended somewhat in the upstream direction by the construction of a bolted mesh-reinforced shotcrete slab on the abutment slope which is continuous with the toe slab. Filters have been placed between the abutment and rockfill downstream of the face slab to prevent migration of fines.

In summary, the foundation for the toe slab requires the same attention to geologic detail as does any dam foundation. Geologic mapping should be done along the excavation to define all potentially adverse conditions and to aid in developing an appropriate plan of remedial treatment. Should some areas require more extensive treatment, this work can proceed concurrently with placement of rockfill without affecting the construction schedule. This flexibility is one of the positive aspects of a concrete face rockfill dam.


### Discussion by N. G. K. Murti,[14] F. ASCE


The writer offers his congratulations to Barry Cooke for the excellent paper.

1. Rockfill dams with an upstream cover of timber planking and/or steel plates belonged to the 19th century. Rockfill dams with concrete face (no timber-nor steel plate) are now edging out other types.

2. A specification that gives little room for rejections is the best. Rockfill dams with a concrete face approach this ideal more closely than any other type. For earthfills, there is no moisture control. Layered and compacted with vibratory rollers, settlements in rockfills are reduced to the minimum. Soft rock, cobbles and gravels can now be used.

3. No suspension of work in the rainy season is required (except in downpours when labor force will not come to work). It can be allowed to be overtopped by floods, if some precautions are taken after studying models.

4. It is ideal in rocky places where acceptable earth for the core is not available except at long leads. Cement is used to the bare minimum. Cement is costly in India and in short supply.

5. Any leakage does not vitiate the stability by disastrous piping, nor the competence of the reservoir to hold water.

6. Under earthquake conditions, its performance surpasses that of other types.

7. It is unfortunate that this type (CFRD) has not yet been introduced

---

[14]Consulting Engr., Vasant Mahal, C Road, Bombay—400 020, India.

in India. Locations appropriate for CFRD Dams are in the westen ghats where the rainfall is heavy—150–250 in. in three or four continuous months and where there is no earth, except a laterite cover (basalt zone below laterite). Many sites have been explored for preliminary designs of a CFRD at Vazarda Dam site. It is proposed to use laterite for the layered and compacted rockfill, and the upstream concrete face and toe mat will rest on the parent rock of basalt.

8. Studies are in hand to use laterite aggregates for the concrete face slab. Deformation data of Cethana, Alto Anchicaya and Areia Dams indicate limits for strains before concrete cracks. If plastic (deformable) concrete could be developed for diaphragm cutoffs, the same can be extrapolated for this exposed cutoff (concrete face). Low shrinkage properties, and also low weathering under alternate wetting and drying up of the slab above the minimum drawdown level of the reservoir are also to be ensured.

9. Deep in the Himalayan seismic belt, with very high rainfall in the north-east, sites suitable for CFRDs are found. CFRDs have an edge over other types.

### Discussion by Ivor L. Pinkerton,[15] F. ASCE

The eighteenth Terzaghi lecture presented by J. Barry Cooke, although entitled "Progress in Rockfill Dams," refers only briefly to earth and rockfill dams and is really a paper on the development of the concrete face rockfill dam from its early beginnings in California. There is little doubt that the high dumping of rockfill used prior to 1960 was the main reason for some of the earlier concrete face rockfill dams having leakage problems. The foundations were always sound non-erodible rock and the high dumped rockfill was usually of large size. For these reasons, the leakages, although often quite high, did not cause failure, but careful remedial measures were required usually by sealing the upstream face to reduce the leakages to acceptable amounts. As Cooke states, these problems led "to a temporary suspension of the construction of the concrete face rockfill dams of more than 300 ft (92 m) in height during the 1960–65 period." Because of this early history of damage to the concrete face often resulting in high leakages, many design engineers were reluctant to regard the concrete face rockfill dam as being technical acceptable. A further reason for this attitude was the excellent performance of earth and rockfill dams which have been successfully constructed to progressively greater heights since they became popular about 1960. This type of dam provided the designer with a rockfill embankment alternative of proven performance.

In his paper, Cooke deals with the improved practices which have been developed from about 1965 onwards in regard to all aspects of the concrete face rockfill dam. These improved practices are now enabling this type of dam to be successfully constructed to increasingly greater heights with excellent performance.

---

[15]Consulting Civ. Engr., Canberra, Australia.

Many concrete face rockfill dams have been constructed in Australia since about 1965, all using rockfill placed in layers and compacted with a smooth drum vibrating roller. Without exception these dams are all giving excellent performance which has justified the adoption of this type of dam. Three dams in particular are worth mentioning as they have made significant contributions to the present state of the art, namely Cethana, Kangaroo Creek and Mangrove Creek. Cethana Dam, completed in 1971, incorporated several innovations in the design including improved face slab joint seals, a low permeability transition zone behind the face slab with a bituminous coating for erosion protection and the elimination of horizontal joints in the face slab. Both Kangaroo Creek and Mangrove Creek Dams were constructed of low strength soft rock for the main rockfill with appropriate drainage layers of good quality rockfill.

More recently, concrete face rockfill dams have been introduced to Asia for the first time with the construction of three important dams: Khao Laem in Thailand, Batang Ai in Sarawak and Cirata in Indonesia. Each of these dams incorporates new features in design which will contribute to the technical advancement of the concrete face rockfill dam.

*Khao Laem Dam*, 115 m high, has foundations of predominately calcareous sandstone and shale interbedded with limestone and containing solution cavities often filled with secondary clay or alluvium. Although the toe slab and upstream rockfill were founded on sound rock, there were some large areas of soft, decalcified rock having low effective modulus under the central and downstream parts of the dam. In these areas, foundation settlements measured during the construction of the dam ranged from 0.8 m–1.4 m. These large settlements caused extensive cracking of the protective shotcrete on the upstream face of the dam. Fortunately, the continuing settlement had ceased by the time the concrete face slab was placed. Other design features of this dam are a diaphragm wall cutoff over a considerable length of the toe slab and a grout gallery. Face slab reinforcement was carried though the vertical joints except those immediately adjacent to the abutments and at an intermediate location where the foundations rise to a high level. The reinforcement was interrupted at the vertical joints in these locations to avoid concentrating all the joint openings at the peripheral joint. Khao Laem has a face slab area of about 140,000 $m^2$ making it among the largest yet constructed. With the reservoir now 3.3 ft (1 m) from full supply level, no measurable leakage can be detected.

*Batang Ai Dam*, 85 m high, on a sound dolerite foundation is curved in alinement being convex upstream. The transitional material behind the face slab consists of 4 parts of river gravel to 1 part of silt giving a well-graded material with a permeability of $10^{-3}$ to $10^{-4}$ cm/s. Placed in 250 mm layers, final compaction on the upstream slope was made using a 1.4 m by 0.9 m vibrating plate compactor mounted on an excavator. This method has the advantage that a small area could be completed and coated with cement mortar to prevent erosion in this high tropical rainfall area. The method resulted in little or no excess concrete being required in the face slab. Filling of the reservoir commenced in November 1984 and is now substantially filled with negligible leakage.

*Cirata Dam*, 125 m high with provision for raising to 140 m, has foundations of breccia/sandstone with some claystone in the upper right abutment. Rockfill will consist of breccia with an andesite drainage zone. Two transition zones are provided behind the face slab with gradings specified to give low permeability and filter action to take care of possible leakage through the face slab joints. Construction of the dam commenced in 1984.

The significant improvements introduced to concrete face rockfill dams in recent years will undoubtedly pave the way for a wider acceptance of this type of dam. It has now been developed sufficiently to give designers confidence that it is an alternative which must be considered. However, as with any other engineering structure, attention to detail both in design and construction is most necessary in order to obtain a high standard of technical performance.

### Discussion by Pietro De Porcellinis[16]

This discussion presents a new type of concrete face dam repair, which has been successfully performed already in two significant cases: Paradela, Portugal and Rouchain, France. The New Exchequer Dam, another example of an early concrete face rockfill dam, is under design for a similar repair. The common problem in these structures has been the excess in leakage, and never questioning the dam's safety, only its economical aspects.

The Paradela Dam had a long history of leakage and repairs (3). Since its completion, cracks in the upstream face and problems with the curtain wall produced important leakage: The curtain wall was quickly repaired by grouting (2), and the concrete face underwent several repairs, e.g., crack sealing, slabs repairs, etc. While studying the dam's leakage problems, the following paradox was observed: The rockfill settlements were quickly reduced with the structure's age (two-thirds occurred in the first five yrs) being actually in the order of 0.4 in. (10 mm)/yr. Despite this fact, leakage rate (even after the most accurate repairs) tended to increase, especially during the last years (Fig. 15). The concrete face was behaving as a fatigued structure, where cracking was not dependent on deformations anymore, but rather on more complex reasons.

The Portuguese EDP company, owner of the structure, decided to study a definitive and permanent program of repairs (3). The complex problem needed a comprehensive solution that could give a definitive remedy to the dam's leaks.

Such a solution was encountered in a continuous sealing membrane covering the whole upstream face and providing th epermanent sealing of every possible future crack of the concrete. A similar coating must have exceptional properties that a normal prefabricated synthetic layer does not provide. The material must resist permanent stresses due to its own weight because of the great length and steepness of the dam's slopes. Additionally, it must be resistant to weathering, winds and waves, and be able to bridge concrete cracks at very high pressures (up to 11

---

[16]Engr. Mgr., New Product Development, Rodio Co., Madrid, Spain.

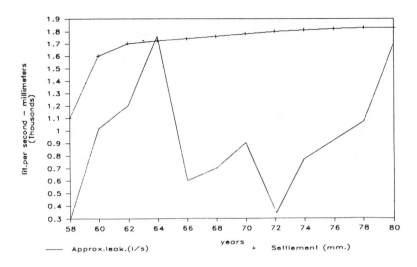

FIG. 15.—Paradela Dam: Approximate Leakage and Settlements

atmospheres, 154 psi). Those conditions were assumed more than ten yrs ago by the author's research team which directed the development of a specialized coating system.

The system is essentially a waterproofing coating produced "in situ" by impregnating a geotextile fabric in a bath of specially developed polymeric solution. The impregnated geotextile is spread with a rolling-out machine over the surface, in strips that overlap. At setting, it glues onto itself, producing a continuous layer, and onto the concrete, giving a sufficient adherence to resist the aforementioned exposure. The installation is possible over very steep slopes (with a special technique it is applicable on vertical surfaces).

The material is conceived as a compound in which the functions of waterproofing and mechanical resistance are separated. The first function is provided by a polymeric substance belonging to the "hydrogel" family, reinforced with special additives. The main features of such materials are as follows:

*Swelling capacity.* When submerged, the polymer swells a limited amount with an osmotic pressure of several atmospheres (its permeability, about $10^{-14}$ cm/s, does not normally depend on the water pressure).

*Permanent adhesive properties* that make it self-gluing.

*Very high weather resistance* due to the hydrogel properties and to the acrylic cross-linked structure.

The reinforcing capacity of the polyester geotextile are well-known today by civil engineers. It is enhanced by the impregnating polymer, especially when the coating is out of water (dryed hydrogel).

This type of solution was approved by the owner and a complete repair program was scheduled for summer 1980. 560,000 sq ft (55,000 m$^2$) of concrete surface were covered between May and September. The concrete face joints, due to the small settlements previewed, were cleaned and filled with deformable mortar; obstacles on the surface were chipped out and smoothed by mortar patching.

Today, Paradela total leakage at complete filling is about 1 cfs (25 L/s). In 1983, the "Direction Departamentale de l'Equipement," Loire, France, decided to repair the concrete face (compacted) rockfill dam of Rouchain. Consultant, Coyne et Bellier of Paris, indicated that the procedure used in Paradela was the most suitable. Works were performed in summer 1983. In 1984, Rouchain reservoir was filled for the first time since its completion in 1976. A minor and located leak appeared at the complex sliding joint provided around the intake tower structure through the upstream face. After repairs performed in 1984, leakage was stablized around 0.4 cfs (10 L/s).

In conclusion, we can state that concrete face rockfill dams, the safest dams conceived by man, may be given a new life, when aging or accidents produce unacceptable leakage through the concrete face.

### APPENDIX.—REFERENCES

32. Gomes Fernandes, L. H., de Oliveira, E., de Vasconcelos Porto, N., "Paradela Concrete Face Dam," *Transactions ASCE*, Vol. 125, 1960, p. 365.
33. Weyermann, W., "Foundation Treatment of Paradela Dam," *Transactions ASCE*, Vol. 125, 1960, p. 419.
34. Comte, C., "Waterproofing of the Paradela Rockfill Dam," *Colloque CEMAGREF*, Paris, France, Feb., 1983.
35. Salembier, M., "Leakage Control Works at the Rouchain Dam," *Water Power and Dam Construction*, Dec., 1984, pp. 15–17.


### Discussion by C. F. Ripley[17]

The paper is timely, informative and authoritative with respect to both the evolution of modern rockfill dams and the current state of the art of concrete face rockfill dams. The author has emphasized that the general progress in rockfill dam technology has been evolutionary. In the writer's opinion, however, there is one aspect of rockfill dam technology in which a trend of change since about 1960 has not been progressive, but has been regressive with impairment of dam safety. This aspect merits attention and has to do with the filter-transition component between the core and downstream shell of thin core rockfill dams.

For earth core type rockfill dams, the primary focus of design and construction considerations is by necessity much different than for the impervious face type, on account of the erodible nature of the earth core material. In the 1940–1960 early development period it had been recognized that cracks through core zones were more common than suspected and that all earth and rockfill dams should incorporate effective defensive measures to render them harmless (37,49). The early earth core rockfill dams commonly had sloping cores and the rockfill was usually dumped and sluiced in high lifts. The dumped rockfill of that era had inter-particle void sizes many times larger than those of the modern compacted rockfill. Likewise, at that time, rockfill deformations were many

[17]Consulting Engr., Victoria, Canada.

times larger. Consequently, the potential for core cracking was great, as was the span of particle sizes between core fines and rockfill that required filtration. Initially with Nantahala Dam in 1940 and followed by 6 other Alcoa dams between 1940 and 1955, Growdon's defensive measures to prevent piping of core material had been shown to be fully effective (three papers by Growdon in Ref. 1). The measures consisted of: A 2 zone filter-transition component on the upstream side of the core; a 3 zone filter-transition component on the downstream side of the core; and careful blanket and curtain grouting of bedrock beneath the core contact area. Processed materials were used in the filter-transition zones; sand minus 1/2 in., gravel 1/2 to 3 in.; and rock 3 to 10 in., respectively, proceeding away from the core. The specified narrow gradation for each of these zones had the practical and essential property that the materials were not susceptible to harmful segregation during dumping and spreading operations. Moreover, the materials used in the sand filter zones were clean and cohesionless, minimizing the susceptibility of the sand to sustain an open crack through which core fines could pass. This "crack stopper" capability of clean cohesionless sand had been proposed by Terzaghi and had been proven to be effective at Brazilian dams where sand chimney drains had prevented piping through homogenous clay dams containing extensive observed transverse cracks (40,50). It was a capability that received particular attention at Kenney Dam, at Terzaghi's insistence (41). The defensive measures used by Growdon became widely used by others for thin core rockfill dams in the 1950s, for both dumped and compacted rockfill, and for both sloping and central core types: Kenney Dam, 1952; Bersimis Dams #1 and #2, 1955; Derbendi Khan, 1957; Brownlee, 1958; Binga, 1960; Trangslet, 1960; all as described in Ref. 1; and Ambukloa (6). Different defensive measures against piping were used at a number of other earth core rockfill dams built in the 1940–1960 period. These ranged from the use of wide central cores with slightly less stringent narrow filter-transition components, to narrow cores with wide transition zones, Notteley, 1942; Watauga, 1948; South Halston, 1950; Kajakai, 1952; Cherry Valley, 1955; all as in Ref. 1. None of these dams has suffered a piping incident or significant leakage to the writer's knowledge.

During the 1960s and 1970s, serious piping incidents within the core occurred at a number of thin earth core rockfill dams, Sir Adam Beck Dykes (48); Hyttejuvet (43,53); Balderhead (Ref. 7, pp. 312–317) (51); Yard's Creek (Ref. 7, pp. 317–324); Matahina (Ref. 7, pp. 308–312); Churchill Falls Dykes (36,38,39,45); Viddalsvatn (42,52). At each of these cases, surface manifestations of piping within the core material took the form of an abrupt increase of seepage discharge laden with sediment, and the appearance of sink holes upstream and/or downstream of the crest. While none of the incidents involved a dam breach, all created a real sense of urgency for remedial action, and all required costly repairs.

The focus of the investigations and analyses in the reports on Hyttejuvet, Balderhead, Yard's Creek, Matahina, Churchill Falls and Viddalsvatn dwelled mostly on transverse core cracks as being the primary problem to be overcome in the prevention of piping. Seepage though joints in the rock foundation was the assigned cause for the Sir Adam Beck Dykes, and a possible assigned cause for Yard's Creek and Churchill Falls. One wonders why this nearly total focus of attention on core and foundation cracks? The obvious and undeniable fact was that the filter and/or transition zones at each dam failed to fulfill their design function as filters; otherwise piping of core material could not have occurred. The filter failures and the causes of them received little attention in the reports, and have not yet received the universal attention that is warranted.

The writer's opinion is that the filter-transition component and/or the foundation treatment at each of these cases involved a significant departure from the proven practice of the 1950s that was initiated by Growdon and adopted by others. What were the departures? The writer believes they were two-fold; first, the use of less stringent particle size gradation requirements for the zones of the filter-transition component, and less stringent treatment of the core-bedrock contact area.

The writer believes that apart from having adequate permeability, there are three basic requirements of the downstream filter zone of a thin core dam that are essential to prevent piping of core material: (1) Appropriate particle size gradation to prevent entry of core fines more than a few millimeters; (2) non-susceptibility to harmful segregation during handling and placement; and (3) "crack-stopper" capability. Experience has shown that a clean sand similar to the Growdon Dams, Kenney Dam and many others, satisfies these requirements. A maximum particle size to eliminate concern for segregation appears to be 3/4 in. with at least 60% finer than #4 sieve. Because the presence of fines imparts undesirable cohesion to a filter sand, the maximum fines content in order to have "crack-stopper" capability appears to be 2% minus #200 sieve, although 0% passing the #100 sieve is preferable. A recent program of severe laboratory tests has indicated the range of sand gradations within these limits that will block entry of the finest core particles (47).

Two basic requirements are considered essential for the transition zones downstream of the filter: (1) Appropriate particle size gradation to prevent entry of particles from the adjacent upstream zone, as well as to prevent migration of particles from the zone itself into the adjacent downstream zone; and (2) non-susceptibility to harmful segregation. A second recent program of severe laboratory tests has confirmed that the conventional ratio, $D_{15}:d_{85} \leq 5$, is a reliable criterion for satisfying requirement 1, (Ref. 46). No definitive studies appear to have been made to determine the limits of particle size range within which harmful segregation during handling and placement of gravel and cobble sizes is prevented. However, ranges of 1/2–3 in., and 3–10 in. have been proven effective in practice for gravel cobble sizes. These performance data suggest that a well-graded transition material of gravel sizes or larger with a $d_{100}:d_{10}$ ratio less than 6 will effectively resist harmful segregation during typical construction operations.

While it is an undeniable fact that many dams have performed satisfactorily with widely graded filter and transition zones, surely the lesson demonstrated by most of the piping incidents previously referenced is that use of widely graded materials incurs a real risk of segregation at random locations where the filter criteria are not satisfied and where piping can occur. The writer believes that segregation is the major culprit to be guarded against, and that the risk of uncontrollable segregation increases as the range of particle sizes becomes wider. The comments on these points by Leps in Ref. 44 are particularly pertinent. The mechanics of eliminating that risk is simple and the additional cost is modest. Surely the processing of filter materials into select sizes for the two or three filter-transition zones of an earth core rockfill dam is as important as the universally accepted processing of aggregates for concrete dams into a larger number of sizes; and surely the costs of processing should be as acceptable.

In conclusion, attention is drawn to the fact that some design changes during the 1960s and 1970s from earlier proven practice for thin earth core rockfill dams have not been progressive, but have resulted in a number of serious piping incidents. There is a need for development of empirical criteria on which to judge the susceptibility of coarse aggregates to harmful segregation.

## Appendix.—References

36. Boivin, R. D., and Seemel, R. N., "Design of the Dykes" and "Construction of the Dykes," 2 papers delivered at CANCOLD Annual Meeting, Quebec City, Canada, 1973.
37. Casagrande, A., "Notes on the Design of Earth Dams," *Journal of the Boston Society of Civil Engineers*, Oct., 1950, pp. 424–429.
38. Chadwick, W. L., Discussion of Q49, *XIII ICOLD*, Vol. V, New Delhi, India, 1979, pp. 410–414.
39. Chadwick, W. L., Discussion of Ref. 38, *XII ICOLD*, Vol. V, Mexico, 1976, pp. 281–282.
40. Hsu, S., "Residual Clay Earth Dams Constructed by Rio Light SA," *2nd Pan Am. Cong. Soil Mech. and Found. Eng.*, Brazil, 1963, Vol. II, pp. 347–363, and discussion by Queiroz, L. A., pp. 679–682.
41. Jomini, H., "The Kenney Dam," *Journal of Engineering Institute of Canada*, Montreal, Canada, Nov., 1954, pp. 6–17.
42. Kjaernsli, B., Discussion of Q42, *XI ICOLD*, Vol. V, Madrid, Spain, 1973, pp. 476–479.
43. Kjaernsli, B., and Torblaa, I., Leakage through Horizontal Cracks in the Core of Hyttejuvet Dam," Public No. 80, Norwegian Geotechnical Institue, Oslo, Norway, 1968.
44. Leps, T., Discussion of Q49, *XIII ICOLD*, Vol. V, New Delhi, India, 1979, pp. 414–415.
45. Seemel, R., and Colwell, C., "Drainage Provisions and Leakage Investigations of the Churchill Falls Dams and Dykes," *XII ICOLD*, Vol. II, Mexico, 1976, Q45, R8.
46. Sherard, J. L., Dunnigan, L. P., and Talbot, J. R., "Basic Properties of Sand and Gravel Filters," *Journal of Geotechnical Engineering*, ASCE, Vol. 110, No. 6, June, 1984.
47. Sherard, J. L., Dunnigan, L. P., and Talbot, J. R., "Filters for Silts and Clays," *Journal of Geotechnical Engineering*, ASCE, Vol. 110, No. 6, June, 1984.
48. Taylor, E. M., "Sir Adam Beck—Niagara Generating Station No. 2 Pumped Storage Reservoir, Observation of Performance," *Eng. Jour. Canada*, Vol. 1, No. 28, Nov., 1963.
49. Terzaghi, K., "Discussion," *Proc. Third Int. Cont. Soil Mech. and Found. Eng.*, Vol. III, Zurich, Switzerland, 1953, p. 217.
50. Vargas, M., and Hsu, S., "The Use of Vertical Core Drains in Brazilian Earth Dams," *X ICOLD*, Montreal, Canada, 1970, Vol. I, O36, R36.
51. Vaughan, P. R., Kluth, D. J., Leonard, M. W., and Pradoura, H. H. M., "Cracking and Erosion of the Rolled Clay Core of Balderhead Dam and the Remedial Works Adopted for its Repair," *X ICOLD*, Montreal, Canada, 1970, Vol. I, Q36, R5.

52. Vestad, H., "Viddalsvatn Dam, A History of Leakage and Investigations," *XII ICOLD*, Vol. II, Mexico, 1976, Q45, R22.
53. Wood, D., Kjaernsli, B., and Hoeg, K., "Thoughts Concerning the Unusual Behaviour of Hyttejuvet Dam," *XII ICOLD*, Mexico, 1976, Vol. II, Q34, R23.

## Discussion by James L. Sherard,[18] F. ASCE

This is an unusually valuable paper describing progress in rockfill dams by an author who has been closely involved with the directions which the progress has taken for more than 40 yrs.

The paper is especially valuable for its treatment of the development of the modern concrete faced rockfill dam (CFRD). Only a few years ago, the CFRD was not generally considered a serious alternative by many specialist engineers because of the earlier history of large settlement (of dumped rockfill) and leakage (through torn waterstops and slab cracks). This situation has changed rapidly, and until now, the CFRD has been generally accepted and is being used with remarkably increasing frequency all over the world.

This current acceptance of the CFRD is due to design improvements which have largely eliminated the settlement and leakage problem, and to the cost and technical advantages of the CFRD over the earth core rockfill dam. To an important degree, it is also due to the influence of the author, who has long been a strong and persuasive advocate of the CFRD and who has been associated with the design and construction of a large majority of these dams. There can be little doubt that the current acceptance and widespread use of the CFRD would not have been reached for many years in the future, if ever, if it had not been for the author's personal consulting practice and remarkable influence.

It is appropriate that this landmark paper was presented as a Terzaghi Lecture. Among his many contributions, Terzaghi put his finger briefly into the developing CFRD practice: His ideas, presented in two discussions of papers in the 1960 ASCE Symposium on Rockfill Dams (54), strongly influenced two major changes, i.e., adoption of a concrete toe slab placed on top of the rock foundation in place of the former concrete filled trench blasted in the rock and the change to compacted rockfill from the former dumped rockfill.

In the remainder of this discussion, the writer presents briefly some opinions generated from his work with the CFRD, on which the author's comments would be valuable.

**The Restrictive Role of Precedent.**—The author correctly emphasizes that the modern CFRD design has evolved step-by-step based primarily on experience with only very little influence from theory. Each design change in a new dam is justified because it is not greatly different from that used on previous successful dams of similar dimensions. This evolutionary design development is comforting but it is also unnecessarily restrictive: When followed strictly, it prevents consideration of large abrupt changes which may be clearly seen to be desirable and perfectly safe and reasonable on the bases of both theory and common sense. The writer believes that this "justified by precedent" design approach has

---
[18] Consulting Engr., San Diego, CA.

unreasonably hindered CFRD development in several main directions, including: (1) Development of a superior perimeter joint; (2) limitation on the maximum permissible height; and (3) prevention of general consideration of spillways over the CFRD.

**Perimeter Joint.**—The current perimeter joint (Fig. 4) with its three separate water stops has been shown to perform adequately under ordinary conditions; however, it has two important technical disadvantages: (1) It is complicated and requires the closest continuous supervision during construction to assure proper installation; and (2) it will not withstand an opening or slab movement in direction normal to the upstream slope line of more than a few inches without rupturing all the water stops. It seems clear that this is not a good detail, especially for higher dams in steep walled canyons where several inches of downstream slab movement relative to the abutment can easily occur. The writer believes it is clear that this joint detail should be abandoned in favor of a new design which will reliably remain watertight for any reasonable movement of the slab relative to the abutment and will be simple to construct. Such a design would be a relatively simple and straightforward engineering task, and the cost would not be significantly greater, but the strong influence of precedent has prevented the abrupt change needed to allow it to be seriously considered or used until now.

**Maximum Allowable Dam Height.**—The writer has participated in recent years on several projects where earth core rockfill dams have been chosen for high dams—90 ft (180 m) or more—although preliminary studies showed the CFRD was substantially less costly: The CFRD was not chosen only because "it would be too large a step from past precedent." The writer believes that the use of precedent in this way is unreasonably restrictive. For compacted rockfill dams up to 820 ft (250 m) height, or higher, we can predict reliably, beyond all question, the maximum possible movements of the concrete slab and we can design joints to remain reliably watertight with complete assurance.

**Spillways on Top of CFRD.**—At many sites, the total project cost could be substantially reduced by putting the spillway over the top of the dam. For the compacted rockfill CFRD, the post-construction settlement of a spillway with control structure on the dam crest and chute on the downstream slope would be negligible. Such a spillway can be designed which is completely safe, but there are only very few examples in recent years. Preliminary designs have been made for spillways over CFRD's, with heights of 328 ft (100 m) and more, for both service and auxiliary spillways, on several projects in the last few years, but were abandoned in final design because of "lack of adequate precedent." The writer believes that this is another unreasonable restriction imposed by the strong influence of precedent which has dominated thinking about the CFRD. Cearly, a compacted rockfill is a superior foundation for a spillway, from all technical points of view, than many of the natural spillway foundations commonly chosen directly adjacent to the dam.

#### APPENDIX.—REFERENCE

54. Terzaghi, K., Discussions of Salt Springs and Lower Bear River Dams and of Wishon and Courtright Dams, *Proceedings*, ASCE, 1960, Part II, pp. 139 and 622–625.

## Discussion by Arthur G. Strassburger,[19] F. ASCE

J. Barry Cooke is to be commended for his excellent summary of the evaluation of rockfill dams as well as his analysis of reasons for the evolution in design and construction. He references the early development of the dumped rockfill dams in the Sierra Nevada, CA, and the contribution of some Pacific Gas & Electric Company (P.G.&E) dams. The writer herein briefly discusses experiences with three P.G.&E dams. Those experiences support at least two of the significant points which the author credits to Terzaghi, i.e., "(3) the benefits of compacted rockfill . . .," and "(4) the potential usability of weaker rock with compaction."

The writer has provided simplified illustrations of settlement for Salt Springs, Lower Bear River No. 1 and Courtright Dams. All three of these were discussed at length by Steele and Cooke (56) and by Cooke (55) in 1960. These three dams were selected because Salt Springs and Courtright are the highest of the P.G.&E Dams and have experienced the greatest settlement and highest maintenance costs. Lower Bear River No. 1, although lower in height, has a generally similar profile and was built, chronologically, between Salt Springs and Courtright. It has required almost no maintenance.

Figs. 16–18 show settlements at the maximum cross sections of the three dams. Fig. 19 shows the comparative profile and height of each dam. Also shown are the approximate areas of greatest settlement. The greater vertical component of settlement, and its high location, experienced by Courtright as compared to the other two dams is noteworthy.

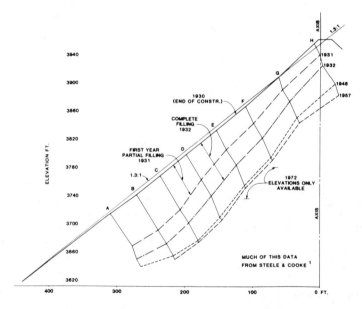

**FIG. 16.—Salt Springs Dam Settlement Sta. 4 + 60 (Maximum Section)**

[19]Consulting Engr., San Rafael, CA; formerly with Pacific Gas & Electric Co.

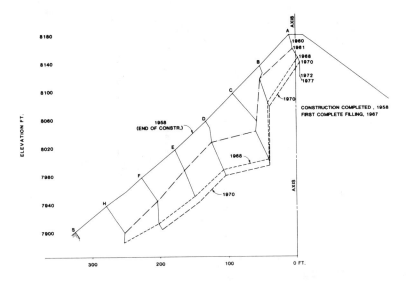

FIG. 17.—Courtright Dam Settlement Sta. 5 + 40 (Maximum Section)

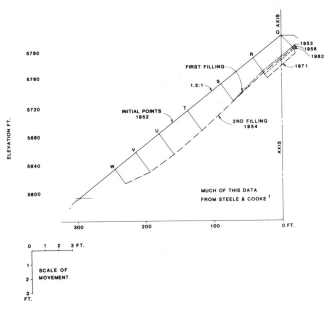

FIG. 18.—Lower Bear River Dam No. 1 Sta. 3 + 80 (Maximum Section)

The writer was associated with the engineering related to the maintenance and safety of P.G.&E's Dams for a number of years. He was intimately involved with repairs and improvements to Courtright in 1968 and 1970. The writer believes there are a number of causes for the "unusual" behavior of Courtright, but foremost are the type of rock used

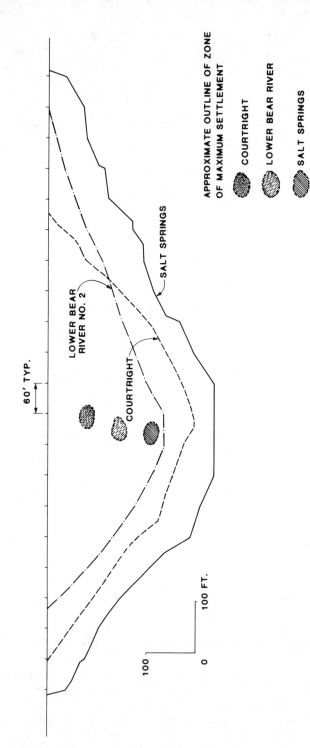

FIG. 19.—Upstream Elevations

in the lower portions of the rockfill, and the profile.

All three dams are constructed of a relatively sound, unweathered Sierra granite. The quarry used for Courtright was highly exfoliated to significant depth. It was this rock, which tended to be rather slabby, which was placed in the bottom of the fill. Naturally, in the dumping process, it tended to land in a "layered" position, and tended to bridge to a greater degree than a more uniform dimensioned angular rock would have. This also resulted in relatively high voids and permitted relatively large settlement. This settlement was, however, somewhat constrained in the lower zones of the dam by the relatively narrow canyon. Thus settlement in the lower zones, as shown by points F and H, is modest considering the quality of rock used. Both Salt Springs and Courtright have at times experienced significant leakage, and both have needed significant concrete face repairs. Both dams, however, have performed safely and adequately. The face repairs to Courtright in 1968 and 1970 were supplemented by a zoned fill placed against the lower portions similar to that shown in Fig. 10 of the author's paper.

The writer concludes that the slabby rock used in Courtright would not have experienced as much post-construction breakage and settlement had it been compacted in accordance with modern technology. The same conclusion also applies to Salt Springs and Bear River, but the differences between the dams particularly emphasize the influence of the slabby rock combined with the lack of compaction.

## Appendix.—References

55. Cooke, J. B., "Rockfill Dams: Wishon and Courtright Concrete Face Dams," *ASCE Transactions*, Part II, Vol. 125, 1960.
56. Steele, I. C., and Cooke, J. B., "Rockfill Dams: Salt Springs and Lower Bear River Concrete Face Dams," *ASCE Transactions*, Part II, Vol. 125, 1960.

### Discussion by William F. Swiger,[20] F. ASCE

Embankment dams are among the oldest engineering structures. Studies indicate that such dams were used for diversion of irrigation water in the Near East in the fifth millenium, B.C. The paper, which gives the history and background of rockfill dam development, is thus a most valuable and welcome addition to the literature on embankment dams.

The highest embankment dams of today are of the rock or gravel fill type. There are a number of these in service, under construction, or in investigation and design which are higher than 500 ft (150 m). As the author states, development of the design and construction of these structures has been evolutionary and seems to the writer to have been guided by the following:

---
[20]Consulting Engr., Buhl, ID.

1. Changes as dictated by observation to correct problems or deficiencies or to improve economy.
2. Development and availability of large and durable specialized construction equipment.
3. Analytical procedures and testing.

As indicated, problems were experienced with many of the early concrete face dams especially at the perimeter joint, crushing failure of the face slab being observed in some. These problems have been greatly reduced by dense compaction of the fill using large vibratory rollers, face design improvements, and a change in zoning. Evaluation of stability of the left bank of the upper reservoir at Cabin Creek led to installing a system of drains. The materials excavated were saprolitic soils containing about 10% of material passing the No. 200 mesh. These were disposed of by placing them in the dead water zone of the reservoir. Thus a low permeability fill was placed over the lower portions of the face slab and perimeter joint. Leakage through the dam is negligible. These observations led to providing similar low permeability fill against the face of several later dams to control seepage at the perimeter joint, e.g., Areia (160 m) and Salvajina (158 m). The author now proposes to carry this design concept further in a dam of unprecedented height for concrete face dams. The sequence is illustrative of the evolution of design.

The author is generally considered the outstanding authority on concrete face embankment dams, and these are discussed at length. However, additional emphasis on certain specifics is considered appropriate.

As stated by the author, not all problems with perimeter joints have been solved. In very steep walled canyons such as Golillas some displacement of the face slab relative to the toe slab has been noted and there has been leakage along this joint. This may be caused in part by problems with obtaining thorough compaction because of limited access and because along such very steep walls, the slab tends to move downstream about parallel to the canyon wall.

Water stops must be carefully detailed to accommodate possible movement of the face slab. Water stops are easily damaged and protection against equipment, tools, rebar, and construction operations should be provided. Further, the water stops should be carefully inspected and, if necessary, repaired just before placing concrete. Very careful workmanship and supervision are necessary in placing concrete to ensure they are properly positioned and properly embedded in concrete.

Modern practice is to place a zone of semipermeable material, commonly termed Zone 1, immediately below the face slab. This reduced problems with raveling during construction and would limit seepage into the dam in the event the cofferdam were overtopped. Suggested gradations for this zone are given. Material underlying Zone 1 must be graded to act as a filter for Zone 1. Also, segregation along the contact must be prevented. For dams subject to significant freezing, it is the writer's opinion that Zone 1 material should be graded so that it would not be subject to ice lens formation (frost heave).

River diversion frequently is one of the most difficult problems of designing a dam. The philosophy and concepts outlined by the author are excellent and worthy of careful consideration.

One of the continuing requests of members of the Geotechnical Division has been for more "practical" papers. This is an outstanding "practical" contribution which will be of value for many years.

### Discussion by H. Taylor,[21] M. ASCE

As the author mentions, Terzaghi has made significant contributions to rockfill dams. One of his most notable contributions was the design of the difficult rockfill/earthfill dam (58) on the Bridge River in British Columbia, now called Terzaghi Dam. This dam has performed well, although in its early years it was troubled with relatively large sinkholes (57).

The writer has been associated with the author as a member of our Advisory Boards for the past 20 yrs and considers the "Eighteenth Terzaghi Lecture—Progress in Rockfill Dams" to be very appropriate, worthwhile, and timely now that designers are studying such dams in the 200 m–300 m height range.

The writer cautions the use of soft (poor quality) rockfill material for high dams with a concrete face. Such rockfill has been successfully used as shell material for high dams with a central impervious core. However, because of the potential for large seepage flows through rockfill from cracks or open joints in a dam with a concrete face, rockfill that contains a large percentage of fine material should be avoided to reduce the risk of internal erosion.

The La Joie Dam on the Bridge River in British Columbia (No. 11 in the author's list of dams on page 1396) is an example of a rockfill dam that has performed well with large seepage flows passing through the rockfill.

The seepage flow through the 87 m high La Joie Dam for a typical one-yr cycle of reservoir operation, when the upstream surface of the dam had a timber face, is shown on Fig. 20. Such peak flows 300 cfs (8.5 m$^3$/s), passed through the hard, durable rockfill each year for some 23 yrs without doing any damage to the rockfill.

In 1973–74, the timber face was removed and a 3–4-in. layer of reinforced shotcrete was placed for the full depth of this dam. The shotcrete has performed well to date and has reduced the seepage to about 20 cfs at full reservoir.

If hard, durable rock is available at a dam site, the designer should consider a rockfill dam with a concrete face as one of his alternatives.

The author and the writer have had many discussions on the use of water when compacting rockfill. The author suggests that water in the order of 20% of volume of rock is beneficial. It has been the writer's finding that water up to 50% of the volume of rockfill placed in each layer prior to compaction is beneficial in reducing subsequent settlement or deformations. The amount of water to be used would depend on the rockfill material. The writer would be inclined to use the higher percentages of water where the rockfill consisted of soft (poor quality) rock

---

[21]B. C. Hydro, Mgr., Geotechnical Dept., Vancouver, Canada.

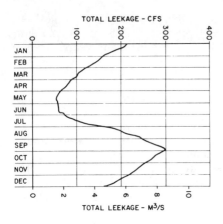 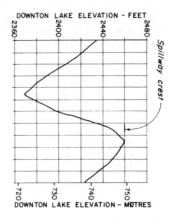

**FIG. 20.—La Joie Dam—Typical Reservoir and Seepage Annual Cycle**

**FIG. 21.—Applying Water to Rockfill Prior to Compaction—Revelstoke Dam**

and contains a relatively large percentage of fines. Fig. 21 shows a watering truck applying water to a rockfill layer at Revelstoke Dam (Columbia River) prior to compaction. For this rockfill material it was specified that water equal to 50% of volume of rockfill is to be applied to each layer.

### Appendix.—References

57. Taylor, H., "Performance of Terzaghi Dam, 1960 to 1969," ISSMFE Conference, Mexico City, Mexico, 1969.

58. Terzaghi, K., and LaCroix, Y., "Mission Dam—an Earth and Rockfill Dam on a Highly Compressible Foundation," *Geotechnique*, Mar., 1964.

## Closure by J. Barry Cooke[22]

The discussions emphasize important points in the paper, provide useful data and views, and raise some questions. They increase the value of the paper.

**Discusser Casinader.**—The example of adapting Winneke Dam, Australia, to a poor foundation, dispersive clay seams and upstream dip slopes on the abutments, is given. The inevitability of occasional fine horizontal cracks in the continuous pour face slabs is reported. Methods of minimizing the cracks, though recognized to be unimportant, are cited.

**Discusser Chadwick.**—Chadwick, former Vice President and Chief Engineer of Southern California Edison Company, reinforces the writer's data on the successful early PG&E Sierra Nevada Dams with Southern California Edison experience with dams as old as 130 yrs with 0.75H:IV downstream slopes; many still in service. His experience with freeze-thaw damage to early Sierra concrete is noted in comparison with no damage to modern concrete with air entrainment. The satisfactory performance of the CFRD face at extremely cold sites is further demonstrated by Cabin Creek Pumped Storage Dam, CO, where shoreline blocks of ice are of 12 ft (3.6 m) dimension and central reservoir sheet is 4 ft (1.3 m) deep.

**Discusser Fetzer.**—The valuable US Army Corps experience, at R. D. Bailey Dam during construction floods, is good to be in the record. The reinforced cofferdam worked when overtopped. The dam without concrete face and with a flood rising 70 ft, resulted in approximately 250 cfs (7 m$^3$/s) leakage which caused negligible settlement of the 210 ft (63 m) high dam. Since Bailey was designed, the grading of the zone under the face has evolved to give a more impervious granular zone, which would have resulted in less leakage. The use of a major zone of poor shale, Zone 2, combined with well-zoned drainage sandstone was economical and unquestionably satisfactory.

**Discusser Fitzpatrick.**—The Tasmania Hydroelectric Commission has been the major contributor to the gaining and dissemination of experience with CFRD's, and also in presenting analyses; they have been a principal contributor to progress. The writer agrees that it is time to extend experience with analyses, but that currently experience is essentially our guide to design. For future higher dams, it is agreed that the area needing instrumentation in new high dams and analyses is in face slab stresses.

**Discusser Fucik.**—Two important dams, the highest of their type at their time in the history of rockfill dams, are referred to. Both established precedents. The 275 ft (84 m) Dix River Dam, constructed in 1925, was precedent for the 320 ft (98 m) Salt Springs Dam in 1932. After 60 yrs, the Dix River Dam of dumped limestone rock and on a limestone

---

[22]Consulting Engr., J. Barry Cooke, Inc., San Rafael, CA.

foundation is still in reliable safe operation. The limestone is a factor in the leakage and settlement; leakage emerges from rock downstream from the dam and settlement is more continuous than for other dumped rockfills due to solution of rock points and edges in contact. Ambuklao, an earth core rockfill, was among the first to use both dumped and compacted rockfill. ("Ambuklao Rockfill Dam, Design and Construction," by M. Fucik and R. F. Edbrooke, *Transactions*, ASCE, Vol. 125, Part I, 1960, pp. 1207–1227.)

**Discussers Hacelas and Ramirez.**—The construction and early performance of the 480 ft (147 m) high Salvajina Dam is summarized by two engineers who closely followed instrumentation, inspection, and design modifications. The dam is on a difficult and variable foundation, and upon filling in 1985, leakage is negligible and settlements nominal. The economical zoning of gravel and required excavation rock is shown and the much higher modulus of compressibility of gravel than of rockfill is documented. The high gravel modulus is consistent with that of the Oroville, W. C. Bennett and Mica earth core gravel shell dams.

**Discusser Houlsby.**—Grouting is usefully emphasized as an important factor for the narrow cut-off slab of the CFRD and the three factors mentioned are important. However, the writer does not agree with all comments on the three factos.

1. *Uplift of toe slab*—At Little Para, on a dip slope, uplift occurred and subsequent work proceeded well with uplift gages and revised grouting procedure. At the sensitive dip slope in siltstone, at Mangrove Creek, uplift gages and grouting control was recognized to be necessary and no uplift occurred. Most rock formations give no uplift problem. The toe slab is the ideal grout cap to get satisfactory grouting near the surface, and good standard practice is to grout through the toe slab.

2. *Surface leaks*—The construction program should never be restricted by the requirement to observe surface leaks. Unseen surface leaks may cause some waste of grout, but a satisfactory grout curtain is obtained.

3. *Safety during grouting*—Some small rocks from the surface zone do roll down the face. The change from 6 in. (15 cm) to 3 in. (7.5 cm) maximum size has improved conditions. Care in face zone placement and simple barrier fencing are means to maintain safety. An advantage of the CFRD is never to allow grouting to affect the contractor's schedule or convenience.

**Discussers Marulanda and Ospina.**—The writer agrees with the comments which usefully point out that, simple and safe as a CFRD is, there are details which require careful attention to minimize seepage. Regarding overtopping, it is the only factor which can be visualized to cause failure of a designed CFRD. Reinforced rockfill, well-known in Australia, is used to a height during construction at which overtopping risk is acceptably low. The development of reinforced rock in Australia is due to the fact that major floods can occur in any month. For a completed dam, reliance is on meteorology, hydrology, and the design flood. Details of reinforcing rockfill for overpour are beyond the scope of this closure; however, rapid and effective current methods have been used at Brogo, Glennies Creek, Boondooma, and Fortuna Dams. The discusser's item

1 refers to yielding foundations, which are agreed to present problems. At and near the toe slab, special design is required. Elsewhere in the foundation, settlement is acceptable, e.g., 1.4 m at axis at Khao Laem Dam.

**Discusser Materon.**—Important points raised are: Unrestricted use of ramps within the rockfill, direct placement of required excavations in the dam, and continuous face slab horizontal reinforcing. Useful experience data on horizontal tension areas at two major dams are given. The discusser correctly suggests that where continuous horizontal steel is to be used, the several vertical joints near the abutments that are subject to opening should be cold joints without reinforcing going through. All the opening should not be concentrated at the perimeter joint. This was done at the major Khao Laem Dam, Thailand, as noted by discusser Pinkerton.

**Discusser Merritt.**—The author agrees with the emphasis placed on geology and the comments on practical means of obtaining an impervious and permanent cutoff where rock conditions are poor and variable. The caution to initially carefully predict conservative toe slab excavation depth in poor rock is important.

**Discusser Murti.**—The points noted enumerate real and useful features of the CFRD. Item 9 suggests a "plastic concrete" possibility for face slab concrete, and on this the author does not agree since plastic concrete, as in diaphragm walls, is about 200 psi and erodible. The present standard concrete of 3,000 psi (20 MPa), or moderately greater, has proven to have adequate flexibility, strength, and durability, and to have minor shrinkage cracks. It is interesting to learn that India is considering Vazarda as a CFRD.

**Discusser Pinkerton.**—The brief discussion contains useful thoughts as well as data on some of the modern Asian dams, Khao Laem, Batang Ai and Cirata. The zone under the face becoming finer graded (Cirata) is a useful trend which: (1) Reduces rocks rolling down the face; (2) reduces excess concrete in permitting a smoother face; and (3) permits sealing a leak by dirty sand. Dirty sand can be maximum #10 or #20 mesh with about 30% minus #200. The interesting use of a vibrating plate compactor at Batang Ai resulted in a smoother surface and higher density than the 10 ton vibratory roller pulled up the face.

**Porcellinis.**—The success of Rodimperm at Paradela and Rouchain has been excellent. The Rodimperm, a geotextile saturated by a latex-polymeric thick mixture, has the unique property of binding (gluing) to the concrete a strong and watertight membrane. The discussion is valuable. Rouchain, constructed in 1974, had soft horizontal joints, which contributed to the damage. Since Las Piedras, Spain in 1966 and Cethana in 1967 no horizontal joints have been used except at Rouchain.

**Ripley.**—The writer is indebted to the discusser for his concise and excellent review of filters for the earth core rockfill dam. Ripley was associated with Terzaghi and Growdon on the thoroughly handled Kenney Dam, referred to in the paper. Such discussion on filters for the earth core rockfill dam was beyond the scope of the paper, but is welcome in discussion in order that the critical feature of the earth core dam be reminded to be core foundation treatment and filters. The thorough list of references is valued as a supplement to the paper. The emphasis on filters is also useful to the CFRD since the zone under the concrete face has now evolved to be a stable and semi-pervious filter zone, to accept construction period floods, to minimize leakage in the event of a face problem, and to enable underwater sealing by placing fine dirty sand.

**Discusser Sherard.**—The writer concurs with the comments on the four points raised and adds the following:

1. *Restructure Roll of Precedent*—There was a practice in dumped rockfill dams of using the same unsatisfactory design details. Precedent on dam design has been discarded to a much greater extent for compacted rockfill dams. As one example, 0.5% reinforcing each way and one slab thickness formula was used on all early dumped CFRD dams. With the greater uniform support of compacted rockfill it was not until Golillas and Areia that reinforcing was reduced to 0.4%, and at Cethana the slab was made substantially thinner. Unquestionably, today, new dams can be designed with the use and extension of present knowledge of design and performance; this, together with new ideas.

2. *Perimeter Joint*—It is agreed that a more simple and reliable joint can be developed. At present, the addition of the extrudable mastic covered by a butyl or PVC thick cover at Areia was so successful that it has been used on subsequent high dams. However, the awkward center waterstop has often been eliminated. The HEC development of the bottom waterstop of copper or stainless steel represented progress in successful performance and construction ease, but it does not take much lateral movement. Further progress is expected.

3. *Maximum Allowable Dam Height*—It is agreed that substantially higher steps in height can be made with the application of the wealth of performance data on existing dams. A basic method of extrapolation is that movements are directly proportional to H squared and indirectly to compression modulus. Other factors are involved. The use of stable filter material under the face, and of earth on the face in the lower face area are conservative steps on high dams. Also, the repair of leakage by applying dirty fine sand gives assurance of leakage control in the event of leakage. This method is particularly useful for high dams since diving is not required.

4. *Spillways on Top of CFRD's*—Yes, precedent has restricted their use. With the success of continuous pour slabs on the water loaded upstream face, certainly a spillway slab will be successful on the downstream slope. With the slab being OK before a flood, the weight of several meters of water won't change conditions. The 45 m fuse plug spillway over Larona Dam, Sulawesi is a current example. Among those presently in design are Crotty Dam in Australia, and Ahning Dam in Malaysia, both dams 280 ft (70 m) high. The designs are in the hands of experienced rockfill dam engineers.

**Discusser Strassburger.**—This discussion is useful in recording the large movements and face crack and joint problems with dumped rockfill dams and with face panels separated by compressible joints. Slabby rock is the cause of the excessive movement for the dumped rockfill at Courtright. The fact that Bear River No. 1 at 245 ft had no trouble established that as the maximum height of a satisfactory dumped rockfill dam. The several higher dams had trouble with face leakage.

**Discusser Swiger.**—This discussion refers to the use of Zone 1 (earthfill) on the lower area of the concrete face. It is a standard procedure in areas below river bed where debris would make face inaccessible for any repair. Its use to higher levels for high dams is a conservative measure, defining high as about 360 ft (110 m). However, with Zone 2 (the zone under the face) developing as a filter, leakage repairs by dirty sand are effective, and the height of Zone 1 is now likely to decrease in future high dams. The further emphasis on the perimeter joint is worthwhile.

**Discusser Taylor.**—The LaJoie experience of 23 yrs of high leakage without significant settlement conforms with a number of experiences with flow through rockfill, normally during construction before the concrete face or the core is placed. Discusser Fitzer gives R. D. Bailey construction flood experience. High flows through rockfill have caused only nominal settlement.

## SUBJECT INDEX

Page number refers to first page of paper.

Adobe, 182
Archaeology, 182

Characteristics, 1
Civil engineering, 124
Columns, supports, 1
Concrete, 361
Construction methods, 182
Construction procedure, 94

Dam construction, 361
Dam design, 361
Dams, earth, 312
Dams, rockfill, 361
Deep foundations, 94
Design criteria, 312
Drainage, 182
Drill holes, 94

Earthquakes, 1
Embankment stability, 222, 312
Embankments, 361
Embedding, 1

Failures, 222
Failures, investigations, 222
Field investigations, 124
Field tests, 1
Foundations, 1

Geological surveys, 124
Geology, 124

Historic sites, 182
History, 182
Holographic interferometry, 1

Laboratory tests, 124
Liquefaction, 312

Mounds, archaeology, 182

Pile bearing capacities, 52
Pile foundations, 52, 222
Pile loading tests, 52
Probability theory, 312
Project management, 124
Project purpose, 124

Resonance, 1
Risk analysis, 312

Shafts, excavations, 94
Shear strength, 94
Site investigation, 124
Slope stability, 222, 312
Sociological factors, 182
Soil properties, 94
Soil-structure interaction, 1
State-of-the-art reviews, 361
Static pile formula, 52
Structural design, 94
Structural settlement, 52
Subsurface investigations, 94

## AUTHOR INDEX

Page number refers to first page of paper.

Biarez, Jean, 88
Bratchell, G.E., 289

Cambefort, H., 297
Casinader, Ranji, 395
Chadwick, W. L., 396
Chaplin, Theodore K., 86
Cooke, J. Barry, 361, 427

DePorcellinis, Pietro, 412

Fetzer, Claude A., 397
Fitzpatrick, M.D., 399
Focht, John A., Jr., 223
Foray, Pierre, 88
Fucik, E.M., 401

Gray, Richard E., 125

Hacelas, Jorge, 401
Houlsby, A. Clive, 404

Isnard, A., 284

Klapperich, Herbert, 358

Lee, Kenneth L., 95
Legget, Robert F., 124, 180
Leonards, Gerald A., 222, 301

Marulanda, A., 405
Mascardi, Claudio A., 299
Materon, Bayardo, 406
Merritt, A.H., 407
Meyerhof, George Geoffrey, 52, 91
Murti, N.G.K., 409

Olson, Roy E., 2
Ospina, C.S., 405

Pender, Michael J., 48
Pilecki, T.J., 284
Pinkerton, Ivor L., 410
Proctor, Richard J., 177

Ramirez, Carlos A., 401
Reese, Lymon C., 94, 122
Richart, F.E., Jr., 1, 51
Ripley, C.F., 414

Savidis, Stavros A., 358
Schmitter, N.J., 219
Schuster, Robert L., 313
Screwvala, Farrokh N., 300
Selig, Ernest T., 362
Sherard, James L., 418
Sowers, George F., 53, 182
Strassburger, Arthur G., 420
Sturm, Ulrich, 358
Sulaiman, Ibrahim H., 120
Swiger, William F., 183, 423

Tannenbaum, Ronald J., 179
Taylor, H., 425

Whitman, Robert V., 312, 359